Land Degradation and Society

Why does land management so often fail to prevent soil erosion, deforestation, salination and flooding? How serious are these problems, and for whom? This book, first published in 1987, sets out to answer these questions, which are still some of the most crucial issues in development today, using an approach called 'regional political ecology'. This approach acknowledges that the reason why land management can fail are extremely varied, and must include a thorough understanding of the changing natural resource base itself, the human response to this, and broader changes in society, of which land managers are a part.

Land Degradation and Society is essential reading for all students of geography, agriculture, social sciences, development studies and related subjects.

Land Degradation and Society

Edited by
Piers Blaikie
Harold Brookfield

Routledge
Taylor & Francis Group

First published in 1987
by Methuen

This edition first published in 2015 by Routledge
2 Park Square, Milton Park, Abingdon, Oxon, OX14 4RN
and by Routledge
711 Third Avenue, New York, NY 10017

Routledge is an imprint of the Taylor & Francis Group, an informa business

Publisher's Note
The publisher has gone to great lengths to ensure the quality of this reprint but points out that some imperfections in the original copies may be apparent.

Disclaimer
The publisher has made every effort to trace copyright holders and welcomes correspondence from those they have been unable to contact.

A Library of Congress record exists under LC control number: 86018062

ISBN 13: 978-1-138-92302-7 (hbk)
ISBN 13: 978-1-315-68536-6 (ebk)
ISBN 13: 978-1-138-92307-2 (pbk)

Land Degradation and Society

Piers Blaikie and Harold Brookfield

with contributions by

Bryant Allen and Robert Crittenden
William Clarke and John Morrison
Narpat Jodha
Judy Messer
Lesley Potter
David Seckler
Vaclav Smil
Mike Stocking

Methuen
London and New York

First published in 1987 by
Methuen & Co. Ltd
11 New Fetter Lane, London EC4P 4EE

Published in the USA by
Methuen & Co.
in association with Methuen, Inc.
29 West 35th Street, New York NY 10001

Printed in Great Britain by Richard Clay Ltd, Bungay, Suffolk

British Library Cataloguing in Publication Data

Blaikie, Piers M.
Land degradation and society. –
(Development studies)
1. Soil degradation
I. Title   II. Brookfield, Harold
III. Series
333.76'13   S623
ISBN 0-416-40140-6
ISBN 0-416-40150-3 Pbk

Library of Congress Cataloging in Publication Data

Blaikie, Piers M.
Land degradation and society.
Bibliography: p.
Includes index.
1. Land use – History.
2. Land use – Case studies.
3. Soil degradation – History.
4. Soil degradation – Case studies.
5. Agricultural conservation – History.
6. Agricultural conservation – Case studies.
I. Brookfield, H. C.   II. Title.
HD 156.B55 1987   333.7313   86-18062
ISBN 0-416-40140-6
ISBN 0-416-40150-3 (pbk.)

Contents

Figures

Tables

Contributors

Bryant Allen Department of Human Geography, Research School of Pacific Studies, Australian National University, GPO Box 4, Canberra, ACT 2601, Australia.

William Clarke School of Social and Economic Development, University of the South Pacific, Laucala Bay, Suva, Fiji.

Robert Crittenden Southern Highlands Province Rural Development Project, PO Box 98, Mendi, Southern Highlands Province, Papua New Guinea.

Narpat Jodha International Crops Research Institute for the Semi-arid Tropics, Patancheru PO, Andhra Pradesh 502324, India.

Judy Messer School of Sociology, University of New South Wales, PO Box 1, Kensington, New South Wales 2033, Australia.

John Morrison Institute of Natural Resources, University of the South Pacific, Laucala Bay, Suva, Fiji.

Lesley Potter Department of Geography, University of Adelaide, GPO Box 498, Adelaide, South Australia 5001, Australia.

David Seckler International School for Agriculture and Resource Development, Colorado State University, Fort Collins, Colorado 80523, USA.

Vaclav Smil Department of Geography, University of Manitoba, Winnipeg, Manitoba R3T 2N2, Canada.

Michael Stocking School of Development Studies, University of East Anglia, Norwich NR4 7TJ, England.

Piers Blaikie School of Development Studies, University of East Anglia, Norwich NR4 7TJ, England.

Harold Brookfield Department of Human Geography, Research School of Pacific Studies, Australian National University, GPO Box 4, Canberra, ACT 2601, Australia.

Acknowledgements

Our first acknowledgement must be to our colleagues in Norwich and Canberra, who have heard a good deal about this book and have made useful if not always gentle comments. Some early draft material was commented on very helpfully by a number of people, most particularly Dr Tim Bayliss-Smith (Cambridge University), Professor Ian Douglas (Manchester University), the late Dr Joe Jennings (Australian National University) and Dr Marc Latham (Director of the International Board for Soil Research and Management, Bangkok); all these provided invaluable suggestions and leads into the literature, as well as encouragement. Ideas and material for specific chapters were provided by Dr Harka Gurung (Kathmandu, Nepal) and M. Jean Vogt (Strasbourg, France). Our wives, Sally and Muriel, provided both tolerance and support, and will be glad the book is now finished. Mary Ann Kernan of Methuen provided more than just editorial control; her friendly insistence is largely responsible for the final form of the book.

The book could not have been written without the strong support of our two institutions, the School of Development Studies, University of East Anglia, and the Research School of Pacific Studies, Australian National University. The latter made it possible for Piers Blaikie to spend three months in Canberra in 1984 and for Harold Brookfield to spend a month in Norwich in 1985, while the former earlier made it possible for Harold Brookfield to spend two months in Norwich in 1982, when the seeds of an idea were first sown. Within our institutions, we particularly thank in Norwich, Liz Goold for research assistance, Yvonne Buckland for word-processing and Barbara Dewing for drawing some of the maps and diagrams. In Canberra, Yvonne Byron provided research assistance and did important editing work on the whole manuscript. She also undertook the mammoth task of collating and checking the references. Carol McKenzie (and earlier also Pauline Falconer) word-processed chapters for us both, and Ian Heyward drew certain of the maps. Carol's and Yvonne Byron's have been the heaviest burdens, and our debt to them is enormous. The mail and telex services, and our two departmental secretaries Yvonne Wilson and Elizabeth Lawrence facilitated collaboration with efficiency and despatch.

We would like to thank the following for permission to reproduce copyright material: J.C. Rodda for Figure 3.1; Catena Verlag and Hans-Rudolf Bork for Figure 7.2.

While the friendly help and tolerance of all the above, and of others who have helped us in numerous smaller ways, have made the task both rapid and enjoyable, we cannot burden any of them with responsibility for the faults which remain. That responsibility belongs to us.

Piers Blaikie
Harold Brookfield

Norwich and Canberra
February 1986

Introduction

If the problems of land degradation could have been solved by research and reports alone, they would have disappeared long ago. It has been forty years since some of the first seminal works on environmental degradation were written (for example, Jacks and Whyte 1939; Osborn 1948; Carson 1962; Commoner 1972), and perhaps ten to fifteen since the high-water mark of environmental movements in the United States and Europe was reached. Even now, the volume of literature and proliferation of national and international institutions flows unabated. In spite of this, whole United Nations agencies and a worldwide environmentalist movement have been unable to make more than a marginal impact upon the prevailing effects of the exploitation of nature for short-term gain. It would be naïve to assume that the root of this state of affairs lies in intellectual failure alone. However, it is argued in this book that much of the literature on land degradation is beset by a fundamental theoretical confusion. Discussants address each others' work but often appear not to discuss the same underlying issues at all. Implicit assumptions about the significance and importance of land degradation remain unexamined. 'Facts', ideologies and beliefs are not identified, and the relevance and accuracy of much of the data base remains in doubt. What people cannot start to agree about is hardly a basis for initiation of change, often of a fundamental, politically sensitive and expensive nature.

This confusion can be traced to three major causes. First, the nature of the debate itself between scientists, commentators and decision-makers in government has hardly been critically examined. Since land degradation is *par excellence* an interdisciplinary issue, a comprehensive theory requires the combination of analytical tools of both the natural and social sciences. Natural scientists have made great strides in understanding and explaining the process of accelerated land degradation, although their work is far from complete. However, there is a need to find ways to bring together natural and social scientists more effectively to address the central question of why 'land managers', as we will call them (e.g. peasants, pastoralists, commercial farmers, state forest departments and so on) are so often unwilling or unable to prevent such accelerated degradation. It will be argued, in our first and final chapters, that the larger part of this task of explanation now lies with social scientists.

Few social scientists have addressed this problem directly. There is a very large literature on land tenure and agrarian structures as impediments to productivity and causes of inequality, but in so far as the environment is considered at all in most of this literature, it is only as a passive background to human interaction. The degree to which this is so is quite remarkable. It runs through almost the whole of the vast literature on agrarian issues in Latin America, for example. Even in the series of reports prepared in the 1960s for the Inter-American Committee for Agricultural Development (CIDA), a major outcome of the Punta del Este conference convened in the aftermath of the Cuban revolution to find means of resolving agrarian discontent without revolution, environmental aspects are statically and briefly described as a basis for the real information on land use and land tenure. Very substantial work on environmental inventory has also been done in Latin America, but the points of connection between the agrarian literature and the environmental literature are so few as almost to be negligible. There has been somewhat closer linkage in Africa, but not to the point of leading toward the sort of integrated analysis which this book seeks. In South Asia, there has been a tremendous outpouring of academic work, as well as political debate at all levels, on the process of agrarian change following the initiation of the Green Revolution in the mid-sixties. Yet there has hardly been *any* serious work, or at least work taken seriously, on what is now regarded as one of the most serious problems faced by the countries of South Asia. It is only most recently that the Centre for Science and Environment (CSE) produced its report on the state of the environment in India (CSE 1982, 1985), and that there has been controversy over the 1985 Forest Bill and the whole issue of conservation, the state's demands upon the environment and the rights of other users. In consequence we have a substantial body of very insightful literature on agrarian problems of transition closely related to political and economic theory, and little until recently on the social, economic and political aspects of *environmental* transition. Perhaps one of the clearest symptoms of this theoretical and methodological failure to combine social and natural science is the constant bewailing of a 'lack of political will' in implementing conservation policies at the international or national level. Ignorance on the part of political leaders and land managers alike, lack of management skills, lack of data collection and monitoring skills are all invoked. Policies are sometimes initiated to put right these alleged causes of non-implementation. Still there is little in the way of successful theories which can explain the paralysis of the state to intervene effectively.

The second cause of confusion arises at a more fundamental and ideological level. There are profound differences of opinion on the overall *significance* of land degradation, which arise from opposing theories of social change. For example, there are such views as that of Simon (1981) who does not see that there is any problem of degradation at all. According to him and other 'technological optimists', the world has successfully managed to feed its rising population and supply its growing industries. Famines and disasters, while certainly unfortunate, can be attributed to bad luck or bad management. Simon and those who think along similar lines (e.g. Kahn *et al.* 1976 or Beckerman 1974) provide telling evidence which throws serious doubt upon the eco-doomsters, at least at the aggregated level of the whole world. More useful land is being added by man's ingenuity, they argue, than is being lost. In any case, we are assured even by many geographers that 'man's mastery over nature' can cope with all problems. Thus, for example, Chorley states that:

> Man's relation to nature is increasingly one of dominance and control, however lovers of nature may deplore it. If the proponents of geography as a scholarly discipline wish to continue to reflect on the relationships between society and nature they cannot afford to adopt models which ignore the glaring probability that this relationship is one which exists between an increasingly numerous, increasingly powerful and progressive, if capricious, master and a large, increasingly vulnerable and spitefully conservative serf. (1973: 157)

Even those who are more prepared to admit that natural forces have by no means been mastered by human technology include many who see the harmful consequences of human interference as no more than 'externalities' of the development process, costs that have to be accepted where they can be afforded and otherwise ignored. It is more important, say some economists (particularly from developing countries as well as some in other places) to ensure economic growth and development first, and only after this has been attained to meet the costs of repair.

The other side of the same coin also applies to natural scientists. Their methods of measurement of land degradation, and the assumptions that are made about the social and political significance of these methods often remain unexamined. The attribution of land degradation to characteristics of soil, geology and climate, and to purely physical constraints tends to leave untheorized where the constraints lie and how far they are social. This task, we would argue, has to combine both physical *and* social theories. It therefore becomes necessary to examine critically the political, social and economic content of seemingly physical and 'apolitical' measures such as the Universal Soil Loss Equation, the 'T' factor and erodibility.

It would be wrong to blame natural scientists any more than social scientists for the failure to ask the right questions about the deeper causes of land degradation. It would not be difficult to present many examples of soil scientists and agronomists writing about degradation in a manner which places all blame on the folly, ignorance or ineptness of the people who actually work the land; indeed, we could cite writings by social scientists which do much the same. This book shows that blame is not so easily placed, and that remedies are not found simply in compelling or persuading the immediate land managers to mend their ways. However, there have been some major changes in scientific perception in recent years. A few years ago, an international and interdisciplinary group undertook a major project on the problems of soil erosion in Tanzania (Rapp, Berry and Temple 1972). They carried out some excellent studies of process both in general and at particular sites, placing special emphasis on varying climatic conditions. A number of 'social' papers was included in a special number of *Geografiska Annaler* in which their results were mainly published, but these concerned principally the progress and failure of land-conservation endeavours by government. The technical papers often included a brief summary of the supposed human factors of causation, but this area was not researched and the whole collection ends rather disappointingly on a mainly 'technofix' note.

Since that time an increasing number of natural scientists studying degradation has come to the view that the problem of human causation is very complex, and that if land management is to be improved then this complexity must be better understood. This has not necessarily encouraged either international or national bodies concerned with land management to include more than a token social science participation in their personnel or at their conferences, and it is interesting that some social scientists who see land degradation as a problem now include parallel token natural scientists in their deliberations – then abuse their contributions for alleged environmental fundamentalism, as one of us has recently witnessed in Australia.

The third source of confusion in the literature of land degradation is related to the others, and arises from a failure to view degradation within a wide historical and geographical framework. Much of the earlier alarmist literature from environmental scientists failed to do this, and has projected a sort of naïve 'edenism' upon the landscapes they saw as irretrievably damaged. What many alarmists would have seen more clearly from a wider perspective was that almost all natural landscapes are being continually modified and parts of them degraded, but this should not necessarily nor automatically engender alarm. Land degradation is a phenomenon not only of our time, and there have occurred 'eco-disasters' (as our generation would describe them) over some thousands of years, and even before human use became a serious contributory factor. By studying these both in physical and social terms, it is easier to evaluate causes and implications today.

To build a wide geographical framework for understanding land degradation is an essential but a very different requirement. Undoubtedly, the circumstances of degradation in any one instance are complex and unique. Therefore analysis must be able to explain the local conjuncture of physical and social processes, as well as to provide a clearly understood basis for generalization about processes worldwide. Mutual unintelligibility between writers frequently arises because of their unexamined assumptions about the societies in which land degradation occurs. However international the experience of writers may be, they frequently make assumptions about the way in which a society operates and changes in discussing the causes and implications of land degradation. The most obvious example has been the export of techniques of measurement of land degradation, conservation methods and policies from the United States to the Third World. Even within the Third World, there are such unbridgeable dissimilarities in the social circumstances of land degradation that causal generalization is usually unwise. These dissimilarities include both the more obvious such as erodibility and erosivity, rural population densities and agricultural technology, but also the political aspirations of peoples and governments and the whole context in which land-management decisions are made at all levels.

This book will argue that there is a fundamental confusion in the literature on land degradation as a whole which arises from an unfulfilled need to examine the social underpinnings of theories implied by both natural and social scientists; it will seek to study land degradation within a wide historical and geographical framework; and it will attempt to develop a methodology which can accommodate detailed local study as well as a basis for theory construction and generalization. How does the book then, in both form and content, aim' to contribute to a better understanding of land degradation? What are the areas which it covers, and what is left aside?

The first few chapters of the book address theoretical and methodological issues. In the first chapter we define our problem and establish a simple but basic terminology; we then discuss some of the methodological problems that need to be faced. The second chapter expands on the first by focusing on the problem of explanation, and includes an extended discussion of certain popular single-factor explanations, specifically those of pressure of population on resources and the conservatism or ignorance of farmers; these are illustrated in a case study of Nepal written by the authors (figure 0.1 shows the location of this and other case studies in the book). The third chapter, contributed by Mike Stocking, reviews the physical problems of definition and measurement, and examines degradation from the standpoint of a soil scientist. In the fourth chapter, the focus is broadened to ask who is the land manager, and how is land management to be defined and described in socially relevant terms. Taking an historical approach we seek to relate the land managers and their decision-making problems to the society and economy within which they find themselves. In chapter 5, of which the major part is contributed by David Seckler (5A), we look at the demand for and supply of land in space and time, and seek a rational way of calculating the costs and benefits of land management. The sixth chapter then looks more specifically at the social relations of land managers in the developing countries, focusing on the colonial period.

Throughout these introductory chapters, an historical perspective is indispensable. The mere fact that there are 'lags' between causation and consequence establishes the need for historical understanding. Society is constantly in transition, and is always in the process of resolving contradictions between humankind and nature and creating new ones. Any social analysis of these requires an historical perspective. Chapter 7 follows the historical path most deeply, through a literature review of the Mediterranean and a closer examination of historical soil erosion in western Europe, most specifically France. One aspect of this chapter is the problem posed by climatic change, potentially important for explanation of historical degradation, and we seek to evaluate the climatic factor along with possible causes of social origin. In chapter 8 the historical theme is developed in an unusual context; Bryant Allen and Robert Crittenden find the causes of land degradation, in an area only recently incorporated within the world economy, in the sort of political economy that developed over a previous period of 300 years following the introduction of a new crop.

In the remaining chapters the historical theme becomes secondary, and we discuss a selection of contemporary problems. Chapter 9 is concerned with the management and mismanagement of areas cleared from tropical rain forest, and converted into grassland or plantations. An introduction which focuses on Southeast Asia is followed by a detailed study of land-management problems in part of Borneo (Kalimantan) by Lesley Potter, and then by a search for understanding of destructive management of the land in Fiji, by William Clarke and John Morrison. Chapter 10 focuses on another specific problem, that of the management of common property resources. The first part examines the connections between degradation and the particular characteristics of common management of resources. The second part is a case study in northwestern India contributed by Narpat Jodha. Chapter 11, of which the major part is contributed by Vaclav Smil, sees continuity of the 'mastery over nature' ethos at the root of the huge degradation problems of China where serious weaknesses of central economic management are discussed. Chapter 12, by contrast, looks at the relations between the farmer, the state and the land in so-called market economies, and is balanced between a review of present management problems in the United States, and a study of the sociology and politics of conservation in Australia, contributed by Judy Messer. Chapter 13 presents our conclusions.

The treatment of land degradation in this book is inevitably selective. Not too much is written about irrigation problems, about either wetland or dryland salinization, about acidification, podsolization and laterization, or about wind erosion. Management and mismanagement of the forest lands themselves are mentioned only in passing, and the theme of desertification is only touched upon in these pages. We say nothing about the problems of the cold lands. In a book which is small relative to its subject matter, we felt it better to focus on certain specific problems in some depth rather than attempt an overview. Our main object is to demonstrate that there are social causes, and that they must be understood if there are to be social solutions; although our frame of reference is global, we have concentrated on themes and areas in which we and our contributors have some experience. There is a fairly strong focus on the southern and eastern fringes of Asia, with only briefer excursions into other parts of the world. It would require another and much larger book to present a full coverage in both thematic and geographical terms.

The book arises out of a workshop held in the Australian National University in February 1984. Piers Blaikie led this workshop, which Harold

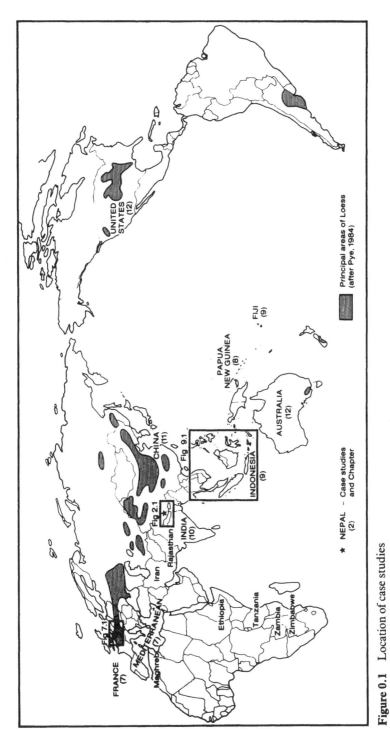

Figure 0.1 Location of case studies

Brookfield organized. Some of the contributors to that workshop are also contributors to this book, but we have also invited a number of others to contribute; these are Stocking, Seckler, Jodha and Smil. Each of us has been responsible for certain parts of the book, and there are some chapters and sections which are almost exclusively written by one or other of us with only minor co-author amendments. Other chapters, and introductory sections, are however truly joint and we have decided that all chapters other than parts by contributors should appear without separate attribution of authorship. Those familiar with our two very different writing styles and theoretical standpoints may be able to identify which parts are by whom, but we suspect that they might make mistakes. Nonetheless the two principal authors have rather different approaches to the subject. The lessons which the authors learnt as the book was written have a wider methodological relevance, and this is discussed in chapter 1, section 9.

Our contributors have had to endure some heavy editing, partly in the interests of length, partly because most of them initially wrote 'stand-alone' chapters which had to be blended into the flow of argument. There is quite a number of insertions by the authors, not separately distinguished. Also there are inevitably some differences in view between contributors, and between them and the authors. The insertions made by the authors in contributors' chapters have preserved these differences and have pointed out where they occur. For example, differences of view between chapters 3 and 5 are quite clear, and the authors think that they add to the strength of the book. There are also explicit differences in emphasis between other chapters and these are signposted clearly.

Specific mention of the various problems with which contributors had to contend would be invidious, but it must be made of Mike Stocking who had to write his chapter twice; a first version, completed in very good time, was lost when a case of research records fell off the back of a truck (literally) in Africa, and was stolen; one hopes that the chapter, and the results of three months' field research that were with it, were of interest to the new owners. We are grateful to all our contributors for their co-operation, even enthusiasm, and for their tolerance of our editorial heavy hands. Our other acknowledgements appear on page xiv.

1 Defining and debating the problem

Piers Blaikie and Harold Brookfield

1 Land degradation and society: initial statements

1.1 Land degradation as a social problem

Land degradation should by definition be a social problem. Purely environmental processes such as leaching and erosion occur with or without human interference, but for these processes to be described as 'degradation' implies social criteria which relate land to its actual or possible uses. Other processes, such as acidification and salinization, are only rarely recognized under natural conditions, at least in an acute form, and have a more directly human origin. The word 'degradation', from its Latin derivation, implies 'reduction to a lower rank'. The 'rank' is in relation to actual or possible uses, and reduction implies a problem for those who use the land. When land becomes degraded, its productivity declines unless steps are taken to restore that productivity and check further losses. In either case, the yield of labour in terms of production is adversely affected. Land degradation, therefore, directly consumes the product of labour, and also consumes capital inputs into production. Other things being equal, the product of work on degraded land is less than that on the same land without degradation.

It may be as well at the outset to face an objection to this statement, to which we return in chapter 5. It may be argued that, if there is abundant land or if losses in productivity can be made up by the provision of chemical fertilizers, degradation is neither an economic nor a social problem. However, this argument can be turned around: without degradation it would not be necessary to move to new land with the attendant costs; without degradation, such large inputs of chemical fertilizers would not be necessary in order to sustain production at constant levels, and efficiency of their use by plants would be greater. Either way, there are both economic and social costs. Also there are secondary costs, such as the nitrification of water supplies, which are purely social in nature in that they affect people and ecological conditions away from the site.

The social significance of degradation has been the subject of a wide variety of views rather than of engaged debate for reasons which are outlined by Blaikie (1985a: 12). We argue that under defined conditions it is a problem of a major order. Decline in the productivity of the land and of labour can be viewed as the 'quiet crisis' which nevertheless erodes the basis of civilization

– to adapt two phrases of Lester Brown (1981). This view claims that the problem is pervasive, often insidious but crucial to the future of humankind. There are elements of environmental fundamentalism in claims of this type – and we examine them in chapter 5 – but they underline the essentially social nature of the problem. Also, there is an important link between the chronic, slow-moving phases of the problem and the acute. When production conditions are adverse, as in a drought, the margin of productivity or of survival for a producer on degraded land is smaller than that of a producer on better managed land. When, as in large parts of Africa in recent years, climatic conditions remain adverse over a long period, farmers on badly degraded land suffer a particularly severe penalty. Land degradation, as well as drought, has been partly responsible for the severity of famine in agricultural areas of Ethiopia and Sudan (Eckholm 1976).

These simple considerations should alone be sufficient to establish land degradation as a problem of social significance. But it is also necessary to our argument to show that land degradation has social causes as well as consequences. While the physical reasons why land becomes degraded belong mainly in the realm of natural science, the reasons why adequate steps are not taken to counter the effects of degradation lie squarely within the realm of social science. Yet the problem of resource deterioration has been curiously neglected by the latter. There have been a few classic texts warning of the problems, such as Malcolm (1938), Jacks and Whyte (1939), Glover (1946), Rounce (1949) and Hyams (1952), but they are rarely cited in recent work. Also, neither classical nor Marxian economics have satisfactorily attacked the methodological problems of studying land degradation, thus depriving social scientists of a developed theoretical base. Seckler points out in chapter 5 that the problems of land degradation are as amenable to economic analysis as any other. But, for a variety of reasons, there has been remarkably little in the way of either empirical or methodological work on the economics of land and water conservation, by contrast with the economics of pollution which has a large literature. The *Journal of Soil and Water Conservation* and the output of some Departments of Agricultural Economics in the United States Midwest are perhaps honourable exceptions.

One of us (Blaikie 1985a) has recently sought to open the issue of degradation of land as a social problem. Essentially, that book built a number of theories to explain different aspects of degradation and conservation, drawn mostly from the standpoint of political economy. The present volume offers a greater diversity of approach, as well as a greater breadth of case-study material. A number of central social issues in land degradation which received only thematic treatment in the earlier book are discussed here in detail. These include the problems of measuring and economically appraising losses, and different institutional arrangements for land management, including common property and private property institutions and the state. More particularly, we also draw on a long and varied historical perspective in order to focus on the reasons why land management fails to be effective.

1.2   Issues of significance

Central to the issues discussed in this volume is the role of the 'land manager'. Land managers may find themselves responding to changes in their social, political and economic circumstances quite independently of changes in the intrinsic properties of the land which they employ. They may be denied access to common resources, or be forced to grow crops by landlords, market or social demand, or by the state. They have to find a strategy with which to meet such pressures, and do this on land which itself changes in nature. The intersection of circumstances and strategies forms our subject matter.

Any interference by humans with the natural processes of soil formation, evolution and erosion has an effect upon these processes, often unforeseen. Leaching, compaction and erosion of the soil, changes in plant cover and hydrological regime, changes in soil and water chemistry all take place naturally in the absence of any human intervention or even presence; in some environments these processes take place quite rapidly under natural conditions. Violent atmospheric events can cause rapid changes in environments empty of people. In some islands of recent geological origin, it can be shown that the soil had been eroded and/or become able to support only a limited biota long before the arrival of people. Yet human interference has modified and usually accelerated all these processes and has created the conditions under which new sets of processes, previously absent or insignificant, come into play. With the exception of the work of bulldozers, explosives, trail-bikes and other tools of malice, all the processes of land degradation occur in nature, but human activity on the land changes the conditions of their operation. The task of land management is to recognize these changes and find some means of bringing them under control.

However, the effect of human interference is not the same at all times and in all places. Human management of the land without leading to degradation is not only possible in a great majority of environments, but has been frequently accomplished in human history. However, the same human skills are not useful and effective in all places; under similar systems of management the productivity of some land is well sustained, while that of other land deteriorates rapidly. The problem is further compounded by the fact that degradation has occurred at one period but not at another on the same land. Agro-technology has not only changed through time, but has also been applied with differing degrees of care and perception.

Human-induced degradation occurs when land is poorly managed, or where natural forces are so powerful that there is no means of management that can check its progress. Some degradation is caused when land that should never have been interfered with is brought into use, but most land now subject to accelerated degradation is capable of more effective management than it receives. Our basic question is why these failures have occurred, and whether or not the problem has been perceived as such by those responsible at the place and time.

Since land degradation has occurred in such a wide variety of social and ecological circumstances, it is clearly futile to search for a uni-causal model of explanation. Equally, there is a number of hypotheses which have useful explanatory power, such as 'population pressure' or the exploitation by people of people, and these are examined in this volume. However, we shall see that while there are many causes where 'population pressure' has contributed to land degradation, in others a marked *decrease* in population densities has led to the same result. Likewise, an onerous burden of taxes, inequitable distribution of landholdings, corvée labour systems and the like, have probably led to declining management on the part of the exploited, but not invariably. On the other hand, there are many examples where very favourable prices for agricultural commodities or for timber have led to accumulation of profits, but also to land degradation. This complexity leads us away from any single theory of land degradation, since there are so many conjunctural factors operating at one place and time. Rather, case-study material and discussion of methodological issues together suggest a general approach to the problem of land degradation to provide an illustrated manual with which readers can approach their own empirical evidence.

2 Definitions of value, capability and degradation

2.1 Choices in defining degradation

As the opening paragraph of this chapter states, the dictionary meaning of degradation is 'reduction to a lower rank'. The term is therefore perceptual and implies at least a 'rank' scale of relative measurement. As a perceptual term, however, it is open to multiple interpretations. To a hunter or herder, the replacement of forest by savanna with a greater capacity to carry ruminants would not be perceived as degradation. Nor would forest replacement by agricultural land be seen as degradation by a colonizing farmer. Usually there are a number of perceptions of physical changes of the biome on the part of actual or potential land-users. Usually, too, there is conflict over the use of land – whether it be between farmers and conservationists, pastoralists and peasants, small farmers and the state, developers and concerned landholders. Since degradation is a perceptual term, it must be expected that there will be a number of definitions in any situation. It is, therefore, essential that the researcher recognizes any such conflict over the use of land and, therefore the definition of degradation. Sometimes the definition is given to the researcher as the 'ruling' one or the state-supported one, in the sense that land should be used in a certain way and degradation is, therefore, defined as reduction in capability to fulfil this demand. Sometimes the researcher will wish to supply other criteria derived from her/his own political and technical viewpoint.

It is of course more usual to employ the language of natural science to describe degradation, from the perspective of the soil scientist or agronomist.

However, the processes are varied and, from a social point of view, their impact may be felt in very different ways. Erosion, especially gully erosion and massive sheet or rill erosion, is very obvious, although the role of human agency may not be. Modification of horizon structure, partial removal of fine particles, pan formation, podsolization, compaction and similar changes are less obvious and have only a more gradual effect on the productivity of the land. Changes in hydrology affect the flow of streams and ground water, affecting storage and the supply of water to livestock and people as well as to the soil. Impoverishment of vegetation, the invasion of weeds and the selective elimination of soil fauna and the larger fauna which live on them affect the whole quality of environment as well as of the land; new environments, such as the Mediterranean *maquis*, may be created and come to be regarded as natural. Among more insidious processes, salinization becomes persistently severe in dry areas and periodically severe where drought is of irregular incidence, where it is seen as a problem mainly at such times. Acidification, on the other hand, affects the rooting depth of plants in a more lasting manner, but its build-up is very slow and it is not at once perceived as a problem.

These physical changes have to be evaluated also in social terms. The first step is to estimate reductions in yields of crops, livestock or useful vegetation resulting from these changes. A useful review can be found in Stocking and Peake (1985). This is a relationship which researchers are only beginning to be able to quantify, and there are many crucial gaps in both our basic understanding and in orders of magnitude under different conditions. The second step is the evaluation of degradation in economic terms. As chapter 5 indicates, there are on-site and external dis-benefits of degradation, now and in the future; however, these are generalized income benefits expressed in money terms. Although these are of obvious and overriding importance in assessing the impact of degradation, they leave unanswered the problem of varying and competing perceptions of degradation. For example, a reduction in income for agriculturalists may result in an increase for herders. Also there are issues of the distribution of losses from degradation between different groups, and access to alternative means of livelihood (e.g. new land) or to new technologies to limit the effects of degradation or to reverse them – all these affect the boundary conditions for accounting the social impact of land degradation.

2.2 The 'value' of land

There is also another issue which should be discussed before proceeding to a definition of land degradation, and this concerns the 'value' of land, which in some way is reduced for the user by degradation. It raises a number of theoretical problems. In none of its forms does the theory of value take adequate account of the 'value' contained in the natural source of all energy in the ecosystem, the sun's energy and of the stored products of that energy, which include the weathered material and nutrients which constitute the soil. Such 'value' cannot be said to be created by labour, does not have a cost of production, and is priced by the market according to a mixed set of utilities, including location, which often ranks higher than quality. Insightful comment on the failure of economics, specifically but not only Marxian economics, to take account of the physical processes underlying production is provided by Alier and Naredo (1982), Alier (1984) and Gutman (1985). While these authors, and the nineteenth-century socialist, Podolinsky, also fail to consider land degradation, they call attention to the failure of economics to consider energy flows or to come to terms with the notions of energy, except in a very imperfect manner.

Marx did, in fact, come somewhat closer to an appreciation of the role of land in production than did most other classical writers. He recognized that:

> Man ... can work only as nature does, that is by changing the form of matter. Nay more, in this work of changing the form he is constantly helped by natural forces ... labour is not the only source of material wealth, of use-values produced by labour. (Marx 1887/1954: 50)

But while there is a recognition of land as the product of natural forces, land – and other natural resources – were considered 'free' inputs into production and did not produce value since it was only labour that was considered to perform this function. On the contrary, it is clear that land may need to be 'paid' a great deal in order to continue to 'exist' at the same quality, as this book seeks to demonstrate. Even modern resource-depletion models fail almost entirely to consider the environment itself as a degradable resource (Hufschmidt *et al.* 1983: 57). It is difficult, therefore, to use the term 'value' in relation to land, and even Robinson's (1963: 46) cop-out in regarding value as a metaphysical concept without empirical meaning does not help; we therefore avoid the term.

2.3 Capability' of land

The term used instead is *capability*. When land is degraded, it suffers a loss of intrinsic qualities or a decline in capability. This term is not one within the economic literature. It is, however, in modern agronomic literature with something like the sense which is required. As a first step towards clarification, degradation is defined as a reduction in the capability of land to satisfy a particular use. If land is transferred from one system of production or use to another, say, from hunter-gathering to agriculture, or from agricultural to urban use, a different set of its intrinsic qualities become relevant and provide the physical basis for capability. Land may be more or less capable in the new context. This is important, because it must not be supposed that deforestation *necessarily* constitutes degradation in a social sense, even though it certainly leads to changes in micro-climate, hydrology and soil. Socially, degradation must relate to capability, and it is only if the degradation process under one system of production has reduced the initial capability of land in a successor system, actual or potential, that degradation is, as it were, carried across the allocation change. In actual practice, this is often the case, since more serious degradation reduces capability for most, if not all, future possible land uses.

2.4 A definition of degradation

We have noted that the effect of human interference need not always be deleterious. It is also possible to restore and improve land, and to create new productive ecosystems of which the outstanding example is the irrigated rice-terrace. The land itself also has its own means of repair: new soil is formed, gullies grass over and become graded; nutrient status is restored under rest. Just as we need to take account both of the interaction between natural processes and human interference in degrading land, so also we must recognize both natural reproduction of capability and of human artifice in assisting this reproduction. Bidwell and Hole (1965) made a useful distinction between 'beneficial' and 'detrimental' effects of human works on the soil. So also should we distinguish between the beneficial and detrimental processes in nature.

Degradation is, therefore, best viewed not as a one-way street, but as a result of forces, or the product of an equation, in which both human and natural forces find a place. We could say that:

Net degradation = (natural degrading processes + human interference) − (natural reproduction + restorative management)

A neat example of the variation of natural reproduction and its impact upon net degradation is provided by a comparative study of Hurni (1983) in which he compares the soil-loss tolerance in the mountains of Ethiopia and the hills of Northern Thailand. In the former case, cultivation has been going on for 2000 years with a fairly low rate of soil loss. However, the cumulative loss and slow rates of natural soil formation have both served to produce very serious land degradation. In Northern Thailand, however, with higher rates of soil loss, the local land-management system has 'compensated' for this and the capability of the land, in which soil formation is more rapid than in Ethiopia, is maintained.

3 The role of land management

3.1 Ways of managing land

With a definition of degradation as a reduction of capability, the role of land management becomes clear. Land management consists of applying known or discovered skills to land use in such a way as to minimize or repair degradation, and ensures that the capability of the land is continued beyond the present crop or other activity, so as to be available for the next. There is no system of land use, anywhere in the world, that does not have agro-technical means with which to achieve or at least approach these ends, provided they are practised in natural environments suitable for their employment.

At the simplest level, rotational grazing and shifting cultivation are effective strategies if well managed, with sufficient land over which they can be applied. Both are 'avoidance' rather than 'control' strategies (Kellman 1974) in that they leave reproduction of capability to natural repair, and avoid the need for intensive inputs on site. Many control strategies are, however, incorporated into modifications of these simple and ancient methods of management; slope control and water control are both employed in association with shifting cultivation, and so also is the addition of fertilizer. Rotational grazing is more easily managed with the addition of fencing and tethering. The major step forward from these strategies in temperate lands was mixed farming, in which both cultivation and grazing were rotated in relation to one another. Thus, perhaps as early as from the eighth century onwards, the two-, three-, and four-field systems of Europe emerged from an essentially shifting-cultivation base. In the humid tropics, mixed rotational farming was less widely suitable and wholly arable technologies evolved, generally involving massive inputs of human labour aided by livestock and their manure to make possible the permanent cultivation of land. Modern technology has added a range of artificial fertilizers, leguminous crops employed in rotation, and the ability to undertake much larger site-management works. We encounter some of this range of practices in the following chapters.

Fundamentally, the land manager's job is to manage natural processes by limiting their degrading consequences, both 'on-site' and 'downstream'. By 'downstream' is meant external effects away from the site, whether actually downstream, downslope or downwind, or effects which undermine the efforts or exacerbate the problems of neighbours, wherever located. The natural processes involved fall into two main groups, the mainly biological/biochemical and the mainly physical. They have a different range of impact, and present different, though related, management problems. The main problems of biological/biochemical management are on-site, though they have important 'downstream' consequences through the movement of mobile ions which can lead to salinization. The basic problem is to cope with the fact that purposive plant growth and removal for use tends to extract mineral and organic elements from the soil faster than they can be reproduced. Natural replacement requires a rest period or the planting of crops and trees which often have a low value in use. Reproduction of the capability of the land itself is usually the secondary objective of farming systems, but it is a vital objective and one that can absorb a great deal of labour.

The natural rate of soil formation varies enormously over the world, from close to zero in a thousand years in parts of Africa and much of Australia, to the formation of a capable solum in as little as ten years on some volcanic ashfalls under humid tropical climates. The impact of the cumulative loss of soil upon crop yields is also probably extremely variable. It has been estimated that a 15 mm loss from an Oxisol in an experiment in Indonesia (Suwardjo and Abyamia 1983, reported in Stocking and Peake 1985) caused a 40 per cent yield reduction, while a mere 2 mm loss from an Ultisol caused a 15 per cent yield reduction. Also, if these results are compared with data from the United States, it appears that the tropical Oxisol suffered a yield reduction ten times that of temperate soils and the tropical Ultisol twenty times that of temperate Ultisols, with similar soil loss. Even if these preliminary data are only approximately correct, they indicate great variation in the manager's task of maintaining land capability.

3.2 Landesque capital

It is important to distinguish between land management in relation to the current crop, the object of which is the production of that crop and the consequences of which are incidental, and purposive land management designed to secure future production. In the nature of things, most of the latter is in the physical area, though if a clearing for shifting cultivation is to last two or three crops, then a part of the labour put into initial clearing creates 'capital' for the second and third crops. The institutional costs of the reorganization of land tenure to make the installation of the three-field system possible in ancient Europe constituted 'capital' which endured for centuries. However, there is a class of works, including stone walls, terraces and such improvements as field drains, water meadows, irrigation systems and regional drainage and reclamation systems which is much more purposive in intent, the specific object of which is to create capital for the future maintenance of land capability. Investments of this nature have a long life and are sometimes described as *landesque capital*, which refers to any investment in land with an anticipated life well beyond that of the present crop, or crop cycle. The creation of landesque capital involves substantial 'saving' of labour and other inputs for future production. There is very little literature on this subject, and what there is suggests that the private benefits to land managers of costly landesque investments are seldom enough over the term of typically perceived discounting rates. We therefore have to supplement these (rather sparse) economic explanations with others to explain why landesque capital is (and was) created at all. As we shall see later in this volume, sheer necessity created by a lack of other options (in order to ensure the survival of the land managers themselves) and particular coercive relations of production are two of the most common explanations, amongst other social and political reasons.

There is a need to be aware that conservation decisions, including the investment in landesque capital, are not often made by individual decision-makers, who will bear all the costs and reap all the benefits. Therefore, one must be able to identify clearly the land manager(s) or hierarchy of land managers, whoever they may be – farmer, developer, landlord, agri-business, manager, government official or whoever. This issue of identification is discussed more fully in chapter 4 but it is enough to say here that managers may have different decision-making environments and different claims or demands upon the same tract of land.

4   Conceptualizing the role of land management

While this larger question of defining the 'land manager' may be deferred until chapter 4, there remains a need to define briefly the task of land management in relation to the natural processes which require to be managed. These are two-fold and concern the role of land management respectively in checking the natural processes of degradation and in aiding the natural processes of repair. What we need to do is to define, simply and unambiguously, the characteristics of the land that is being managed in such a way that will specify the nature of the land-management task.

4.1   Sensitivity and resilience of land

There are two qualitative terms which are useful in describing the quality of land systems (soil, water, vegetation) and these are *sensitivity* and *resilience*. A number of other terms have been used, including 'susceptibility' and 'fragility' (Winiger 1983; Glaser 1983), but some of these are loaded terms. The first term chosen here is sensitivity and it refers to the degree to which a given land system undergoes changes due to natural forces, following human interference. The term used here refers to sensitivity to erosion as well as to other forms of damage, such as the accumulation of mobile ions (which can give rise to salinization).

The second group of land characteristics of importance in land management concerns the ability of land to reproduce its capability after interference, and the measure of need for human artifice toward that end. This restoration of capital in the form of organic matter, nutrients and soil structure occurs naturally under forest or grassland fallow, as Nye and Greenland (1960) demonstrated in a manner that is still relevant. It occurs, however, at very different rates in different situations, while the depletion under cultivation which creates the need for restoration also takes place at very different rates. Certain ecosystems offer high initial productivity but this is rapidly depleted; in others, productivity is better sustained under repeated use. This property of standing up to, or absorbing the effects of interference, is only partly correlated with what is loosely termed 'fertility' of the soil.

Broadly following Holling, we propose to term this property *resilience*. Holling wrote of the resilience of a natural system where 'resilience is a property that allows a system to absorb and *utilize* (or even benefit from) change' (1978: 11). Where resilience is high, it requires a major disturbance to overcome the limits to qualitative change in a system and allow it to be transformed rapidly into another condition. Also, resilience is independent of the quantitative primary productivity of the site, be it small or great.

It will be apparent that, where a site is highly resilient and also insensitive to the forces of damage, the task of land management is relatively easy. Many wetlands, even though they require some initial drainage and may be liable to occasional flood, have both these properties, as do alluvial plains in humid climates. It may be for this reason that, as recent research has established, most early agriculture in southern Europe and the Middle East, and perhaps elsewhere also, was on moist land; it was fixed-plot cultivation on land easy to manage, from which there has been subsequent differentiation into various forms of wetland and dryland farming (Sherratt 1980, 1981). Even shifting cultivation, adapted to land of low resilience, is seen as a subsequent development in this argument.

Usually the resilience of land has limits, and the task of land management becomes one of supplementing natural resilience with devices such as land- and crop-rotation, manuring and fertilization, the planting of legumes and a range of tillage and land-preparation methods, many of which are also linked in part to the control of sensitivity to damage. It is a part of our argument to show that almost all land other than the most infertile or least capable, least resilient and most sensitive, can be managed at some level of production wherever there is water and a sufficient growing period. Recent research even in the Amazon basin has shown that only about 3 per cent of its soils are incapable of management in some form, despite the acidity and low fertility of 75 per cent of the remainder (Sanchez *et al.* 1982; Wade and Sanchez 1983). The cost of management may, however, be very high whether in terms of labour or material inputs.

To summarize the 2 × 2 table of characteristics of land and the implications of land use and management:

(a) a land system of low sensitivity and high resilience only suffers degradation under conditions of very poor land management and persistent practices which remove soil, increase compaction, salinity, etc.;

(b) a land system of high sensitivity and high resilience suffers degradation easily but responds well to land management designed to aid reproduction of capability;

(c) a land system of low sensitivity and low resilience is initially resistant to degradation but, once thresholds are passed, it is very difficult for any system of land management to restore capability;

(d) a land system of high sensitivity and low resilience easily degrades, does not respond to land management, and should not be interfered with in any major way by human agency, except (paradoxically) where major works create the landesque capital of a wholly new agro-ecosystem. The comparison between the impact of soil loss on productivity in temperate and tropical soils indicates that the latter tend, as a class, to have a relatively high sensitivity and low resilience and, hence, present more difficult management problems.

Two examples may serve to illustrate further the implications of different degrees of sensitivity and resilience for land. The first concerns the middle hills of the Nepal Himalaya, where some of the world's worst induced erosion is said to be taking place (e.g. Eckholm 1976). It is now established that the Tibetan plateau has been uplifted some 1000 m over the past 100,000 years (Ives 1981). Over the whole period, this is a mean rate of 1 cm/year. The Himalayan face has been uplifted at a lower rate, creating high natural erodibility as the slope becomes steeper, but an estimate of current uplift in the middle hills is 1 mm/year (Iwata, Sharma and Yamanaka 1984). In a small catchment in central Nepal, Caine and Mool (1982) calculate an annual lowering rate from mass wasting of 1.2 cm/year, while Williams (1977, cited in Carson 1985) calculates total denudation rates in four large catchments ranging from 0.51 to 2.56 mm/year. Regional uplift and regional degradation are natural processes, and the effect of terracing for agriculture has often been to check natural surface erosion rates, though with no significant effect on the more sporadic and localized mass wasting processes (Carson 1985). The management of such terrain presents enormous problems. This is an example of land with high sensitivity and of variable resilience. We return to this example in chapter 2, section 5.

The second example is from the lowlands of western and central Europe, which would seem on *prima facie* grounds to present a much less sensitive environment, with geological stability, a climate of low erosivity, and low relief. However, the whole region is mantled by a loess-type periglacial *limon*, of low permeability and, in the absence of management designed to ensure such permeability, has been subject to substantial erosion, leading to the redeposition of colluvial material. Discussed further in chapter 7, this region has been shown to have quite high sensitivity and to be subject to episodic damage. Sensitivity is not always readily explained and the less obvious it is, the greater perhaps the danger that a relaxation of management might lead to damage. However, under better management, the land system was able to reproduce its capability and even to increase it as a result of the degree of its resilience.

4.2 A summary

At the outset, the problem was posed as the search for social causes within the interaction between natural and human causes of degradation. In order to undertake this task, a definition of degradation was needed, which is a loss of capability to satisfy the demands made upon it. These may be competing, hypothetical or future ones, and it is important to specify against which of them a loss of capability is being measured. Noting that most processes of both damage and reproduction are natural but that their operation is greatly influenced by human interference and artifice, the problem can be summarized by an equation in which degradation becomes a net function, both of human and natural forces, both of damage and repair.

This led us to the consideration of the role of land management and of the importance of the land manager – an issue which we sidestepped at this stage. In order to define the task of land management and the means of encapsulating the work of natural scientists, simple terms were required. *Sensitivity* to physical and other forms of damage and *resilience* of the site characteristics in the face of use were identified as the two most relevant characteristics.

5 Relationships between society and land degradation

Having defined our key terms, the next task is to outline the main characteristics of the relationship between land degradation and society, and then to draw conclusions about an appropriate method of analysis. We identify three main characteristics: the interactive effects of degradation and society through time; the crucial considerations of geographical scale and the scale of social and economic organization; and the contradictions between social and environmental changes through time.

5.1 Interactive effects

As in many complex issues of social or physical change, there is a reflexive and two-way relationship between land degradation and society. To take the similar case of population growth and development, for example, rapid population growth can, under certain conditions, adversely affect economic development and the living standards of the majority of the population unless the economy can be expanded at a comparable or greater rate. Conversely, however, many aspects of poverty lead couples to have large families, and thus encourage a high population growth rate. In the same way, land degradation can undermine and frustrate economic development, while low levels of economic development can in turn have a strong causal impact on the incidence of land degradation. Blaikie (1985a: 117) offers examples of 'desperate ecocide' by peasants and pastoralists under extreme pressure to survive, and chapter 2, section 4 in this book gives a further illustration.

These interactive effects also take place through time. A period of rapid degradation may reduce the range of options over the possible uses to which land can be put in the future, unless there is effective repair. The future history of the affected region therefore takes a different course. This simple observation is somewhat complicated when establishing the impact of such land degradation upon the future history of the relevant people who use, or would have used, the land. The problem revolves around the convenient word 'relevant'. First of all, land degradation can affect, presumably adversely, the options of people living in the afflicted area, and future generations. However, if these future generations have the option of migrating elsewhere the issue becomes hypothetical. If, on the other hand, they do not have this option – perhaps because of national barriers as in the case of the Sotho of Lesotho, if the option of working in the South African gold mines is closed in the future – then the impact of degradation of a region on the present population becomes a very real question for analysis. This issue is one of 'option values' which is discussed in chapter 5.

5.2   Interaction and scale

The scale issue is crucial to the definition of land management because it focuses on the boundary problem of decision-making and of allocating costs and benefits. One person's degradation is another's accumulation, and this is equally true of uphill and downhill positions of a slope, regions, nations and even continents. For example, the 'hollow frontier' of Brazil in the early twentieth century, and that of the United States in the nineteenth, might be said to have contributed to the process of accumulation and the development of infrastructure on a national scale in the form of railways, roads and services. The fact of degradation on the settlement frontier had its effect on future options there, but the immediate effect of extracting short-term profits from the land was beneficial in the national context.

On a smaller scale, the physical transfer of fertility via riverborne silt and dissolved minerals, or by deliberate transportation of organic or mineral fertilizer from one place to another, makes it necessary to develop a more sophisticated set of criteria with which to analyse the impact of land degradation in one area upon the wider society. The exceptional case of Nauru has particular point here. The removal of rock phosphate from Nauru since 1900 has destroyed the agricultural capability of the island, which was never high, in the interests of overcoming phosphate deficiency in the soils of Australia and New Zealand. Latterly, the Nauruans have received good compensation for this loss, which they have invested mainly in the Australian economy, and on the proceeds of which they now largely live.

5.3   Contradictions between social and environmental change

The third aspect for debate concerns the possible contradictions between the criteria used for land degradation, and those for beneficial social change, or 'development', through time. An increase in cash incomes through commercial cropping and ranching can yield a temporary increase in rural incomes, maybe even over several generations, but can lead to degradation through lack of attention to management of the land, and hence to subsequent income reduction. Examples of this contradiction are legion. With the development of synthetic fertilizers, and their manufacture in larger and larger quantities, it can be argued that those pioneers who put profit first and good land management second made the right decisions, since the deleterious consequences of their actions are now masked by inputs of industrial origin. Moreover, while it may be that the modern oil-based fertilizers will not always be available, and more certainly will not be available so cheaply as oil resources finally approach exhaustion, the optimists would maintain that substitutes will be invented as the need arises. It is impossible to refute this argument, other than by pointing to the lower long-term cost of adopting management strategies which rely more upon natural processes of regeneration and repair.

6 The approach adopted in this book

6.1 Demands made by the society/land degradation relationship upon the method of analysis

Three characteristics of the relationship between land degradation and society have been identified: the importance of interactive and feedback effects through time, the importance of scale considerations, and the contradictions between social and environmental changes through time. These have to be recognized as placing difficult demands upon the way in which land degradation and society is studied.

One of the chief demands is a great deal of data, and there immediately arise technical problems of definition, measurement and availability (these are discussed in chapter 3). The second set of data problems involves the relationship between physical changes in soil and vegetation and declines in the productivity of the land (e.g. crop yields, livestock production). Again, this is partly a technical exercise, and much of the biophysical modelling of these relationships is beset by enormous uncertainties and errors (Amos 1982), but it is also an exercise which must try to distinguish the impact of physical changes in soil and vegetation from the impact of other purely socioeconomic changes in the circumstances of the land manager. Thirdly, there are difficult problems in the quantification of flows between people and regions. These derive from several distinct sources: the problem of conversion of flows of qualitatively different types to a common measure where energy, nutrients, available calories for human consumption, and market or shadow prices are only sometimes interchangeable; more abstract theoretical problems of incorporating the 'value' of resources found in nature (see page 12); and lastly the 'unit of account' problem discussed on page 14.

Wide degrees of error can therefore be made in the assessment of the important causes and rate of degradation and the reduction in capability of land. The ambiguity is compounded by the scantiness of data on farming and pastoral practices. Over long periods particularly, the causes of degradation usually involve social and economic changes which are difficult to measure, even if it is possible to reconstruct qualitative processes (see chapters 9B and 10B). If, for example, it is suggested that onerous rates of taxation and rents were responsible for heavy-handed and exploitative management of the soil, the challenge is to 'prove it'. A rigorous explanation linking the cause and effect would also have to predict that a reduction in rates of taxation and rents would reduce exploitation of the soil. This account of the problems should not be a charter for sloppy reasoning and inadequate empirical verification, but it does indicate that the extent of rigour in any analysis is as much a matter of circumstance as it is of necessity.

What then is our response to these demands for data which probably cannot be met? Presented with these problems it looks as if the task of explanation outruns the prospect of empirical verification. Part of the response is an adaptation and development of the ideas of Thompson and Warburton (1985a, b) who suggest ways of 'getting to grips with uncertainty'. The first element in our approach is to accept 'plural perceptions, plural problem definitions, plural expectations and plural rationalities' (Thompson and Warburton 1985a: 123). There *are* competing social definitions of land degradation, and therefore the challenge of moving away from a single 'scientific' definition and measurement must be taken up. This means we must put the land manager 'centre stage' in the explanation, and learn from the land managers' perceptions of their problems. Thus land becomes a 'resource-in-use', inextricably related to the people and society that uses it. It also means that we avoid single hypothesis explanations of degradation (and these are critically reviewed in the next chapter). Degradation at one place and time will be conjunctural and complex. There *are* patterns that repeat themselves in human–environment relations, but their modelling can only be partial at best. Case-study material therefore becomes crucial, and is a dominant feature in this book. But it is easy to lapse into a mere recording of unique events full of 'emic' data, which are difficult to relate to each other. Therefore an approach is suggested which allows for complexity, uncertainty and great variety, and one which takes as its point of entry those data which are beset with *least* uncertainty – the direct relationship between the land-user and manager and the land itself.

The other response to uncertainty leads us in a different direction, but one which is not contradictory. This is to try and improve our means of measuring and evaluating land degradation. If outside institutions are to make any contribution to the reduction of land degradation and of the incomes of people who rely on the land for their livelihoods, they will have to know *if* there is a problem and how great it is. Therefore, reliable methods of measurement of land degradation are crucial. Of course data are not reliable, they are *constructed*, and considerable attention in this book is devoted to their ideological nature, but this does not detract from the necessity to improve techniques of measurement. To this end chapter 3 explores the problems and prospects. Also, we need a methodology to evaluate the importance of land degradation in economic terms and a contribution to this is offered in chapter

5. First of all, the theoretical basis of the approach to land degradation and society is outlined in the next section.

6.2 The approach of 'regional political ecology'

The complexity of these relationships demands an approach which can encompass interactive effects, the contribution of different geographical scales and hierarchies of socioeconomic organizations (e.g. person, household, village, region, state, world) and the contradictions between social and environmental changes through time. Our approach can be described as *regional political ecology*. The adjective 'regional' is important because it is necessary to take account of environmental variability and the spatial variations in resilience and sensitivity of the land, as different demands are put on the 'land through time. The word 'regional' also implies the incorporation of environmental considerations into theories of regional growth and decline.

The circumstances in which land managers operate in their decision-making over land use and management can be considered in the context of core–periphery relations. Location-specific studies of the settlement frontiers of Brazil, the United States and Southeast Asia, as well as of agricultural decision-making in economically declining areas, provide considerable evidence for suggesting that declining regional economies provide an important context for lack of initiative and investment of labour and capital in managing land. Chapter 6 gives examples from hill and mountain areas of this link between the status of regional decline and the circumstances of decision-making in land management. Chapter 7 on the other hand provides evidence from eighteenth-century France to show that both the downswing and the upswing in a rural economy can almost equally press on the welfare and freedom of those who occupy the most vulnerable position in the social order.

The phrase 'political ecology' combines the concerns of ecology and a broadly defined political economy. Together this encompasses the constantly shifting dialectic between society and land-based resources, and also within classes and groups within society itself.

We also derive from political economy a concern with the role of the state. The state commonly tends to lend its power to dominant groups and classes, and thus may reinforce the tendency for accumulation by these dominant groups and marginalization of the losers, through such actions as taxation, food policy, land tenure policy and the allocation of resources. The agrarian history of Europe provides abundant examples (Abel 1980; Kriedte 1983). Very recent work on the relationship between cumulative soil losses and crop and livestock yields has shown a negative exponential relationship (Stocking and Peake 1985; Hufschmidt *et al.* 1983: 146) which strongly encourages the state to allocate resources to protect productive and still capable land, rather than to repair already degraded land which has fallen to a low level of productivity. Such a trend may be accentuated by the need of dominant groups to protect the source of major commercial crops. The allocation of state-controlled resources in rural development therefore usually disfavours the physical and social margin. This is shown for Latin America by Posner and MacPherson (1982) and for Nepal by Blaikie, Cameron and Seddon (1980). It may be added that the efforts of international agencies have hitherto tended to concentrate in the same direction, notwithstanding contrary statements of policy. These ideas are developed in an introductory fashion in section 7 of this chapter.

Extended examples of regional political ecology which consciously uses theoretical material from the core–periphery model, applied theories of the state, and the ecology of agricultural systems, are offered in chapter 2, section 4 and chapter 6, section 6. In the latter, it is hypothesized that many areas of the Third World suffer from a set of related symptoms which combine the results of land degradation, political and economic peripheralization, stagnant production, outmigration and poverty. However, there are clearly important variations in the politico-economic and physical histories of peripheral areas. Some areas, especially in hills and mountains, have avoided colonization and have preserved elements of ancient culture and social structure, such as segmented tribal organization and unformalized rules of land tenure. Other areas and their people have been intensively colonized and have attracted metropolitan capital into plantations, large farms and ranches, but are limited by sensitive and unresilient environments of a different type altogether. The distinction between these two is clearly drawn in chapter 6 and again in chapter 10.

However, there were and still are political economies which predate the world capitalist system, or remain only loosely articulated with it in modern times. Today, post-1945 Albania is an example and historically the Asian and tropical-American empires grew and differentiated on the basis mainly of internal division of labour and trade, with only peripheral dependence on external exchange. Such writers as Chevalier (1963) and Borah and Cook (1963) have shown how a class structure had evolved in central Mexico under the Aztec empire, how this was reflected in the management of land and the exaction of tribute, and how remoter groups brought under Aztec rule were incorporated into this system in a peripheral relationship. Degradation and erosion were substantial (Cook 1949). In this volume the more remarkable – because little stratified – case of the highlands of Papua New Guinea is analysed in chapter 8; here a political economy based on surplus production for competitive prestation evolved in the 300 or so years before there was any direct contact with the world political economic system, and a significant degree of land degradation was brought about under that isolated system.

In chapter 7 we undertake a more specific historical inquiry into the conditions of degradation in the past. We seek to explain how and why erosion of a type generally associated with sub-humid areas of southern Europe came to prevail in quite large parts of central and western Europe in the past, reaching a peak, at least in France, in the eighteenth century. Finding the evidence to favour a preponderantly human causation, it is hypothesized that pressures on the peasantry came to be translated into inadequate management of the land. Landlords, the emergent bourgeoisie and the state all contributed to these pressures. This historical example, and other historical material in this book, are introduced for a very specific set of reasons. Not only was the early-modern condition of the peasant and working classes in the west comparable with, or worse than, that of their modern counterparts in the Third World, but the pressures on them assumed a severity rarely encountered today. The historical examples thus provide something of an 'extreme' case of our thesis that damage to the land and damage to certain classes in society are interrelated. Moreover, they also provide long-term depth of material that is not generally available to us in the Third World or in countries of recent European settlements, and hence provide both an illustration of political ecology in time depth, and also a corrective to facile conclusions that might otherwise be drawn from the examination only of contemporary problems.

7 The margin and marginality

The approach of regional political ecology makes considerable use of various models and ideas surrounding the concept of the margin and marginality, and in the last substantive section in this chapter we turn to defining them, and to relating them to land degradation. There are three rather different although related uses of the term in neo-classical economics, in ecology and in political economy. In the following sections each of the three uses are examined, and then in section 7.4 they are brought together, and the reader will recognize that we have returned once again to the ground of regional political ecology.

7.1 The economic concept of the margin

The concept of the marginal unit of a factor of production, that last unit which when brought into use yields exactly its own cost and no more, is implicit in the classical theory of rent. Ricardo (1951) developed the theory of rent in regard to qualities of land; when all land of the first, and by definition uniform, quality has been brought into production, and land of the second quality is then employed, the cost of production on the latter will be higher than on the first. For this to be possible the price must rise, and so all land of the first quality will receive an unearned income in consequence of the incorporation of the second; the unearned income of labour inputs on the land is rent. If land is more intensively cultivated, the law of diminishing returns will apply. Hence the schedule of production will form a parabola, so that at the optimum ratio land and labour will both be utilized fully, and beyond this point there is a shortage of the forces of natural growth relative to the input of labour. Further increases in demand will therefore make it necessary to bring in new and inferior land, and the last land to be brought into use, or to be intensified, will just repay the cost of production and no more; this is the margin.

Von Thunen (Hall 1966) noted that beyond the optimum point of intensification it is a combination of constant land and increasing labour that becomes less productive, so that it is the additional units of labour that will in fact earn less. Gossen (as cited in Heimann 1945) noted that the value of any given unit of a quantity, wherever produced, is appraised like the marginal or last unit and thus has the same utility, and showed that the value of any individual unit produced must be equal to the marginal utility. The marginal unit is therefore that whose marginal cost is equal to the marginal utility, and if we are writing of land qualities, then this unit is the marginal land (Heimann 1945: 186–7). Add to this Von Thunen's arguments about the effects of intensification as the margin is approached on the distribution of returns to the factors of production, and we also have a link with the political-economy view of the margin which is developed below.

7.2   The ecological concept of the margin

In principle, at least, the ecological concept of the margin is comparable with the neo-classical one. For a given plant, or association of plants such as a forest, the marginal unit of land is that where natural conditions will just permit the plant to survive. However, an ecological view cannot avoid the question of environmental variability, so that we have to define the margin in terms of expected adverse conditions, recognizing that in some years plants can grow well beyond their 'secure' domain. This being so, a marginal environment for plants is better interpreted as the area or zone within which there is expected killing stress, but over which a plant or plant association can expand when that stress is absent. The same concept applies to marginal habitats for wildlife, and by extension also to crops and livestock.

Discussion of the 'ecological margin' does not always follow this logical approach. Perhaps it is better to be more restrictive and to define the term by extrapolation of the neo-classical definition to take account of environmental variability. The Sahel, for example, is thus defined as a marginal zone within which droughts of great severity and length can be expected. Discussion of the 'advance' of the desert margin into this zone (Stebbing 1935; Rapp 1976) means essentially that its marginality is becoming accentuated as human interference assists natural forces in the elimination or pauperization of plant communities, and makes their re-establishment in good years less likely.

Ecological marginality need not relate only to 'natural' conditions. Agro-ecosystems created by people immediately acquire a new set of relevant environmental variables. In all irrigated land, the availability and the quality of water become paramount. A clear example of ecological marginality in the context of created agro-ecosystems is provided by the annually reconstructed fields made in the gravelled beds of rivers in parts of the Mountain Province of the Philippines, while another is the gardens fed with human manure that were until recently encountered on embanked portions of the sea beach around the inlets which penetrate the New Territories of Hong Kong. Both were economically better than marginal, otherwise they would not have been constructed, but both were ecologically marginal at grave risk from storm and flood. We illustrate the more complex example of the *sawah*-rice terrace later.

7.3 The political-economic concept of marginality

The political-economy approach concerns the effect on people as well as on their productive activities of on-going changes within society at local and global levels. Use of the term in this context has arisen in the Latin American literature, where it was used to describe the sort of process described by many writers from Mariategui (1971) onward, and pithily summed up by Stavenhagen:

> The channeling of capital, raw materials, abundant foods, and manual labour coming from the backward zones permits the rapid development of these poles or focal points of growth, and condemns the supplying areas to an increasing stagnation and underdevelopment. (1969: 108)

At about the same time, Casanova (1970: 123) wrote of the 'marginal masses' who are outside the political system of Mexico, and of the 'marginal population' which is disorganized, uninformed and which can make demands only 'in the traditional forms of supplication, petition and complaint'. The term was quickly adopted (Parra 1972) to refer to a whole class of people who are excluded from employment, services, participation in decision-making, opportunity and secure housing (Brett 1973). Gaining wider currency, 'marginalization' has been used in the feminist literature to describe the exclusion of women from productive employment (Hartmann 1976; Young and Moser 1981), and in being widened to this and other contexts has perhaps lost something of the force contained in the original Latin American formulation.

7.4 The relation between three concepts of the margin

Writing of Kenya, Wisner (1976) wrote of *marginals* created by colonialism and capitalism who, in the process of social allocation of space, were quite literally pushed into *marginal places*. However, socio-political and ecological or economic marginality are not necessarily correlated in this way. 'Marginalized' peasants can, and do, occupy smallholdings on highly fertile land, while ecologically marginal land that is also near marginal in the neo-classical sense can, if a holder has enough of it, offer the basis for a highly profitable commercial operation. Much of northern Australia is ecologically marginal, but while most of it would be sub-marginal for commercial agriculture as has repeatedly been shown, it can support very profitable pastoral operations when coarsely divided into properties and chains of properties the size of small European countries. However, the Aboriginal people dispossessed of their land and now working on these estates share none of this affluence, and have been marginalized within the new relations of production.

If we control the comparison within a single mode or system of production, however, a relationship can more readily be established. An Asian rice-growing community has land sharply differentiated by fertility and hydrology, and its upland areas are sensitive under interference. When *sawah*-rice terraces are created, these new agro-ecosystems differ greatly in their ecological security. If they are on unstable slopes, the terrace walls may collapse. Some are difficult to supply with water in dry years, while others lose water readily by seepage. Under a high population density all the land capable of *sawah*-rice production has been taken up and converted; some of this is ecologically marginal even though economically secure in most years. Great differences in rent are yielded by the *sawah*-rice parcels. Some farmers without or with insufficient *sawah*, take up dry land for swidden cultivation on the ecologically marginal slopes, where they get good short-term returns of dry crops, but are at risk from erosion and loss of fertility. Those who are most marginalized in the socioeconomic sense have no land, and are forced to seek casual work from others. This is a hypothetical example, but is not unlike an upland West Java village (*kampung*) studied by members of the International Rice Research Institute (IRRI). They conclude

> As growth of population presses hard on limited land resources under constant technology, cultivation frontiers are expanded to more marginal land and greater amounts of labor applied per unit of cultivated land; the cost of food production increases and food prices rise; in the long end (*sic*), laborers' income will decrease to a subsistence minimum barely sufficient to maintain stationary population and all the surplus will be captured by landlords as increased land rent. This is exactly what has occurred in the *Kampung*. (Kikuchi *et al*. 1980: 15)

It will be useful to summarize some of the postulated and demonstrated relationships. To clarify, we identify the three concepts of marginality as economic (EN), ecological (EC) and politicoeconomic (PE) in what follows.

Land managers can become marginalized (PE) through the imposition of taxes, corvée labour and other relations of surplus extraction. The responses they make may be reflected in land use and in investment decisions over the preservation of productivity of their land. Adversity of this sort can produce innovations which raise productivity – to pay for the extraction of surpluses – as well as safeguard future productivity. However, more extreme marginalization (PE), often involving a whole number of readjustments particularly a loss of labour power (through war, conscription or emigration), has frequently led to changes in land use and the inability to keep up longer-term investments in soil and water conservation (e.g. repair of terraces and cleaning of irrigation and drainage ditches). The land then becomes economically marginal (EN) and the result is a decline in capability and marginality (EC) of the agro-ecosystem.

Spatial marginalization (PE) may also accompany these changes. Dominant classes may gain control and use more fertile land and force others to use more marginal land (EN). The attempts of the latter to make a living with reduced resources have often led to land degradation. Marginal land (EC) which has a high sensitivity and low resilience to even skilful or light interference by land managers can attract land uses, for this reason, which permanently damage the capability of the land. Here the emphasis rests not only upon the socially imposed marginality (PE) of the land manager, but also upon the intrinsic marginality (EC) of the land itself. Commercial ranching in the Australian interior is a prime example. If land degradation comes about as a result of either commercial exploitation or socially induced marginalization (PE) of land managers, a vicious circle of increasing impoverishment and further marginalization (EC) of land and land managers (EN) can sometimes result. Hence land degradation is both a result of *and* a cause of social marginalization (PE). It can accentuate the physical marginality (EC) of land by reducing its present capability, and marginalize (EN) it for present alternative uses. Much of Ethiopia, the Sahel region as a whole, and other areas of low resilience find themselves in this position.

8 Degradation, hazards and the environmental paradigm

The approach of regional political ecology taken in this book is compatible with the new directions in hazards and disaster research. Both approaches share an historical and a dynamic approach to human–environment relations. Nature is seen to be in constant flux, and measurement must constantly be updated (see also chapter 3). Also, nature is not universally nor statically defined; resources 'become' resources when people define them as such (Blaikie 1985c). The multiple definitions of natural resources and degradation by three groups of land-users and three government departments in the Indonesian case study (chapter 9B, section 3.3) are a good illustration amongst others in this book. Both approaches emphasize underlying social order rather than capricious nature in the explanation of calamitous events:

> causes, internal features and consequences (of natural disaster) are *not* explained by conditions or behaviour peculiar to calamitous events. Rather they are seen to depend upon the ongoing social order, its everyday relations to the habitat and the larger historical circumstances that shape or frustrate these matters. (Hewitt 1983: 25)

The three concluding chapters in Hewitt's book provide the basis of this 'alternative approach', linking the ongoing social order to hazardous events. Susman, O'Keefe and Wisner (1983) build on the work of O'Keefe (1975) and of Wisner (1976) who for a decade have linked disasters to processes of marginalization and proletarianization. The trigger events which start disasters or catastrophes have explanatory linkages with land degradation because both arise from the conjunction of physical and social processes. Sayer urged that we must start with the essential and necessary unity of society and nature, and that 'to start in the conventional manner with ... a separation followed by a listing of interactions would be to prejudice every other aspect of the exposition' (1980: 22). Approving this view, Watts goes on to argue that 'the subject matter of human ecology is accordingly *inner-*actions with nature' (1983: 234). This formulation is close to the idea of a 'resource-in-use' used earlier in this chapter. Also shared with the alternative approach in hazards research is an avoidance of relegating natural processes to a mere context or backdrop to hazards or degradation. Some radical literature has tended to do this and to imply that studies of climatic change in the Sahel, for example, are no more than a smokescreen and decoy to cover the tracks of the 'real' culprit – capitalism. It is vital to understand (as accurately as data, measurement and modelling will allow) the natural forces which create a variable management task to which decision-making, sub,ect to political economic conditions of choice, has to respond.

9   The social scientist's contribution: the need for open minds

We set out initially to write this book from position papers which adopted respectively Marxist and behavioural approaches, in each case with qualifications. What happened instead was something unforeseen: large areas of agreement emerged between the two authors, and several of the contributors also. While a more abstract (and no doubt rigorous) analysis of the two positions would undoubtedly expose fundamental contradictions, there is a broad area within which the explanation of land degradation can draw upon similar themes. There is something to be said for declaring a truce on the more abstract structural differences in the interpretation of social change, however important these differences may be, if it allows cross-fertilization of approaches. There are certainly fundamental contradictions between the 'human adaptation', neo-classical and various Marxist approaches, to take these three only. However, they share the objectives of understanding and problem solving, and of bringing about change in the situation, albeit in different degrees and in different ways. While there are epistemological reasons why Marxists have not been too interested in 'decision-making' models, there is nothing inherently revisionist in building them. Likewise, there is no betrayal of the profession of neo-classical economics in trying to pursue the quantification of costs and benefits of degradation and conservation into the realms of politics and unquantifiable conjecture (as done in chapter 5). Nor is there any reason why the study of human behaviour should fail to take advantage of the insights of theory about economic rationality or disregard the contradictions inherent in all social change and social formation.

There is a need for open minds, too, in the use of quantification and model building. There is an extraordinary schism between two self-perceived epistomological camps, the one which measures, creates its own data and uses others' in model building, and the other which calls itself 'radical' and eschews analysis of this sort as positivist, and the data as ideologically tainted and reductionist. Whilst this book amply shows that data do not simply exist but rather are constructed, it also argues strongly for technically better *and* more ideologically aware measurement of process, costs and benefits. Quantitative modelling of resources-in-use and land managers themselves need not be mindless number crunching. Nor need a central concern for the social meaning of degradation and for conscious ideological choice in explanation be dismissed as biased and not 'real' science.

Open minds assist in clarifying and sharing objectives. There are many blocks to open minds: the criteria for excellence and promotion differ between various practitioners (academics of different disciplines, consultants, administrators, politicians); there can be interdisciplinary rivalry between different academic departments (particularly between natural and social science); and more specific epistomological differences, mainly about the domain and status of proof in discourse and research. Land degradation and society, because of its complex and multidisciplinary nature, and its theoretical and practical elements, encounters most of these blocks.

If these blocks are not removed, the issue of land degradation will remain shrouded in controversy, uncertainty and incomprehension. What people cannot understand, they tend to avoid; what is unclear, people cannot decide upon. So it is with policy-makers and land degradation. While solutions will be as multiple as the causes of land degradation, the general approach outlined here aims to unify but through an appreciation of plurality of purpose and flexibility in explanation. For the discipline of geography at least, Carl Sauer put the problem and challenge perfectly more than forty-five years ago:

> Surely nothing could be more geographic than critical studies of the wastage of surface and soil as expressions of abusive land occupation. On the one hand are the pathological physical processes; on the other, the cultural causes are to be studied. Next come the effects of continued wastage on survival of population and economy, with increasing tendency to degenerative alterations or replacement. Finally, there is the question of recovery or rehabilitation ... Geographers have given strangely little attention to man as a geomorphologic agent.... The theme was clearly indicated as a formal problem of geography three-quarters of a century ago by Marsh. Geographers have long given lecture courses on conservation of natural resources and considered the evils of soil erosion. But what have they done as investigators in the field, which may actually lie at the

doorsteps of their classrooms? Is the answer that soil students should study sheet wastage, geomorphologists gullies, agricultural economists failing agriculture, rural sociologists failing populations, and the geographer prepare lectures on what others investigate? (Sauer 18–19)

10 Summary and conclusion

All aspects of the relationship between land degradation and society are both social and physical – a commonplace statement that is self-evidently true, but not trivial. It means that degradation is perceptual and socially defined. There may well be competing perceptions and these can be put into the context of the political economy as a whole, in which different classes and groups perceive and use land and its resources in different ways. Our four central terms – land management, land degradation, resilience and sensitivity – are all defined in a social context, and with explicit reference to ongoing processes of social change. There are extremely severe problems of data availability and of verification and proof. The approach taken in this book must respond to this problem of uncertainty and does so by seeking a point of entry where uncertainty is least, at the point of the land manager. The land manager is then 'contented', and her or his actions explained within a set of dynamic human–environment relationships which we call regional political ecology. The various definitions of the margin and marginality are central to this approach.

It will be obvious that we avoid an ethical and fundamentalist approach to land degradation. The definition of degradation, and whether it is 'bad' or not are both related to the people who use land. The field of interest in this book does not include difficult environmental-ethical questions such as the extinction of endangered species, or conflicts between national parks and other human uses of the biome, where ethical judgements assume greater importance. The approach taken here is that land degradation is judged in terms of the altered benefits and costs that accrue to people at the time and in the future.

2 Approaches to the study of land degradation

Piers Blaikie and Harold Brookfield

1 Chains of explanation

We have described our approach to the explanation of land degradation in any specific area as regional political ecology, and essentially the approach follows a chain of explanation. It starts with the land managers and their direct relations with the land (crop rotations, fuelwood use, stocking densities, capital investments and so on). Then the next link concerns their relations with each other, other land users, and groups in the wider society who affect them in any way, which in turn determines land management. The state and the world economy consitute the last links in the chain. Clearly then, explanations will be highly conjunctural, although relying on theoretical bases drawn from natural and social science. In this context we examine the major 'single hypothesis' approaches to land degradation. After all, there *are* discernible patterns of social change and land degradation, and models which would claim a degree of universality. The first is the explanation of land degradation in terms of population pressure and is the main concern of a number of influential theories. The second is very much more limited and different in character and explains degradation in terms of maladaptions and ignorance of land managers themselves – the problem lies uniquely with them. This chapter examines each of these models in turn in sections 2 and 3, and finishes with a case study which illustrates their performances.

2 Population and land degradation

2.1 Attribution and generalization

Any attempt to find the cause of land degradation is somewhat akin to a 'whodunnit', except that no criminal will ultimately confess, and Hercule Poirot is unable to assemble the suspects on a Nile steamer or in the dining car of a snowbound Orient Express for the final confrontation. The analogy is an apt one. Murders are generally easier to identify than land degradation; but guilt is often shared in different degrees between different people (e.g. assassin, accessory, etc.), as in each case of land degradation. However, any *general* statement about the causes of land degradation is of a very different order from the usual 'whodunnit', except perhaps in the case of the Orient Express, where *all* the suspects were found guilty! For the purpose of the analysis of land degradation such a statement may be true but is not very useful. At the other extreme, single general hypotheses of guilt do not get us very far either.

Perhaps the most common such hypothesis is that which attributes degradation to pressure of population on resources (which we shall henceforth call PPR), and therefore to growth in the numbers and density of population on the face of the earth. It does indeed seem self-evident that growth in numbers will cause land to be used more heavily; and that as the *per capita* area of arable and grazing land grows smaller, the sheer necessity of production will force farmers to use land in disregard of the long-term consequences. Yet, very severe degradation can occur in the total absence of PPR. Periods of population decline have often been periods of severe damage to the land.

2.2   Malthus rides again

Only a minority, among whom we are not included, would continue in the 1980s to believe that the rapid increase in the earth's population, an increase now heavily concentrated in the less-developed countries, is no matter for serious concern. A steady decline in the amount of arable land per head of the world's population means that this land will be required to produce more in order to provide the food and industrial raw materials needed by a population anticipated to grow by at least half its 1975 number during the limited space of the last quarter of this century. In the less-developed countries it is projected that the *per capita* area of arable land will have declined from just under 0.5 ha in 1955 to under 0.2 ha by the year 2000 (Council on Environmental Quality 1982: 403).

In modelling the effects on the land of population growth, however, the assumptions are based on inadequately measured and understood current trends often stated with great conviction.

Eckholm, for example, writes:

> Whatever the root causes of suicidal land treatment and rapid population growth ... and the causes of both are numerous and complex ... in nearly every instance the rise in human numbers is the immediate catalyst of deteriorating food-production systems. (1976: 18)

Ehrlich and Ehrlich more baldly say that 'an area must be considered overpopulated ... if the activities of the population are leading to a steady deterioration of the environment' (1970: 201). Yet they go on to say that while Australia may be underpopulated 'the "frontier philosophy" is more rampant in Australia than in the United States in terms of environmental deterioration and agricultural overexploitation'.

The Ehrlichs' confusion is helpful, for it at once challenges their own simplification. The terms 'underpopulation' and 'overpopulation', like PPR itself, imply that there must exist a critical population density at which none of these conditions obtain. This critical population is often described as the 'carrying capacity' of the land, a notion which applies to human populations a principle that is well-established among animal populations. Once animal numbers exceed the available food resources, they undergo a severe decline through mortality or other Malthusian checks. At least in so far as human populations depend on their land for their livelihood, similar conditions should apply. Or so the argument runs.

Efforts to employ the carrying-capacity concept continue to be made, notwithstanding criticisms. A notable effort is that made on a pan-tropical scale by the Food and Agriculture Organization of the United Nations (FAO) in association with the International Institute for Applied Systems Analysis (FAO 1982; Higgins *et al*. 1984). Here countries were subdivided on the basis of the major soil types and the length-of-growing season periods. Calculations of production of a range of crops were made under conditions of 'low input', 'intermediate input' and 'high input' – these being in all cases conditions of modern technology and fertilizers, not of labour. Assessment was made of country-level ability to supply the projected year-2000 population, with encouraging results except in the arid regions of Africa and Bangladesh, provided that 'high levels of farming technology' are adopted. This is an interesting exercise, but its assumptions carry a degree of unreality, and the assumption of access by the majority of the agrarian population to high levels of modern farming technology is perhaps the most unreal of all.

If carrying capacity changes with each turn in the course of socioeconomic evolution, each new technological input or new crop introduction, and can also vary markedly according to the bounty or otherwise of rainfall in a given year, of what use is the concept? Writing of Malthus' original essay, Peacock has argued that 'like any other theory, the theory of population must be regarded as a conditional hypothesis' (in Glass 1953: 66).

2.3   Malthus unhorsed?: The Boserup hypothesis

The objections to the PPR explanation of land degradation introduced above do not constitute a refutation. An alternative hypothesis was required, and was provided by Boserup (1965, 1981). Boserup proposed that the neo-Malthusian view of population capacity as dependent on resources and the state of technology was erroneous. She gathered evidence to show that output from a given area responds far more generously to additional inputs of labour than the neo-Malthusians suppose, even under pre-industrial conditions. She then proposed that 'the growth of population is a major determinant of technological change in agriculture' (Boserup 1965: 56). Population becomes the independent variable, and the dependent variables become agro-technology, the intensiveness of labour inputs and hence the capacity of the system to support people.

Boserup's conclusions were policy conclusions for development strategy, made more explicit in a later summary (Boserup 1970). As such, however, they were largely disregarded. The prevailing view among development scientists of whatever background was that the modern biogenetic, chemical and organizational agricultural revolution, rather than labour intensification, would solve the problems if they were to be solved. A more gloomy view, closer to the neo-Malthusian approach, would be that of Cassen:

> While the Boserup theory may have some validity in the broad sweep of history ... there is no reason to believe the argument is of general validity in today's developing countries. Cases of the opposite effects are not hard to find: over-exploitation of land, overgrazing of pasture, man-made erosion and so forth (1976: 807).

There is an important ambiguity in Boserup's hypothesis which concerns the innovation process itself. Her model may be likened to a toothpaste tube – population growth applies pressure on the tube, and somehow, in an undefined way, squeezes out agricultural innovation at the other end. However, as Cassen points out above, there are many contrary examples. What appears at the other end of the tube is often not innovation but degradation. Why? One of a variety of explanations provided in this book is the lack of access to productive resources on the part of the cultivator, and this is intimately linked to the class nature of most land management (see also chapter 6).

That is not to deny that PPR *is* often an important and reinforcing link in reducing this access to sectors of an agrarian population. This argument has already been made by one of the authors (Blaikie 1985a: 18, 107), and will only briefly be recapitulated here. There is a wide variety of land management practices which can be adopted and there is a good deal of variation in the amount of resources required for each practice on the part of the individual cultivator or pastoralist. Agronomic methods of management are usually less demanding than soil conservation works, but even the former require some spare capacity of labour, nutrients, land or capital. Crop residue incorporation is one effective conservation technique but requires that the farmer can 'do without' residues which may be important for other purposes (e.g. fuel, roofing for houses or fodder for livestock). Cover crops require extra labour and seed, while a soil-conserving reorganization of crop planting times may demand a reallocation or increase in labour demand. These observations apply *a fortiori* to chemical fertilizers, tree planting or the construction of grassed waterways. Also, state-sponsored research and development in conservation requiring relatively plentiful and locally available resources tend to be neglected in favour of more paying concerns such as the development of commercial crops on large farms on uneroded lands (Beets 1982; Belshaw 1979; Richards 1985). Thus PPR may well produce degradation and fail to produce agricultural innovation. Nonetheless, a lack of access to productive resources must itself not be promoted as a rival single hypothesis.

By attempting to isolate population as a single causal variable Boserup comes close to the very neo-Malthusians whom she criticizes. For, while neo-Malthusians believe that there are ultimate limits to the capacity of the land to support population without famine, damage or both, Boserup merely converts these limits into launching pads without successfully demonstrating that this conversion can always be made in all environments, or can continue indefinitely. Indeed, the evidence of diminishing returns, stagnation in rural wages and increased hours of work in both of Boserup's books, and especially in the second, suggests that Malthus climbs slowly back into the saddle as the Boserup sequence advances. At least within the domain of pre-industrial and early industrial agriculture which is her preferred ground, Boserup emerges more as a corrector of Malthus than as his refutor.

Cassen (1976) has drawn attention to the impact of population growth in some countries (notably in South Asia) upon the composition of resource costs required to produce basic consumption goods. The argument runs as follows. As long as population growth creates a demand in basic consumption items which can be met by the application of labour alone to land, the impact of population growth is not deleterious and may well run the course which Boserup argues (the creation of landesque capital, clearing and cultivation of new land and so on). Mao Zedong's adage of 'with every mouth comes a pair of hands' may hold. However, when increased aggregate demand for consumption items has to be met by expensive purchased inputs using scarce foreign exchange, then in a sense these are 'wasted' in keeping alive a growing population without increasing surpluses and savings. In an important sense, the reassuring model of Boserup is also stood on its head, and PPR may again assert degrading tendencies upon the land. Whether the 'Green Revolution' restores Boserup rather than Malthus to a standing position, and if so for how long, is a matter of contemporary controversy.

2.4 Innovation and intensification

The question really comes back to land management in the course of achieving production. We must once again pause to clarify the meaning of terms. The term 'intensification' is much used in literature about the Boserup model to mean the adoption of production systems which gain more output, averaged over time, from a given unit of land. Thus, elimination of the fallow year from the European three-field system and its replacement by fodder crops would be intensification. So also, however, would be the addition of labour inputs to wet rice cultivation to squeeze more and more production from the same field, as described by Geertz (1963).

Properly speaking, however, intensification means the addition of inputs up to, or beyond, the economic margin where application of further inputs will not increase total productivity; in the case of agriculture these are measured against constant land, and in pre-industrial agriculture we mean mainly inputs of labour, plus livestock. When, however, there is a change in the manner in which factors of production are used the inputs are applied in qualitatively new ways; a new curve of intensification is created. Such qualitative changes are innovations.

Brookfield (1984a) sought to distinguish between innovation and simple production intensification by means of the example of the West Indian sugar industry under slavery and subsequently. A production system which involved intensive land management with slave labour, and one effect of which may have been to save the production base from rapid degradation, was introduced for gain by seventeenth-century entrepreneurs. Requiring heavy inputs of labour, from 4000 to 8700 person-hours/year/ha, for both production and management, this system created a high density of population, the effect of which, even for a century after slavery was abolished, was to inhibit the innovation of new farming practices. Labour was cheap and abundant; furthermore, a large population required to be supported (Goveia 1969). Only the modern introduction of cane-breeding and fertilizers finally increased production per worker and so created a labour shortage. The old situation had endured some 200 years, with the effect that:

> Population-based theory is turned on its head in this case: given the social conditions of production, pressure of population on resources became a disincentive to innovate. Intensification and innovation became alternatives in a situation in which the means to innovate existed and were known, but conditions produced by intensification led the landholders to resist adoption. (Brookfield 1984a: 32)

In section 4 of this chapter we shall examine material from Nepal, where a high PPR is interpreted as both the cause of degradation, and also as the means by which degradation is managed and contained. Without anticipating this examination, it is clear that PPR can be seen both as creating a need to exploit resources in environmentally sensitive areas in such a way as to expose them to damage, and also as providing the means of a labour-intensive management system which seeks to contain the consequences. There is no reason, even on *prima facie* grounds, why both cannot be true.

2.5 Taking a position on population and degradation

The growth of human numbers, and the growth of numbers among their livestock, can undoubtedly create stress. It requires the extension of interference into new areas, and the subjection of these areas to the high levels of damage that follow initial interference. It requires the occupation of sites of lower resilience and higher sensitivity, for which existing management practices may be inadequate. Since the expansion is likely to be carried out largely by those displaced from older areas by poverty, or by other pressures of social or political origin, the new land has to be managed by those with the fewest resources to devote or divert to its management. Here we see the three definitions of margin and marginalization (economic, ecological and political economic) combining in a downward spiral. An increase in damage to the land is an inevitable consequence, at least for a time, and PPR and lack of access to the means to innovate go hand in hand.

High PPR may also create stresses within existing systems with well-tried management practices. As the margin of subsistence grows narrower, so the pressure to maximize short-term production will grow stronger. The need to innovate will grow, but the means with which to innovate will be lacking. Wealthier landholders whose own resources are not gravely threatened by the 'downstream' effects of degradation on the land of their poorer neighbours, may welcome the growing abundance of cheap labour and see no need to embark on larger innovations which might be of benefit to all. While they may have the means to innovate, they may not see it as being to their advantage to do so. Grinding poverty is a poor environment for good management, and a favourable one for degradation. But grinding poverty is not only brought about by PPR, though it may well be exacerbated by it, as in the following case study of Nepal.

Except in the presence of a situation such as outlined above, however, a high PPR also provides abundant labour with which to undertake intensive management. Where known and tried innovations are available, but require a high labour input, they are more likely to be both undertaken and maintained under conditions of high population density. The long-enduring ecosystem stability in Roman and Byzantine Palestine, and after that in Lebanon, noted in chapter 7, existed under conditions of very high rural population density. Most of modern Java, despite the serious erosion that takes place in headwater areas and on land of high environmental sensitivity that is unsuited to irrigated terracing, exemplifies the high productivity obtainable under intensive management with extremely high densities of population. The peril is that such systems require abundance of labour not only for their establishment but also for their maintenance. If some of that labour is withdrawn, as by an increase in off-farm employment opportunities, or by emigration, or by the demands on male labour generated by the state for corvée work, or for war, the consequences can be disastrous. The created system itself is one of high sensitivity, although it is resilient as long as the necessary inputs for its maintenance are available.

Where land is abundant, the need to conserve it may not be apparent. It is only after major damage has occurred, as in North America in the 1930s, that the need to halt and reverse damage comes to be perceived. Shifting cultivators in the humid tropics are often regarded as a prime example of destructive land users because of their practices, made possible by low levels of PPR. This may be to malign them. None the less, some of the major modern examples of land degradation are in areas of low to medium population density, rather than in areas of high density.

For the present, then, we adopt an open approach to the relation of population pressure to land degradation. Degradation can occur under rising PPR, under declining PPR, and without PPR. We do not accept that population pressure leads inevitably to land degradation, even though it may almost inevitably lead to extreme poverty when it occurs in underdeveloped, mainly rural, countries. The question of why management fails, or breaks down, is not answered so simply. Population is certainly one factor in the situation, and the present rapid growth of rural populations in many parts of the world makes it, in association with other causes, a critical factor. But 'in association with other causes' is the essential part of that statement, for the other causes can themselves be sufficient. PPR is something that can operate on both sides, contributing to degradation, and aiding management and repair. In general and theoretical terms, then, Hercule Poirot remains with no proven case against his prime suspect. Unlike the situation in a 'whodunnit', however, this has become obvious at an early stage in the narrative.

3   Behavioural questions and their context

3.1   Them and us

For the natural or physical scientist who diagnoses the immediate causes of a specific problem of land degradation, the debate about PPR is of use only in so far as (s)he seeks general explanation; his or her more immediate concern is with the means of introducing or enforcing protective measures, and (s)he is often aware of considerable resistance on the part of farmers and rural communities in general. Indeed it has been claimed by Blaikie (1985a) that most conservation policies introduced by governments fail, although there are significant exceptions. Major schemes with substantial funding for compensation to cultivators, and/or with the political means of mass mobilization, have a better chance of achieving results. Models for this exist even in the nineteenth century, as, for example, in major efforts to introduce large-scale land management changes in the French Alps in the 1860s (Henin 1979). In modern China, the efforts of thousands of workers could be mobilized to plant shelter-belts against wind erosion, fix dunes and terrace hillsides in the loess regions, though apparently without significant success, as Smil argues in chapter 11. But the lesser task of obtaining the co-operation of farmers in works of protection and drainage, often involving restrictions on land use and the creation of some landesque capital at the farm level, is seldom achieved without a great deal of persuasion and example, and not infrequently fails.

For a long time it was the fashion to decry the 'stupidity' or the 'conservatism' or the 'uncaring idleness' of such farmers, or to stress their 'ignorance'. A minority of farmers, like a minority in any walk of life, are certainly stupid and many more are conservative, but conservatism does not necessarily arise only from an unwillingness to change. Where there is a known set of practices and behavioural responses, it is thus much easier for the farmer to adhere to an established pattern than to make changes, as Kirkby (1973) showed in Mexico. Changes may only be forced when a major 'discontinuity' becomes apparent; yields have fallen alarmingly; more land has been lost than can simply be rationalized away; the alternative to the risk of doing something new is the seeming certainty of losing everything.

However, where farmers are peasants, in contact with a larger economy, and are subjected to pressures to change from government, different conditions apply. Bailey (1966) wrote a classic interpretation of the 'peasant view of the bad life' which is particularly relevant to the assumption by 'outsiders' that peasants are traditional, conservative and stupid, when these outsiders are confronted by yet another failure to bring about effective land management. To the peasant the government is seen as the 'enemy' and its representatives have a completely different cognitive map from that of the peasant. Intervention by government to bring about 'better' land management is frequently met by suspicion and non-comprehension. In the multiplex and non-specialized relationships of a peasant society, specialized interventions such as forest protection or pastoral regulations seem incomprehensible and quite incompatible with the moral economy of peasant society. Often the peasant views the future as the 'round of time' rather than the 'arrow of time'. The farmer allocates resources on the assumption that next year will be, more or less, like this year and is seen against a round of time – so many years before an ox is replaced, two or three years before the house needs rethatching, and so on. Not so a conservation officer, who must persuade the target-group of tangible benefits through innovations, and try to set in motion a definite change within a specific time period.

An alternative explanation for conservatism of land managers is an economic one, and rests on the often-observed risk-aversion behaviour of peasant and other farmers. Living and working in an environment of risk and uncertainty in which it is not possible to predict with confidence that such-and-such a set of inputs will yield such-and-such a return, farmers are reluctant to embark on new practices which might increase risk. Poorer farmers, it is argued, adopt what Lipton (1968) termed a 'survival algorithm'. Lacking the resources to weather failure, they both suffer greater risk and are more inclined to behave in a risk-averting manner than wealthier farmers. Innovative behaviour involves risk and uncertainty. The rich farmers are better able to bear its risks, and so stand to gain more from its benefits.

'Ignorance', 'stupidity' and 'conservatism' imply that there *is* a choice, but people are too ignorant, stupid and conservative to make the right one: that provided by governments or international 'experts'. Economic constraints caused by uncertainty and risk, compounded by onerous relations of production, perhaps too by PPR, provide a more clearly defined economic map of what is possible and what is not. Unfeasible government plans can all too easily be laid at the door of unappreciative farmers and pastoralists. At least economic explanations give some rationality to peasant behaviour and partially take the lid from the black box of ignorance and conservatism.

3.2 Ignorance and perception

Ignorance in a non-pejorative sense is another matter altogether. Ignorance of subtle changes in the quality of the soil is not only possible, but widespread. There is a number of degrading processes which show little immediate effect until a threshold of resilience is past, or until some untoward event exposes an increase in sensitivity. Leaching and pan-formation within the soil may operate in this way, as may changes in structure and chemical status. A striking example has recently emerged in eastern Australia where the planting of clover to upgrade pastures by nitrogen fixation has greatly improved capability since it began some fifty years ago. Surplus nitrogen gradually acidifies the soil in depth, releasing plant-toxic trace elements (aluminium and manganese); plants root to shallower depth to avoid them and become more sensitive to drought. The resilience of the system has a threshold beyond which acidification causes capability to suffer a sharp decline (Williams 1980; Bromfield et al. 1983; CSIRO 1985). Not only is none of this perceptible to the land manager until the limits of resilience are reached, but it has not even been perceived by soil scientists until lately.

It is equally possible for land managers to remain quite ignorant of the effects of low rates of erosion, where these exceed still lower rates of soil formation, until a critical level of accumulated loss is reached, and this can even take centuries (Hurni and Messerli 1981; Hurni 1983). The effect of loss of organic matter in restricting the ability of crops to respond to inorganic fertilizers is not likely to become apparent until the latter are applied. Both ignorance of degradation and its perception are functions of the rate and accumulated degree of degradation, as well as of the intelligence of the land manager.

A succession of good years can delay the perception of degradation to a critical extent, or at least facilitate an optimistic ignorance of the real consequences of observable changes. Without doubt, this has been an important factor in the 'degradation crises' that have struck many areas of new settlement during the past century. Newly settled farmers and graziers, who have used the natural 'capital' of long periods of 'rest' on land never before ploughed or grazed since the present soil–vegetation complexes were formed, had little means of knowing that the land they worked was of high sensitivity and low resilience until disaster – usually a drought – exposed the consequences of a period of heavy use. Writing just before the disasters of the 1930s, Webb (1931) showed how the perception of the Great Plains of the United States evolved rapidly during the nineteenth century, to the extent of a belief that occupation of the land increased rainfall and could ban the spectre of drought. In Australia, Meinig (1962), Heathcote (1965, 1969) and Williams (1974, 1979) have shown how the hazards of the semi-arid regions were first ignored, then harshly recognized and later only partly accepted.

Given the apparent success of remedies and adaptations a false sense of security could quickly become re-established. However, 'ignorance' shades from real to wilful, and may arise in part from other causes, such as a strong market imperative, a need to occupy new land for cash cropping or because of PPR elsewhere, or an ethos which believes that 'man' (and here we use 'man' rather than 'people') can and even must 'master nature'. This ethos, strikingly analysed by Passmore (1974), is at the root of much of the 'ignorance' so strongly complained of by those who work for a better management of the land. We present a stark illustration in chapter 11.

Still further removed from involuntary ignorance is the speculative abuse of land by commercial ranching and farming corporations and individuals, and by logging contractors. Here ignorance is not an accurate term since whether these land users know of the costs they inflict on future users of the land, and on present users elsewhere, through externalizing their costs, is beside the point. Wholesale disappearance of forests is the result (Plumwood and Routley 1982; Myers 1985), while the devastation of large tracts of agricultural land through now-abandoned commercial enterprises (Dinham and Hines 1983) is undoubtedly the result of calculated human agency and not of ignorance, nor stupidity.

4 The erosion problem in crowded Nepal – crisis of environment or crisis of explanation?

4.1 The environmental situation

Nepal is a classic area for the study of land degradation; we have already referred to it in chapter 1, and shall do so again at several points in this book. With an immensely varied environment, including the world's highest mountains, a strip of the Gangetic plain, and the high-altitude desert of the trans-Himalaya, Nepal is among the world's least 'developed' countries with a high and rising density of population on its limited areas of arable land (see figure 2.1). Rural population densities reach over 1500 per km^2 of cultivated land, or 15 to the ha, and there are districts in the middle hills with even higher densities. This is similar to densities in central Java and is 50 per cent higher than Bangladesh (Strout 1983).

In the main agricultural areas of the middle hills or *Pahad* almost all arable land is terraced. Irrigated land (*khet*), whether fed by rivers or from springs, grows rice and winter crops of wheat and potato wherever possible; dry land (*pakho*) is also terraced, but with a slope from almost level to as much as 25° or occasionally more. These dry lands grow mostly maize, often with finger millet as a relay crop which is transplanted from a seed bed under the growing maize. Unterraced land, used mainly for pasture (*charan*) but with a few cultivated patches, now occupies only the steepest slopes and ridgetops, but was in former times much more extensive and the site of shifting and semi-permanent cultivation among the mixed forests (*Quercus* spp., *Castanopsis* spp., *Pinus roxburghii*), and *Rhodedendron arboreum* that used to

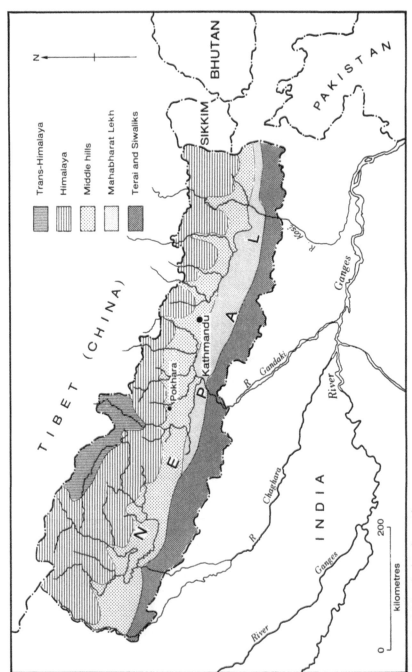

Figure 2.1 Nepal, showing main topographic belts

cover these hills (Burkill 1910; Mahat 1985). Cultivation and the collection of firewood were not the only reasons for clearance; charcoal-making for the metal industries of Nepal continued to be a major cause of depredation until early in the present century, though this has now diminished, and trees are also lopped to provide leaf-fodder for livestock (Bajracharya 1983). Livestock is a very important element in Nepalese farming systems (Axinn and Axinn 1983), and the numbers of livestock per human inhabitant are among the highest in the developing world.

The growth in population has eliminated the forest from large areas and has replaced it by cultivation. Commentators agree that virtually all land capable of being terraced has now been taken up in the middle hills (e.g. Caplan 1970: 6; Mahat 1985), so that forest boundaries are not now much in retreat. The degraded condition of the remaining forests is indeed now attributed to historical rather than to current practices (Mahat 1985; Nepal–Australia Forestry Project 1985). Yet this relative stability is only recent. In 1928 the government policy of replacing forest by human cultivation wherever possible was praised by one observer, who felt that

> this policy must be pursued for many years before there need be the slightest grounds for fearing that sufficient forest will not remain. For in the temperate zone [the middle hills] it is certain that cultivation can never occupy more than one-third of the total area.... Perhaps in the valley of Katmandu (*sic*) and its vicinity a condition has been reached in which it would be wise to call a halt.... But elsewhere the day on which restriction of cultivation need become a question for consideration is still far off. (Collier 1928/1976: 253)

From ancient times the State owned all land, and as it was the principal source of revenue to rulers no cultivable land should be allowed to lie idle (Stiller 1975; Regmi 1976). These principles were sustained and developed by the rulers of unified Nepal after 1768 so that the first king himself directed that all land convertible into fields should be reclaimed, and if homesteads were built on such land they should be moved (Regmi 1978). Peasants paid the state half the produce of the land in tax, or rent, and later this was paid to officials and others who received grants of land (and its income) in lieu of salary or as reward for service. Even so, only the irrigated *khet* land seems to have had firm definition until modern times, and in a village near Pokhara, Macfarlane (1976: 52, 87) noted that the *pakho* land was not shared out individually until about 1940; before that 'patches of jungle were slashed and burnt by those who had the labour'. By the late 1960s, however, only a few rocky and steep patches remained to be cleared; 'the limits of maize cultivation had been reached'. On the other hand, there have been no significant changes in areas of agricultural land and forest since at least 1900 in a densely peopled area at Thokarpa, east of Kathmandu, despite large population increases (Mahat 1985: 232).

Exposure of the slopes has serious consequences for runoff under the torrential monsoon rain, especially at the beginning of the monsoon when there is little ground cover after the long dry season. Heavy erosion of the sloping dry terraces is a result, especially as the risers of these terraces are cut back each dry season to add new mineral phosphorus and soil to the terraces, which are thus widened and progressively flattened through time; each season a proportion of these terraces is rendered useless, and the land reverts to uncropped *pakho*; each season, however, new or reclaimed terraces are created on the *pakho* land.[1] Slope-wash, gullying and especially mass movement occur on all slopes, and mass movement also affects the lower lying *khet* terraces, which become overcharged with water and saturated in depth. In occasional flash floods, the best valley bottom *khet* is sometimes washed out by heavily laden streams which widen their channels. As much as 50 per cent of the eroded material may be carried into the lowlands, causing wide channel braiding, choking rivers and adding up to a metre of soil per 20 years to adjacent fields; sediment loads as high as 25,000 ppm are regularly recorded in the major rivers.

The distribution and nature of erosion are, however, important. Dry terraces may lose only 0.4–1.6 mm/year, with corresponding losses of organic matter, nitrogen, phosphorus and potassium, and some of this is redeposited downslope, especially in irrigated terraces which can gain soil and nutrients. Much higher losses are experienced under degraded forest without surface protection from small shrubs, and on grazing land. In one small watershed gross estimated losses ranged from only 2 t/ha/year (tonnes per hectare per year) from the irrigated terraces to 20 t/ha/year from grazing land. However, while top-soil lost from this 63 ha catchment *and* carried beyond its bounds totalled only 220 t (3.86 t/ha from 57 ha of the area), the total loss was 1320 t/year, the balance being derived from mass wasting on the remaining 6 ha (Carson 1985). Caine and Mool (1982) calculated annual mass wasting losses to be as high as 13 t/ha from another catchment but derived from only 1 per cent of its area. Moreover, much of the landslip damage is repaired, and much land that appeared irreparably damaged when mapped in 1979–80 (Kienholz *et al.* 1983) was totally reclaimed under terraces, wet or dry, three or four years later (Ives 1984; 1985). Our own observations, using the 'geomorphic damage map' of the Ives and Messerli (1981) Mountain Hazards Mapping Project in the field, confirm that specific areas mapped as debris flow in 1979–80 were again wholly terraced in 1984 (Brookfield 1984b).

The importance of accelerated erosion in Nepal has been the subject of a great deal of uncertainty (Thompson and Warburton 1985a, b). A variety of

[1] Most of the modern literature cited in this chapter refers to dry terraces as *bari*, thus distinguished from the irrigated *khet* and the unterraced *pakho*. The Nepalese geographer Dr Harka Gurung (personal communication) advises that this is incorrect, being based on the writings of Kathmandu scholars who are unfamiliar with the languages of most of the farmers. *Bari* are infield dry terraces, heavily manured and mulched and often fenced; *pakho* is outfield, whether terraced or not. Gurung's advice is followed in this chapter. The same usage also seems to be followed by Regmi (1971).

commentators have been convinced that Nepal is an ecological catastrophe. A single doom-laden quotation from a wide choice will suffice:

> Population growth in the context of a traditional agrarian technology is forcing farmers onto even steeper slopes, slopes unfit for sustained farming even with the astonishingly elaborate terracing practised there. Meanwhile, villagers must roam farther and farther from their homes to gather fodder and firewood, thus surrounding villages with a widening circle of denuded hillsides. (Eckholm 1976: 77)

However, a detailed and quantified analysis of the extent of degradation has produced a more qualified interpretation. A recent FAO study stated that the results of their analysis suggest that past descriptions of conditions in Nepal have exaggerated the erosion problems, and go on to say that the 3 per cent figure of the total land surface classified under poor and very poor watershed conditions is not a serious erosion problem (FAO 1980: 5). Interestingly, they attribute part of this possible misconception to the effects of road-biased rural tourism, according with Chambers' (1984) apposite diagnosis of misconceptions of foreign and urban-based observers. There is an often-visited panorama which almost all visitors see as they travel west from Kathmandu by road towards Pokhara, at the point of leaving the Kathmandu Valley. From the lip of the valley westward and northward lies a rather desolate scene of considerable erosion, deforestation and degraded pastures. The higher parts of this same area were clad in heavily lopped *Quercus semecarpifolia* scrub when Kirkpatrick traversed it in 1793 (Burkill 1910). This scene finds its way into numerous articles and publications on the ecological crisis of Nepal, but is not representative of the country as a whole. Indeed, much of the agricultural land is in surprisingly good condition (FAO 1980: 7). The worst land erosion is on shifting cultivation land in the Mahabharat Lekh, and in the arid trans-Himalaya, where mass wasting and wind erosion carry away all soil from steep, unprotected slopes. In general, it seems that *charan* land including abandoned cultivation patches and pastures suffer from most serious erosion, particularly sheet erosion and loss of topsoil. *Pakho* land is the next most seriously affected followed by *khet* (Carson 1985: 7). However, mass wasting and catastrophic events such as large-scale slope failures are probably *not* due to human interference (Carson 1985; Ramsay 1985). Hence the most noticeable and dramatic forms of erosion, remarked on by rural tourists, are probably not caused by Nepalese farmers at all.

4.2 How Nepalese farmers cope

At a superficial level, the Nepalese problem has appeared to many observers, local as well as foreign, to be a classic example of the effect of PPR, specifically through deforestation and the dry terracing of steeply sloping land in a sensitive environment. Eckholm (1976) suggests that Nepal will slide away into the Ganges by the year 2000, and has no doubt that PPR is the principal culprit. A secondary culprit, however, is the Nepalese farmer of whom the Asian Development Bank (1982 II: 34) was highly critical. The practice of building unbunded, outward-sloping dry terraces was attacked with particular severity, and inward-sloping bench terraces were urged as a means of checking runoff. Water management was also regarded as primitive, a view shared by two local agricultural scientists (Nepali and Regmi 1981) who wrote that 'the technology of water management is scanty if not absent'. Under pressure of declining resources *per capita* cattle holdings are declining, and moreover the technology of composting manure is primitive, so that up to 52 per cent of the nitrogen and up to 80 per cent of the phosphorus are oxidized. Over 90 per cent of the fodder consumed by the animals does no more than keep them alive, leaving only 10 per cent for yield of 'economic products' (Asian Development Bank 1982). All this is regarded as very inefficient.

Others differ, at least to parts of this deluge of complaint. Axinn and Axinn (1983) analyse the flows of energy within the farm system, and stress the vital significance of 'keeping the animals alive' to plough the land and manure it, both 'economic products' of their existence. Ives (1985) notes that dry terraces slope outward to avoid waterlogging of dry crops, and to prevent accumulation and penetration of water which would cause landslipping; the absence of a bund is a deliberate measure to *ensure* runoff. Like ourselves, he is impressed by the skill of Nepalese levelling, terrace construction and water management, which includes extensive systems designed to enlarge the command area of irrigation flows. Gurung (personal communication) argues that it is the capital needed to build irrigation systems that is 'scanty', not the technology. Against the view of the critics is the glowing tribute paid by Cool:

> Personal observation suggests that it may require up to twenty years to fully transform an afforested hillside into a relatively stable irrigated terrace. The enterprise is marked with difficulty, setbacks and occasional failure. Yet what stands out is the skill and energy that goes into their design and execution and how successful the hill farmer is in maintaining and improving his terraced fields year after year, generation after generation. Flooding, landslips, goats and cattle, and occasional earthquakes are taken in stride. With only hand tools and simple bullock-drawn ploughs, but with enormous fortitude, the mountain farmer rebuilds, reploughs, reseeds and survives. (1983: 7)

Much greater detail concerning the manner in which farmers of the Kolpu Khola area manage the specific hazards of their environment is provided by Johnson, Olson and Manandhar (1982) in a paper written within the 'natural-hazards school' paradigm, but illuminated by use of ethnoscientific method (Conklin 1954) and by a degree of social awareness not common among this school of research. Methods of maintenance, repair and damage reduction are described, both those used by individual farmers and those known to them but beyond their means. The problems are well understood, although the supernatural forms part of the folk explanation of sudden and unpredictable terrace collapse. Maintenance is a regular and time-consuming task. When damage actually occurs the problem becomes one of resources. Thus:

> Farmers evaluate the options and, often, must choose the less effective one which is, however, the one within their means. Timing is a crucial factor in this decision. Constraints and limited resources may lead the farmer to postpone taking preventive measures in the face of warning signs. This may result in rapid deterioration or destruction of the endangered field, or it may allow time for the accumulation of resources for complete repair. (Johnson, Olson and Manandhar 1982: 84)

All farmers, however, are willing to experience temporary loss, even of long duration, in order to reduce the risk of greater loss. Thus farmers may cut irrigation off from endangered *khet* land and use it as lower yielding dry terrace, or even let it lie waste until consolidation is achieved. Even more drastic is the deliberate diversion of erosive flows of water on to land threatened with slumping to wash it away before new terraces are built to entrap new soil and rebuild irrigated fields, which then take some years to recover full capability. On the other hand, loose temporary terraces are sometimes made on the *pakho* land in order to obtain a little extra production at extreme risk. Differences in resources between farmers may lead to differences in net damage suffered by the poor and the wealthy, so that 'the overall effect of "random" landslides and floods may result in increased disparities between rich and poor' (Johnson, Olson and Manandhar 1982: 188).

The larger socioeconomic problems which underlie both differences in the quality of land management, and in the impact of damage, thus emerge. Some of these are discussed in a wider context in chapter 6, and a detailed study is offered by Blaikie, Cameron and Seddon (1980). Bajracharaya (1983) also places deforestation in such a context. Gurung (1982) argues forcefully that 'the basic problem of the Himalayan region ... is not ecological but the low level of development'. However, development of some types is not necessarily beneficial to the land. Khanal (1981) offers convincing demonstration of the effects of labour migration on land management in another part of the middle hills. While 90 per cent of the population of Nepal still depends on the land for its major sustenance, by one source only some 64 per cent of rural income is farm income (National Planning Commission, Nepal 1978), though by another the proportion is 85 per cent, 60 per cent from crops and 25 per cent from livestock (Nepal Rastra Bank, cited in Nepal–Australia Forestry Project 1985). Already some two-thirds of the agricultural production of Nepal comes from the lowlands (or *terai*) along the Indian border, where only one-third of the people live; a major shift in population is taking place (Goldstein, Ross and Schuler 1983), and the effect of a substantial reduction in labour inputs into the intensive farming and land-management systems of the middle hills could be more devastating than the increase in PPR that has taken place in historical and recent times. Some shortage of labourers and livestock hands is already felt in some areas (Gurung personal communication).

There is a further question of significance to be derived from the Nepalese case. It has been widely assumed that severe land degradation is a recent problem, but this is mainly because the problem has only been recognized since foreigners began to enter freely and move around the country after 1950. Logically, one would suppose that earlier shifting cultivation and wood cutting for charcoal-making on a large scale would have done more damage than the intensive terracing that has covered the same hills under rising PPR in modern times. Maize and potatoes were introduced into Nepal in the eighteenth century (Regmi 1978), and the effect of these crops, which offer poor ground cover in the early stages of growth, must surely have been to increase erosion before terracing was widely established.

Certainly, the massive transportation of material from the Himalayas and the middle hills into the Ganges plain is not new. The Kosi river, which drains eastern Nepal, has shifted its course 120 km to the west (i.e. up-Ganges) since 1736, and flooding has been experienced since ancient times (Carson 1985). The contribution of deforestation is disputed in the light of this evidence, and in any case forest destruction in Nepal has been in progress for several centuries (Mahat 1985). There is some slender evidence that the Ganges itself may have become more shoal-ridden during the first half of the nineteenth century, necessitating the replacement of early steamboats by shallower draught vessels (Headrick 1981: 22) but catastrophic events (such as glacial or debris dam-bursts) have been bringing vast quantities of sediment down into the lowlands for millennia.

In these circumstances, the effect of rising PPR in Nepal appears double-edged. Terracing has probably diminished erosion; deforestation and the creation of larger areas of grazing land have probably increased it. It is by no means clear that the balance has been towards greater degradation. Nor is it clear that the pessimist's vision, caricatured by Ives as desertification of the Himalayas, 'the devastation of the Ganges and Brahmaputra plains and a major opportunity for Dutch polder engineers in the Bay of Bengal' (1985: 428) will come about under present practices. 'Development' of a type which would involve major reduction of available rural labour in the middle hills, or the introduction of machinery and of large but untried works conceived by engineers without understanding of the dynamic montane environment, might however be more conducive to this result.

4.3 The 'chain of explanation' in the middle hills

The precise role of PPR still needs to be explored further: in terms of the interpretation of whether accelerated erosion *is* a major and widespread phenomenon; whether it has recently become worse; and also in terms of the causal relationship between population growth and erosion. A more comprehensive frame of analysis must be provided or, to use the metaphor with which this chapter started, the chain of explanation must be followed back to broader socioeconomic links.

One of the most fundamental relationships in evaluating the impact of PPR is that between population growth, extension of cultivation and the productivity of land and labour. Between 1970–1 and 1980–1 the increase in population was greater than the increase in cropped area in both highlands and lowlands (Gurung forthcoming), and in the highlands most of the increase was due to extension of double-cropping particularly involving wheat cultivation on *khet* (Mahat 1985). However, inadequate data indicate that food-crop production declined by 0.5 per cent between 1970–2 and 1980–2 in the middle hills but increased by 9.6 per cent in the lowlands, against area increases of 11.6 and 15.1 per cent respectively (Gurung forthcoming). Forest clearance in the *terai* has now reached the limits of good land, and a weakening situation of irrigated rice production probably reflects the extension of cultivation on to less fertile land.

Mahat (1985) describes the double-cropping, inter-cropping and relay-cropping systems in use in a part of the middle hills east of Kathmandu. Livestock manure is the basis of production, and in the hills there is one 'livestock unit' per head of population, mainly cattle and buffalo (41 and 22 per cent at weightings of 1.0 and 1.5) plus goats and pigs for meat. However, in Nepal as a whole, the annual increase in cattle population between 1966–7 and 1979–80 was only 0.12 per cent while the buffalo population declined by 1.3 per cent (Rajbhandari and Shah 1981). The consequences for production are probably better illustrated by a representative interview with a farmer than by unreliable statistics:

> Some thirty years ago we still produced enough grain to allow us to exchange surplus for necessary daily goods, which we could not get from our farming. Of the grains harvested, one third was exchanged.... While the good farmers who have enough cattle and do very intensive cultivation can still increase their yields, this is not the general trend. In a *khet* (irrigated) field where we sowed 4 *mana* of seed we used to get 1 *muri* of paddy; now we need an area with 8 *mana* of seed to get 1 *muri*. Our wheat used to have big ears and long halms and we filled six baskets a day, nowadays it is sometimes only one or twó. In many houses there is no longer enough food. For some, the harvest grains are sufficient for only three to four months a year. (Banister and Thapa 1980: 90)

The small size of corn-cobs on stacks outside houses in the middle hills is a matter of observation, though whether or not any trend is present is impossible to determine except by farmers' recall.

Intensive questionnaire work in west-central Nepal undertaken by Blaikie, Cameron and Seddon in 1973–4 and 1980 (Blaikie, Cameron and Seddon 1979, 1980) indicated that farmers were in fact well aware of decline in yields on old-established fields, particularly dry fields which supported maize and millet. The main problem is one of a reduced availability of plant nutrients, which come predominantly from composting of forest products and involve a 'transference of fertility' from the forest to arable land (Blaikie 1985b). Cattle are stall-fed much of the time and are fed forest-litter, tree loppings and hand-cut grass, all of which are gathered daily. The manure, together with leaf material from animal bedding, is then applied to fields. As the population rises, the increased demand for food crops is met by heavier lopping, which thins and destroys the forest. Mahat's (1985) informants at Thokarpa, east of Kathmandu, held that the greatest forest degradation occurred between 1951 and 1962 when there was a partial breakdown in central government administration, but that controls have been restored under the local government system established in the latter year.

Fuel needs also reduce forest cover. Indeed, wood fuel demands exceed supply by 2.3:1 in the middle hills and by 4:1 in the drier far west. One source estimates that all accessible forests will be eliminated within twenty years (Asian Development Bank 1982: 12), but other information cited here sheds doubt on this estimate and indeed all estimates are fraught with uncertainty (Thompson and Warburton 1985a). Crop yields decline when the forest-to-arable ratio is upset, and whatever the direct effect of deforestation on erosion it certainly has an effect on the capability of the arable lands. The inability of most Nepalese households to make good this 'energy crisis' by importing chemical fertilizers, and kerosene or other alternative fuel for cooking is an essential part of the explanation.

So we must pursue the question of PPR and degradation to the next link in the chain of explanation. The problem, as Harka Gurung argues, could be conceived as a lack of development. Simply, incomes are so low that few can *afford* chemical fertilizer, or any capital investment other than that generated entirely by their own labour. Half the households of a rural household survey undertaken in 1973–4 (Blaikie, Cameron and Seddon 1980) sold less than Rs. 250 (*c.* £10) of agricultural, pastoral and handicraft produce per year. Most budgets were balanced to within a few rupees, or were in deficit (Blaikie in Seddon, Blaikie and Cameron 1979: 69). Also chemical fertilizers do not produce a satisfactory return on unirrigated crops and, therefore, are not usually applied to summer maize and millet, but are largely limited to winter wheat, and winter maize at low altitudes, to industrial crops and sugar cane in the *terai*, and to vegetable growing by larger farmers in the Kathmandu Valley. Thus, as traditional forms of energy use start to fail because of population growth and degradation, the peasantry is, and becomes more, unable to transform itself and substitute new and imported forms of energy for agriculture, livestock rearing and fuel. It also becomes less and less able to harness water for more irrigation, especially as the replacement of local smelting of iron and copper by imported metal has reduced the population of miners whose winter off-season work used to be the building of irrigation systems for villagers (H. Gurung personal communication).

However, rural households *have* responded to PPR in other ways. They have migrated out of the hills altogether (so that the *terai* now holds over 40 per cent of the population of Nepal, according to the 1981 Census). They continue to extend cultivation and upgrade land (from dry *pakho* to terraced *pakho* and to *khet*). As the Land Resources Mapping Project has shown, a number of new crop rotations have been introduced, particularly on south-facing slopes and lower elevations, involving relay-cropping, inter-cropping, the selective introduction of quick maturing varieties, zero-tillage cultivation allowing broadcast wheat to be sown before the harvesting of *padi*-rice in *khet* fields, and so on. However, the endeavours and ingenuity of Nepalese farmers can in many areas of the middle hills and mountains barely cope with the increasing challenges being heaped upon them. However, lest the erroneous impression arises that PPR *is* the ultimate culprit, the question must then to be asked, *why* is the majority of the Nepalese rural population so poor?

The last link in the chain of explanation is vital, but it is only explored in outline here to avoid anticipating the argument of chapter 6. Nepal has had a long history of political independence, but also of quite important economic relations with British and later independent India. The independent state, however, taxed the farmers heavily and placed heavy demands on their labour for corvée work; in some areas landlordism developed as state officials were allocated areas of land and people to exploit in lieu of salary (Regmi 1971, 1978). Until very recent times the state remained unchanged in its antique and quasi-feudal form and kept out modernizing reforms in education, forms of representative government as well as productive capitalism in industry and agriculture (Blaikie, Cameron and Seddon 1980). Nepal remains landlocked and dependent upon India, both politically and economically. Such surpluses as there are in Nepal tend to be used in merchanting, smuggling, real estate and speculative purchase of land. Attempts at manufacture are undermined by cheaper products from India, and the 'leaky' frontier allows grain from the surplus-producing *terai* to flow south to India rather than to the food-deficient hills. The state itself has had great difficulty in outgrowing its quasi-feudal and extractive nature to meet the daunting challenges of development at the present time (Shaha 1975; Blaikie and Seddon 1978).

4.4 Crisis in explanation?

The explanation starts with changes in the intrinsic properties of the Nepalese landscape and ends with the problems of the Nepalese state and its relations with India. Each link of the explanation is firmly closed around the next, but initially the direction is away from the environment and towards more general 'development problems' of the agrarian economy and the Nepalese state. It leads back to land degradation because a vicious circle of links has been created, leading from degradation to underdevelopment and back again to degradation.

The palliation or gradual reversal of the problem of poverty requires a series of 'techno-political' decisions. However, there is a considerable degree of indeterminacy between the 'levers of policy' and intended results. For example, we cannot state with certainty that more commercial economic opportunities for Nepalese farmers (e.g. growing fruit and other crops for which they have a comparative advantage) would *necessarily* improve land management and reduce degradation. As the discussion of the land-management decision-making process in chapter 4 will suggest, there are many *other* exits from the vicious circle just described, aside from more effective land management. The more the explanation is linked to the social and political economic context, the more unpredictable and less direct the impact of environmental policy becomes. On the other hand the more directly environmental and technical the explanation, the less it is able to account for human agency in the management of land. The knack in explanation must lie in the ability to grasp a few strategic variables that both relate closely together in a causal manner, and which are relatively sensitive to change. In that way the most promising policy variables and paths of social change can be identified.

In the Nepalese case, the need to distinguish between the natural processes of a high-energy environment and the effects of human interference – and further to distinguish between the harmful and the positive elements of that interference – is of paramount importance. As we shall see again and again in this book, not enough is known of the modern history of the land surface to establish this distinction, although there is certainly enough evidence to discard the generalizations of even a decade ago. What is certain is that the management efforts of the farmers of the middle hills are under stress of such an order as to threaten the basis of their livelihood. More specifically, the stresses are felt most keenly on the overgrazed forest-fringe areas and on the upland dry-terraced fields. Here, poverty is the basic cause of poor management, and the consequence of poor management is deepening poverty.

3 Measuring land degradation

Mike Stocking

1 Introduction

1.1 The context

It would be good to believe that science is fact and that measurement is right. Indeed, so tempting is the thought of the neutrality of science (and the objectivity of measurement) that many who should know better – scientists for example – believe it, and those not in a position to judge believe it too. The white-coat syndrome is a powerful force, and nowhere is this more true than in the presentation of results of experiments and programmes of measurement. Measurement, however, is not an isolated process. First, somebody has to decide to do the measurement; set a working hypothesis for the measurement to test; choose a set of techniques; arrange a sampling programme and people to do the sampling; analyse the results and use judgement in the interpretation of these results; and decide how those results should be presented, and to whom. Then there is the recipient of the measurement who puts the data into context (or rejects them entirely) and who has to make value judgements as to the worth and applicability of the information. Finally, there is the end-user of the measurement; the person who makes the decisions, who bases a course of action on the results so presented. All these people have their preconceptions, misconceptions and ideologies. Therefore, measurement is never neutral, never a pure service for science or policy. To quote Weatherall: 'Man, as a scientist, is inescapably part of any experiment he conducts' (1968: 159).

1.2 What to measure and how?

Hanson (1973) describes the problem well when he notes that, when looking at a single object, scientists do not begin their enquiries from the same data, do not take the same observations and do not even see the same thing. We could begin, for example, with a case in Zimbabwe where 'common sense' and visual 'evidence' proved a poor guide to diagnosis of the problem of erosion in an area dissected by deep gullies. Gullies by their very nature are obvious features – you can break your neck falling into one! It is natural then that, where gullies exist, soil conservationists will attempt to tackle the gully erosion problem by check structures, stone-filled gabions, brush dams or whatever. This was the case in a heavily gullied part of Central Mashonoland, Zimbabwe; yet the structures all failed or had no measurable effect on the progress of erosion. A five-year monitoring programme on sediment sources and erosion rates revealed why. The gullies contributed only 13 per cent of the sediment; the rest came from inter-gully sheet erosion (Stocking 1978). Unlike gullies, sheet erosion is insidious and difficult to detect unless the observer has the eyes and experience to see it. As in this case, sheet erosion can be extremely serious, whereas the gullies dominate the observer's perception but are only symptoms of the wider malaise, the degraded state of the whole catchment. Conservation efforts were, therefore, addressing the wrong problem, the wrong part of the catchment, and the symptoms rather than the disease.

More than this can happen when measurement leads to explanation of a process and to an associated response. If one variable changes in association with another it is very easy to assume cause and effect, especially where the results seem to confirm one's prejudices. Again in Africa we may take another example, from the eastern highlands of Zimbabwe. Iron Age hut sites and pit structures occur extensively and are thought to originate from occupation by the Barwe-Tonga peoples who had been pushed out of the Zambezi Valley by the Manyika up to about the sixteenth century. Pit structures were the sites where the Tonga kept their cattle for protection, and usually the pits were encircled by huts. It is possible to gain a good picture of the density of human and cattle occupation of the area in the African Iron Age. Also evident is that those areas with more pit structures tend to have large gullies and evidence of past and present high rates of erosion. Does this mean then that the land-use activities of the Tonga caused the erosion? It is a tempting proposition, and given our knowledge of the effect of overgrazing on land degradation, it is not unreasonable to conclude that these Iron Age people degraded their land. Indeed, this is the explanation subscribed to generally in Zimbabwe.

However, consider an alternative: the gullies predate Tonga settlement and are an expression of geomorphological instability of parts of this landscape where the soil and slope conditions are conducive to the formation of gullies. Perhaps the Tonga were attracted to these parts because of the better protection afforded by the gullies against marauding tribes, and the better grasses and more open vegetation of the areas that are gullied. In other words, maybe the settlements and the gullies are not directly related, but instead each is related to other (unmeasured) variables. Under such an alternative explanation, it would be wrong to blame the Tonga for the erosion; they may merely have been rationally reacting to their environment and choosing the very areas where erosion also happened to be high. Neither explanation is proven and each would need detailed comparative research for verification.

In land degradation research, data collection itself is probably the major source of error. Some measurements are notoriously difficult. Suspended sediment loads, for example, not only vary with discharge but can vary even between equal discharges. A preoccupation with the implications of the data can, as Walling (1978) suggests, lead to failure to place confidence limits on the results.

Much data are also very poorly presented, and are practically unusable. Learned tomes line the shelves of planning offices but fail to be useful to those who work there. For example, land-resource surveys have been justifiably criticized on the grounds that they often collect facts about the landscape with little reference to the use of those facts and the relevance of the data (Moss 1978; Young 1978). Many are technically superb but are bulky, indigestible documents containing data of only dubious value. To argue that such studies are inventories of what is there, archival material waiting to be used, is to ignore that this costly luxury becomes outdated very soon. Land degradation is dynamic; it changes in response to a host of environmental variables, each of which may also change. Measurement has to have a purpose and trying to use measurements originally designed for another purpose is like wearing somebody else's suit – it may cover the body but rarely does it fit.

2 Problems of measuring land degradation

2.1 Two technical and methodological problems

First, there is the frequency/magnitude probem: how can a measurement overcome the vast difference between slow, continual changes on the one hand, and massive, infrequent changes on the other hand? Secondly, there is the scale problem: can measurements be taken at one scale, for example, in a laboratory or on a field plot, and then be extrapolated to whole fields, catchments, regions, continents or the world? These two problems are essentially technical ones of measurement.

The following problems are technical to a degree but concern the interface between the measurement and the user, and the extent that supposed factual statements on land degradation can be misinterpreted and abused. The first concerns data reliability and 'truth'. How can different interpretations arise from apparently the same data or the same objectives of measurement? There are also interpolation and extrapolation problems. What do experimental results really mean and can they be extended to unmeasured conditions with any confidence that they are technically correct and appropriate to local needs? Lastly, there arises the problem of relevance. Is the research currently being conducted in land degradation relevant to the needs of society?

2.2 The frequency/magnitude problem

Processes of land degradation occur at varying rates and with varying degrees of severity. Acidification of a soil occurs almost continually and imperceptibly, accelerating during periods of heavy rainfall and decelerating when evaporation exceeds leaching. Individual soil-creep and sheet-wash events may number more than twenty per season, the majority being minor but still representing at least 50 per cent of the total magnitude of the processes. At the other end of the scale, landslips are major catastrophic events, unlikely to be recorded more than once in a lifetime. Against such variation in rates and frequency of process, the measurement of variables of land degradation must concern itself with the frequency of observation, the spacing and regularity of observations and the overall sampling frame in time. To illustrate the quantitative importance of carefully considering the frequency/magnitude problem in degradation studies, take the case of the measurement of the suspended sediment load for the River Creedy in Devon, UK, reported by Walling and Webb (1981). They had available a series of hourly concentration data over seven years, representing as near to a continuous record as feasible. From this data base, they compared the accuracy and reliability of a range of sampling strategies that are in common use for estimating suspended sediment loads in streams against the actual load calculated from the continuous records. The sampling strategies interpolate sediment load on the assumption that the values obtained from instantaneous samples are representative of the intervening time periods between measurements. Depending on the sampling interval and the interpolation procedure, the calculated sediment load was found to be as little as 20 per cent of the actual load.

Why should there be such a discrepancy? The major problem is the extreme variability of the parameter being measured. Concentrations of sediment in excess of 100 mg/litre occurred for 5 per cent of the time, and in excess of 1000 mg/litre only 0.05 per cent of the time. The validity of sampling strategies that have many low concentration events and few, if any, high concentration events, thus missing the vital high magnitude occurrence, must be brought into question.

Considered in terms of the frequency/magnitude problem, the crudest level of measurement has been called 'incidental observation' by Thornes and Brunsden (1977). These are events which were recorded because an observer just happened to be there or because it was regarded as an unusual happening. Measurements from 'incidental observation' would also include fixed-time observations, the best example of which in land degradation studies is the use of periodic photography. In northern Ethiopia, Virgo and Munro (1978) used 1965 and 1974 photography to calculate that gully encroachment was of the order of 5–10 m/year. In order to calculate rates of gully erosion in Zimbabwe, the present author used a combination of time-lapse aerial photography, ground-based observations, District Commissioner's reports back to 1920 and interviews with elders in the local community to establish the position of gullies at memorable times in recent history (Stocking 1978). Such methods of incidental observation are crude, uncontrolled and subject to error. They can also be misleading in that two observations separated by a long period of time will spread out the importance of the extreme event, allocating the magnitude of that event equally to every year. Nevertheless, they can be valuable.

A second order of observation is the 'controlled observation' that has regular spacing in time but which attempts to measure irregular occurrences. Gardner's (1969) study of slope movements in the Canadian Rockies showed that true creep and subsidence are nearly continuous; rolling and sliding of rock debris are discontinuous and partly seasonal; while avalanching is definitely seasonal. Processes which are irregular in time and very rapid in occurrence – wind erosion, for example – need closely spaced observations over as long a time period as possible. As Thornes and Brunsden state 'The basic irony is that we need a lower density of sampling for regular events than for the very infrequent ones ... unfortunately in practice the reverse is usually all that we can afford' (1977: 57).

2.3 Scale problems

Can we make measurements of land degradation at one scale and from them infer rates of degradation at other scales? It would be convenient, for example, to measure soil erosion in a few representative locations, then calculate catchment erosion rates, and finally compute regional, country or even continental rates of denudation. The prospect is appealing; indeed, it is so appealing that several authors have attempted just such an extrapolation with strange results. Are they justified?

The short answer is 'no'. Rates of erosion measured at the field scale will, when extended to the scale of the catchment, grossly overestimate the total amount of sediment leaving the catchment. Field erosion rates are usually based on the results of experiments carried out on plots of a fraction of a hectare that are effectively isolated by solid boundaries. On a real slope soil movement is more complex and much is redeposited at some point, only to be moved again on another occasion. Therefore, the complete removal of one 'unit' of soil may require several storm events over an extended period of time, but on the experimental soil-loss plot it would have needed only one such event. Taking the argument a step further, soil is not only redeposited as accumulations of sediment on a slope, but it may also be stored in gullies and on stream banks. This leads to even greater overestimation of sediment output from a catchment from field plot measurements.

The importance of scale and the difficulties of making comparisons between measurements conducted at various scales are highlighted by Millington (1981) in a soil-erosion and sediment-yield study on two small drainage basins in Sierre Leone. Measurements were taken by erosion pins, literally, measurement at a point, along transects through areas of cultivation, bush regrowth and forest, by small plot techniques, and by monitoring basin sediment yields. There was not only a far higher basin sediment yield computed from the pin and plot measurements than actual sediment yield, but also a poor correlation between the erosion pins and the small plots. Each measurement has different aerial scales of operation and different temporal scales, thus making it very difficult to make absolute statements of sediment yield from the catchment without specifying the method of measurement.

2.4 Data reliability and 'truth'

Most of the examples discussed so far concern problems inherent in techniques and sampling, and in the comparability of measurements conducted by different methods. Faced with contradictory measurements which are you to believe? The one that proves your preconceptions? Or the most complicated and apparently technically superior measurement? Or the one that gives the neatest, cheapest or most satisfying solutions?

This problem may be illustrated by comparing world denudation rates as given on the global maps of suspended-sediment yield produced by Strakhov (1967) and Fournier (1960), and the estimates of regional erosion produced by Corbel (1964). Apparently consistent data bases give totally different patterns of erosion and orders of magnitude (Selby 1982; Walling 1984). Corbel (1964) summarized total denudation data for eight climatic zones in three humidity categories and two relief types. From this it appears that erosion is highest in the mountains and that there is a trend for the highest rates to be coincident with higher precipitation and temperate and cold climates, and lower rates in all tropical areas. Rates of denudation according to Corbel are of the order of 4000 t/km²/year in the most extreme environments, but only 0.2 t/km²/year on arid tropical plains. Strakhov's map gives sediment yields of 50 to 100 t/km²/year for most of the inhabited part of Africa (1967) (Figure 3.1). Based on an extrapolation of sediment yields of sixty river basins, Strakhov distinguished primarily between a temperate moist belt in the northern hemisphere and the humid zones of the tropics and subtropics. Fournier (1960) used seventy-eight drainage basins up to 1,000,000 km² in size, of which none are in Africa, and by means of correlating sediment yield with a climatic parameter (p^2/P, where p is rainfall of the rainiest month; and P is mean annual rainfall), he came up with totally different sediment yields. Not only are they of an order greater than Strakhov's but they also exhibit a different pattern from both Strakhov's map and Corbel's estimates. Walling's map (1984) (Figure 3.1) is different again but it is wisely accompanied by the comment that, 'it should not be viewed as a definitive map' (Walling 1984: 271).

2.5 Problem of relevance

Measurements have to perform a service. They have no intrinsic value unless they are used. To be used, they must be in a form capable of use and appropriate to the user. This is especially vital in the field of land degradation where measurement is the needed input to a whole range of decisions, from evaluating optimal uses of land, through the design and implementation of

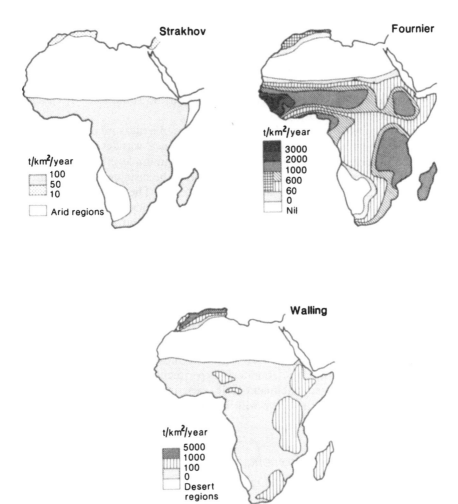

Figure 3.1 Maps of suspended sediment yields within Africa (from the work of Strakhov 1967, Fournier 1960 and Walling 1984)

conservation schemes, to the planning of rehabilitation projects for degraded areas.

Tell a farmer he has one hundred Bubnoff Units of erosion on his field, he might mistake the comment to be referring to a new breed of cattle – he certainly would never guess that his soil was eroding down by 100 mm per 1000 years (about 1.3 t/ha/year). Erosion rates are usually specified in t/ha/year or in mm lowering of the ground surface. Neither measure has any real meaning to a farmer or a planner; neither says how serious the erosion is or whether it can safely be ignored. Ten tonnes of soil eroded from a hectare may seem a lot and, indeed, may be very serious on some soils. It could, however, be inconsequential on a deep loess.

In an attempt to overcome the meaninglessness of a measure of soil loss in weight per unit area, the concept of 'soil loss tolerance' was developed by the US Soil Conservation Service to describe 'the maximum level of soil erosion that will permit a high level of crop productivity to be sustained economically and indefinitely' (Wischmeier and Smith 1978: 2). The tolerance level or T-value is an accepted part of the process of making land-use decisions in the United States but recently it has come under close scrutiny and criticism (Cook 1982; Schertz 1983), principally because the assigned values for T have no physical or research basis, and they have been accused of being too low by some (i.e. too restrictive on farmers) or too high (i.e. allowing too much soil loss). See Stocking (1984a) for a fuller discussion of T-values.

3 Types of measurement of land degradation

It would be impossible to review all aspects of measurement of degradation here. The main types are laboratory experiments, field measurements, field-scale experiments, catchment-scale experiments, and broader scale survey procedures. Of these, we will briefly review some of the advantages and disadvantages of each.

3.1 Simple field measurements for erosion

Because of their simplicity, it is easy to decry field measurements as being inaccurate, simplistic, subject to random error and hedged by too many assumptions. However, they take easily observable phenomena in the field in order to make rapid, direct and uncomplicated assessments of the rate of erosion. Tree root exposure and pedestal development have been used by several workers (e.g. Lamarche 1968; Rapp, Murray-Rust, Christiansson and Berry 1972; Dunne, Dietrich and Brunengo 1978), as well as more recently in a regional survey of Tanzania. Using the fact that many trees and bushes can be dated, it is possible, by measuring the difference in height between the present surface and the old ground surface, to calculate a mean rate of erosion per year. Thus for the Shinyanga Region of Tanzania it was found that over the preceding twenty years the mean erosion rate was 22.4 t/ha/year; between twenty and thirty years ago it was 10.5 t/ha/year; and between thirty and ninety years ago it was only 1.4 t/ha/year (Stocking 1984b).

Of the simple field measurements in use, erosion pins are the commonest. Pioneered in the mid-1950s by Schumm (1956), they are wooden or metal stakes driven into the ground to act as a fixed reference against which the lowering of the ground level may be monitored.

Such simple field measurements will never be fully respectable amongst the more technologically minded of the scientific fraternity, and quite valid criticisms can be made of the accuracy of measurements so obtained. However, the very strength of field measurements lies in the possibility of taking large numbers of them cheaply, with only semi-skilled technical assistance, and giving results that are probably more meaningful and visually impressive to the farmer and the extension worker than some super-sophisticated experimental facility at a distant research station. There would appear to be a lot of scope for development in this area of measurement of land degradation because increasingly it is the user of the land who must be convinced that degradation is a problem.

3.2 Runoff and erosion plots

Erosion plots were first used in the 1940s in the United States and their form and pattern remain largely unchanged today. Soil loss and runoff is collected at the foot of the plot in a trough and led away to a first, second and sometimes third collecting tank. The larger the plot area, the more soil loss and runoff will be collected. Depending on the infiltration characteristics of the soil and the local climate, plots above a certain size become unwieldy.

Once in operation it is normal to carry out standard treatments, attempting to hold most factors constant while varying only one that is of interest; for example, planting density, or direction of ploughing. In this way, over a number of years, it may be possible to isolate the influence of management factors, and thereby make recommendations as to appropriate practices that minimize soil loss.

As with all experiments, major areas of uncertainty remain with erosion plots. Inputs of erosive rainfall cannot be controlled. Conclusions based on a near-drought year are often overturned in a good rainfall year. Also, chance variations affecting an erosion plot can give spurious results which are often not identified because of the costs of replicates. Further, sampling techniques are by no means standardized. Strictly speaking the measurements are not absolute field-erosion rates; they are comparative, and they bring out the sometimes remarkable differences in soil loss between treatments. No viable alternative exists for erosion plots as the major data input to predictive equations of soil loss.

3.3   Prediction equations of soil loss

In agricultural development and the planning of soil conservation, it is often desirable to have an estimate of the erosion that would occur if a particular crop or treatment were to be applied to an area of land. Such estimates ideally should be part of the planning process, allowing a farmer to exercise personal choice in farming practices within specified boundaries of tolerable erosion. A farmer may choose then between, say, a perennial crop giving good year-round cover and no physical conservation measures, or a crop with poor cover characteristics and stringent conservation measures. Provided erosion does not exceed a predetermined level and provided erosion can be predicted from unmeasured and unexperimented conditions, then rational decisions about land use can be made. These are major provisos which will be examined below.

Soil-loss prediction techniques have been developed from the impetus of plot experiments conducted in the United States. With some forty-nine separate research sites, all producing an ever increasing body of data on soil loss under a variety of conditions, it became imperative that some means be available for gathering, collating and putting into a usable form the vast mass of information. A National Runoff and Soil Loss Data Center was established in 1954 at Purdue which processed some 10,000 plot-years of information into what became known as the Universal Soil Loss Equation (USLE). Refinements over the years have expanded the use of the USLE and added further detail (e.g. a soil erodibility nomograph: Wischmeier and Mannering 1969).

As its *Handbook* states, 'the USLE is an erosion model designed to predict the long time average soil losses in runoff from specific field areas in specified cropping and management systems' (Wischmeier and Smith 1978: 3). The form of the equation is given in most basic texts and will not be repeated here, but essentially it splits the erosion process into six major factors, uses variables to express each of these factors, and provides tables and design equations to calculate appropriate numerical values for the variables (and this is further discussed in chapter 5A, section 3). The end-product of the equation is a prediction of soil loss in tonnes/acre. At the very heart of the USLE is the data upon which the variables are calculated, data based upon regression equations derived from experimental field plots.

It is a fact that the USLE has served the US Soil Conservation Service admirably and has enabled farmers and conservationists to make informed predictions of the consequences of changed land use. However, its very attractiveness in being able to be computed easily has led to gross misuse of the equation. So worried was the equation's founder, Walter Wischmeier, that an article on the 'Use and misuse of the Universal Soil Loss Equation' had to be published (Wischmeier 1976). In it, Wischmeier warns of the extrapolation of equation factor values into unmeasured areas and severely cautions against the use of the USLE for storm calculations of soil loss and for catchment erosion.

As the most widely used and misused predictive measurement of erosion, it is timely to review again some of the problems surrounding the Universal Soil Loss Equation. First, there is controversy over its appellation 'universal'. The equation itself is universal only in the sense that it lumps together six factors, five of which are ratio measures. What are not universal are the equation's factor values: the values derived from experiments conducted predominantly in the Midwest United States on a specific range of crops, soils, farming practices and climate. The USLE is, therefore, a wholly empirical equation and is merely the vehicle for assembling and processing local data and regression equations based on that data. To expect the equation and its factor values to work in, say, the humid tropics is like expecting a plant which happily functions on the soil, air and water of the temperate plains of North America to be transplanted successfully to the steamy jungles of Borneo. It will not work, despite the fact that the plants of Borneo also function on soil, air and water. The equation *can* be made to work anywhere provided there is a sufficient local data base. (See chapter 5 for an attempt to supply that data base.) Consider that there are some 20,000 plot-years of data for the relatively uniform farming conditions of the Midwest United States, and still they are not satisfied, and it boggles the mind (and the purse) to start calculating the number of experiments that would be required in tropical farming systems with their variety of intercrops, cultural techniques and environments. Developing countries simply could not afford it.

Lastly, the USLE takes an historical soil-loss record to make predictions of future events. What happens as agricultural practices evolve through changes in technology, alterations in producer prices, shifts in input costs or whatever? The historical record becomes obsolete. The data base will always be a minimum of ten years out of date; i.e. the length of time it takes to create it. Whether that is serious is a matter of judgement and of the local environment.

These are fundamental problems with erosion measurements processed through a predictive equation of the USLE type. The record of difficulties and anomalies in the use of USLE is slowly increasing (e.g. Vanelslande *et al.* 1984, who found that erodibility values on a range of tropical soils were hopelessly miscalculated). Yet research workers continue to apply it in countries with only the sketchiest of data bases: Brazil (Leprun 1981), West Africa (Roose 1976), India (Singh, Babu and Chandra 1985) to name but a few. Perhaps it is justified to employ the USLE for a restricted range of conditions upon which local information is available, but to advocate it for general use is surely wrong.

Alternative techniques do exist. A model for local use, where data are limited and maximum flexibility is required, is the Soil Loss Estimator for

Southern Africa (SLEMSA) (Elwell and Stocking 1982). SLEMSA suffers some of the inadequacies of the USLE but claims to be within the grasp of many data-poor countries and combines several factors into what are termed 'rational' parameters; for example, the seasonal amount of kinetic energy that a growing crop intercepts – an expression of the interaction of vegetation, rainfall and some management practices such as planting date, which has been demonstrated to be of vital significance in the seasonal tropical rainfall regime of Africa. Looking to the future, process-based models (e.g. Foster, Meyer and Onstad 1977) will become more important because they will have the ability to interpolate and extrapolate to unmeasured conditions and they should have far more generalized applicability than the data-hungry models of today.

3.4 The FAO's Soil Degradation Methodology – an example of the broad-scale survey

In 1975, the United Nations Environmental Programme (UNEP), UNESCO and FAO initiated a major project to develop a methodology for measuring soil degradation with, as a first objective, a pilot programme to map North Africa and the Near and Middle East at a scale of 1:5 million giving present soil degradation rates and the present state of the soil, and the soil degradation risks. As the only undertaking of its kind attempting to measure soil degradation, this project is worth highlighting if only to exemplify the very real practical difficulties of combining a diverse set of processes into one document. The most substantive statement of the methodology and presentation of provisional maps is given in FAO (1979).

Six categories of soil degradation are recognized: water erosion, wind erosion, excess of salts, chemical degradation, physical degradation and biological degradation.

Clearly, many of these processes interact but for the purposes of the model of soil degradation they are taken independently of each other. Chemical, physical and biological degradation will often occur simultaneously, and water erosion will usually always result in loss in structural stability of the soil and physical degradation: these are but two examples. Nevertheless to make the methodology workable they are treated separately, recognizing that, although each will reduce soil productivity, they do so in very different ways which cannot be isolated at the present state of our knowledge. Space forbids any further detail on this approach which is a fascinating exercise in systematic number crunching. To their credit, the designers of the methodology eventually lump the detailed calculations into broad class categories which help to hide some of the technical inadequacies in the original data.

Nevertheless, the methodology suffers weaknesses. First, it is very ambitious in taking many processes and trying to present them together in one uniform format. Can one impose uniformity on non-uniform processes with different scales of operation and different relative magnitudes? Secondly, the parametric approach gives equality and independence to each process where inequality and interrelationships could well dominate. In reality some processes in some areas will be more important in limiting productivity than others. In addition, it is conceivable that the coincidence of two processes could be far more devastating than the sum effect of the two processes. For example, water erosion and high leaching could lead to severe aluminium toxicity and almost total nutrient depletion. Thirdly, information requirements to work the methodology are high and many areas would not even have potential evapo-transpiration records. One is forced further away from the direct measure of interest (such as t/ha of soil loss) to proxy measures, the validity of which is unproven in the environment concerned. It is a dangerous measurement procedure liable to multiplicative errors. Lastly, it is a complicated and daunting procedure.

It is easy to criticize the Soil Degradation Methodology but hard to suggest alternatives. Perhaps its major strength is not in the numbers and categories that come out at the end, but in the way in which it focuses attention on the range of degradation processes and at least ensures that each is considered and estimated. It is a pioneering attempt that one suspects may not be repeated very often but it shows an admirable degree of interdisciplinarity which is also all too often lacking in scientific measurement.

4 The case of erosion-induced loss of soil productivity

Intuitively many people recognize that soil erosion is an evil. Siltation of reservoirs is often cited as a major result of erosion, and it is known that soil conditions and crop growth become worse as erosion progresses. In early conservation writings the dangers of soil erosion to soil fertility, agricultural productivity and farmers' livelihoods were often spelt out (e.g. Huxley 1937; Jacks and Whyte 1939), and the Dust Bowl conditions in North America made it obvious that degradation was serious. But the impact of this evidence (and other) of the severe consequences of erosion was only temporary and slight. The reasons? – lack of firm, quantitative evidence, plenty of emotional rhetoric and a notable absence of measurement.

Evidence for loss in productivity comes from many sources, including worldwide land productivity trends, sometimes offset by the application of new inputs, but which are frequently shown to be giving declining marginal returns through time (Brown 1978), the impoverishment of shifting agricultural systems, the proven relationship between erosion and crop yields and the declining nutrient levels in eroded soils.

However, all these lines of evidence are indirect, suggesting that there is a link but not conclusively proving it in a manner which is both demonstrable to decision-makers and convincing to ordinary farmers. The evidence and associated measurements suffer from at least four problems.

First, productivity and erosion are not independent, and do not change discretely in isolation of other factors. As one changes, so does the other, and because of the multivariate characteristics of each, so do many other factors change. Evidence can be confused and measurements are ambiguous. Secondly, there are other factors responsible for yield decline and productivity losses, and, although these factors may themselves be related to erosion, they are distinctly separate processes. Thirdly, erosion rates by themselves are poor indicators of loss in productivity (e.g. Larson, Pierce and Dowdy 1983). The very measure on which there is considerable data in the form of t/ha of soil loss is poorly representative of productivity decline. Some soils may suffer much erosion and be relatively unaffected, while others need only a very small quantity of soil loss to decline dramatically in yield levels. Finally, technology tends to mask the decline in land productivity (see Krauss and Allmaras 1982).

Therefore, we have what is, on the one hand, a pressing problem that urgently needs measurement and data in order to kindle a response from persons who may be able to halt the erosion-induced loss in soil productivity; and on the other hand we have confusing and ambiguous snippets of evidence which find a ready ear in the conservationally converted but a deafening silence amongst those who have least to gain in the near future in heeding the message, i.e. commercial farmers, agribusiness and developed countries. Where is the fault? At least part of the blame can be attached to agricultural and related research, and to the lack of adequate measurements.

In a recent review of the erosion–productivity relationship (Stocking 1984a), four major problems were identified with regard to existing investigations and experimental work. First, there is a general lack of innovation in techniques of experimentation. The same tired approaches feature again and again – runoff plots, research station trials, etc. Secondly, there is a failure to relate experimental work to practical uses in extension, economic planning and further research development. Measurements are mute, and, therefore, any notion that the results should speak for themselves must be a resounding failure. Thirdly, United States research has dominated the whole field of soil erosion and yield losses with a consequent stultifying effect elsewhere and an unfortunate and totally mistaken assumption that the results can be transferred to the developing tropics. There is, in fact, evidence that yield losses under temperate conditions are at least an order of magnitude less than under tropical conditions with the same degree of soil loss and equivalent soils (Stocking and Peake 1986). The conditions causing an innocuous yield loss in the United States might well be a disaster on a tropical soil. Finally, most erosion research worldwide goes towards collecting soil loss data from experimental field plots. The automatic assumption is that soil loss is a good and direct proxy for productivity loss. As already discussed above, it is not. Narrow disciplinarity goes so far that on some research stations agronomic experiments are conducted measuring yield and other vegetative parameters, while on plots close by soil loss is being measured but not yields. It is not surprising, then, that measurements of practical value are rarely forthcoming.

5. Conclusion

Measurement has a strangely ambivalent role. It is expected that the measurement will produce the right answers to and inspiring insights into problems such as soil erosion, salinization or sedimentation. Yet, at the same time, measurements are necessarily preconditioned by the experiences, preconceptions and prejudices of the observer. There is, therefore, an inertia which prevents the perceptive analysis and innovative answer. New measurements of land degradation are largely prisoners of their historical development with a preconditioning that precludes flexibility in dealing with a changed world and worsening degradation. Science, though, has always progressed in a series of lurches, but it does seem that measurement and experimentation in environmental-cum-socioeconomic topics such as land degradation lag far behind. Maybe we are due for a big lurch in degradation studies. There could be no more appropriate field than the whole question of declining productivity, food security, economics of soil conservation and the social consequences of allowing soils to erode.

4 Decision-making in land management

Piers Blaikie and Harold Brookfield

1 Scale: spatial and temporal

1.1 The significance of scale

The importance of the explicit treatment of scale in the study of land degradation and society has already been mentioned in chapters 1 and 3. In this chapter the issue is discussed more fully. There are three reasons why the scale implications of any analysis of this subject should be clearly specified. First, the geographical scale of the 'unit of account' (e.g. field, farm, areas of more than 10° slopes, watershed, nation) is usually tied closely with the decision-making process over land use, conservation and degradation. Central to much of the social analysis of land degradation are the questions, 'Who decides to conserve, and how?' There is seldom a neat one-to-one correspondence of geographical scale and 'level' of decision-making. It is tempting to equate decision-making at the farm level with the family decision-making unit or farm manager; the common pasture with the village; and the national territory with the national government. This equation is sometimes valid, but it is often not so easy.

Secondly, any evaluation on the part of either analysts or land users themselves must demarcate very clearly how the costs and benefits of degradation and conservation are to be accounted. The 'upstream–downstream' effect is important where either beneficial silt or costly floods or layers of gravel are the benefits or costs from upstream erosion (see also chapter 5A). However, the assumption commonly made that the downstream benefits of erosion usually equal or outrun the costs upstream must be resisted. Even at the level of an individual field, there can be spatial displacements of costs and benefits. It may take the form of removal of soil and nutrients from the top of a field, and their redeposition at the bottom. Land mangement may take the form of physically transferring soil back to the top of the field. At the level of a whole farm, degradation of one part may be offset by enhancement of another. A simple illustration is the removal of leafy matter from woodlots in order to feed livestock which supply manure to the fields. On a local, or subregional scale, soil removed by erosion from a steep hillside may find lodgement on gentler slopes below, where it can more readily be used. However, decision-making with regard to the use of woodlot and arable land is tied together by the farming household which exercises its choice in both areas and types of land use. In other cases costs and benefits flow to different land users altogether.

At the opposite or subcontinental scale, the loss of soil and dissolved nutrients from Ethiopia has for millennia enriched the soils of Egypt. A more accountable situation occurs where accelerated erosion has a detrimental impact outside the area of origin, as maybe in the case of Nepal (although this is in dispute). If, for a moment we assume economic rationality and co-operation to be dominant in international relations, it could be argued that India should consider paying for part of Nepal's conservation effort since it will be India's population in the Gangetic plain which may be the most numerous beneficiaries in terms of reduced flooding and an improved supply of irrigation water. Also the high mobility of certain ions provokes salinization of wetlands and even drylands at a considerable distance from areas in which this mobility has been augmented by forest clearance. While the 'geographical transfer of value' may thus acquire a new dimension, it is very evident that we must take care to define the scale at which we are working if the social causes and consequences of degradation are to be described adequately. Indeed, the distribution of costs and benefits is of crucial importance to the land managers themselves, and relates to the first point above. As a comprehensive review by Amos (1982) shows, it is very difficult to justify many soil-conserving practices for the individual farmer on economic grounds except on very productive land which is not already damaged but which is threatened with large reductions in capability. Therefore, conservation measures, particularly those which involve costly land-management works, frequently require other and more centralized decision-making and additional sources of funds or resources other than those which can be raised at the local level. It also frequently requires the co-operation of numerous private property owners (everyone has to agree to take a collapsing hillside of *sawah* terraces out of irrigated cultivation, for example), and in cases of common property the co-operation of members of common property management institutions.

Thirdly, it is important to understand the implications of scale, because the scale at which the analysis is pitched tends to affect the type of explanation given to land degradation. This aspect has been discussed by Blaikie (1985a: 80), and only a brief mention is made here. Obviously, the variations of slope, soil, and moisture availability and vegetation over even a few metres have a primary impact upon decision-making in land management for the land users themselves in terms of cultivation techniques, cropping strategies, types of terracing or bunding, and the type or extent of repair work needed when faced with a soil slip or other less acute event. When governments become involved, land-use planning formalizes these variations and plans for them in terms of mapping and legislation or securing agreement from the land users themselves. In cartographic terms, this scale may be 1:5000 in the former case, or for governments interested in planning for watershed management 1:50,000. Here the focus is primarily upon physical variations of land capability, the sensitivity and resilience of the land and upon the detailed relationships between land-use practice and the biome. At this scale, the explanation is largely a physical and technical one provided by the natural sciences.

However, as one proceeds along the chain of explanation, there are also other considerations which move the analyst away from the detailed scale and physically specific focus, and away from those who make the direct land-management decisions. Settlement histories and the spatial unfolding of political economy are often important in explaining the variations of the social relations of production, and of important cultural aspects of agriculture and pastoralism, which impinge upon the range of choices land managers have. The occurrence of spatial marginalization of producers is a case in point and is explored in detail in chapter 6, section 3. In these cases, their decisions are constrained because they have been confined to smaller and less fertile areas by other more powerful groups. Pastoralists may no longer have access to land now irrigated and forming part of a rural development project or a plantation. The settlement histories of many parts of Africa have resulted in tribes of very different social organization and cultivating ability inhabiting neighbouring and physiographically similar environments. Yet, an analysis of the decision-making environment requires a broadening of geographical scale from the local and immediate variations of people and the land, to 'regional' concerns such as these.

Lastly, there are non-place-based or non-location-specific networks of economic, social and political relations acting directly and indirectly upon land managers. These relations therefore usually do not have a geographical expression at all but nonetheless have important repercussions upon the decisions of the land manager. Land tenure and reform, taxation, prices of inputs and outputs of the agricultural system, soil conservation programmes or logging concessions – all these can be relevant factors in the process of agrarian change for land managers. Some of these come about as the result of conscious decisions made at the state level, particularly so in centrally managed economies. However, these networks too may produce geographical patterns of degradation on effective land management, and as such are an essential consideration in regional political ecology.

1.2   The temporal dimension

Scale is measured not only in spatial terms, but also in time. The origins of degradation may lie in deforestation one or two generations ago, the immediate effect of which was the creation of arable land or an increase in the carrying capacity of livestock. These beneficial effects may have endured for years before erosion, changes in hydrology or soil structure, the movement of mobile ions, and the invasion of new biota begin to wreak real damage on the capability of the land. In many cases of degradation there are substantial lag effects measured over a period ranging from months to centuries. Although there are dramatic and classic examples of newly cleared land that is already losing centimetres of soil per year, together with most of its carbon and nitrogen, and in which trace elements are quickly assuming toxic levels, they represent only a part of the whole spectrum of degradation. We encounter such clear examples later, particularly on recently cleared land in Fiji (chapter 9), but we also encounter other examples where a less dramatic degradation has taken one or two centuries to become visible as patterns on the land (chapter 8B).

Not only may degradation as a whole be fast or slow, but so also may the processes contained within it. The discussion of human versus natural causation of erosion in historical times, parts of which we touch on in chapter 7, has been bedevilled by some confusion between the debatable effects of long-term climatic variability and the more demonstrable effects of short-term events such as storms and snowmelt. Both may be significant, but in different ways. An Australian puzzle illustrates an aspect of the problem. It has been observed that while there is abundant evidence of erosional landforms and soil loss over large parts of the continent, little of the resultant material seems to be reaching the sea to supply the beaches, many of which are being starved (E.C.F. Bird personal communication). R.J. Wasson (personal communication) suggests an answer. Whereas the fine products move rapidly, the coarser products of erosion move much more slowly and many of them have not yet proceeded far downstream; they are choking river channels which spread widely in response to a massive increase in bedload, yet depositional jetties even in sheltered waters at the mouths of the same rivers are not currently being augmented. It will be centuries before much of the material eroded from the catchments over the past 200 years reaches the sea. Yet in other parts of the world, in Madagascar for example, a much more rapid sequence seems clearly evident.

The search for social causes of degradation must also extend backwards in time; moreover, some of the degradation that might arise from present failings in land management will not become evident in our time, or even the time of several more generations. Where there are major time lags in the physical erosion processes, there must also be time lags in social causation. But while slow processes should yield abundant time for adaptation, and undoubtedly have done so in many parts of the world, their very slowness carries with it dangers if the damage is not perceived. The example of the contrast between degradation in Ethiopia and northern Thailand, developed by Hurni (1983) and presented in chapter 1, provides a clear case in point; Ethiopia is less sensitive to erosion than northern Thailand, but is also now less resilient, so that long and slow erosion works a more deadly effect. From our present point of view, however, we need to note that while the causation of current erosion in northern Thailand can be sought in the practices of the current generation, the causation of the Ethiopian problem needs to be sought in the cumulative agricultural practices over a period of centuries. It is so also in many other areas which are now shown to be heavily damaged; if processes are slow it is an error to attribute the damage only to current practices.

Generally, the largest spatial scale also involves the longest time scale in the complementary achievement of transformation in two widely separated regions. Within a single region, however, Mixtec farmers in the Nochixtlan valley of Mexico have adapted erosion processes to enhance the capability of other land. Erosion gullies from the sensitive and unresilient uplands have been fed on to terraced valley fields, so that over about 1000 years the main valley floors have been widened from about 1.5 to 3.0 km (Kirkby 1972; Whyte 1977). In Barbados, three centuries of sugar cultivation have modified the soil so greatly through marling and manuring that it is itself 'almost an artefact' (Saint 1934) but at the same time have permitted the movement of soil from now bare coral-limestone slopes into solution hollows where it sometimes has a depth of many metres. None of this has happened quickly and the whole history of land use has to be taken into account in explaining the present situation.

The definition of degradation advanced in chapter 1 (page 7) is essentially relevant only to a single site or small area, over a limited period of time. It does not incorporate 'imports' which are the 'exports' of another site, nor does it allow for the effect of long time-lags. The creation of landesque capital in a bygone time – the restorative management or prospective management of an earlier generation – forms part of the human aid to reproduction of capability, but it is not the work of the present managers of the land. Almost any long-used site will have experienced several different systems of management, each of which will have had its beneficial or detrimental effect. Such temporal externalities affect the nature of the present land manager's task, and provide a set of conditions that are not within the control of the present generation.

1.3 Nested scales of explanation

Clearly there is no 'correct' scale for an investigation of land managers and their decisions, but there is an appropriate one for answering different questions. Frequently a comprehensive enquiry into land management will require an approach which employs a nested set of scales: local and site specific where individuals or small groups make the relevant decisions; the regional scale involving more generalized patterns of physiographic variation, types of land use, and property relations and settlement history; the national scale in which the particular form of class relations give the economic, political and administrative context for land-management decisions; and the international scale, which, in the most general manner, involves almost every element in the world economy, particularly through the commoditization of land, labour and agricultural production. How those pervasive processes of agrarian change impinge upon land managers at the local level is to some extent the result of a mediation through the state structure and the various interests which use and work through it. Thus, any particular enquiry about the circumstances of land degradation may well have to ask and answer questions on a number of scales which fit inside each other like a set of Chinese boxes.

An important distinction must be made between decision-making *per se*, and those factors which merely affect the range of choices of the decision-maker. For example, a family farm may be the decision-making unit, and it will have the freedom to manage the land as it wishes, and to choose between different options. However, the parameters of choice are usually controlled by the decisions of others. A landlord may set a rental for a piece of land, a government may fix prices of inputs or outputs and so on. These others make decisions too, but these decisions only indirectly affect land-use and management decisions. It is therefore very common to find that *direct* decision-making is frequently local, for example, the manager of a sugar plantation or the peasant farm household, but many of the parameters of choice are determined by others, for instance locally by a landlord, centrally by corporate management of a group of plantations, or nationally by government parastatal boards.

In summary, the significance of scale for the study of land management rests upon: (a) a clear understanding about who, at what level, makes land-management decisions; (b) a statement about the unit of account for assessing costs and benefits of these decisions; (c) an appreciation that different scales of study tend to focus on different elements of land degradation; and (d) a proper understanding of the time-lags involved in both physical and social processes.

In the next section, a formal decision-making model for land use and management is presented. One of the most underspecified aspects of decisions to use and manage land is the decision-making process itself. The model below indicates clearly that there can be many *different* options taken by the direct land user and manager which do *not* involve land conservation and intensification of production at all. Indeed, quite the opposite in many instances, and this situation frequently produces outcomes which run counter to those posed by Boserup and discussed in chapter 2, section 2 and following sections.

1.4 Decision-making and land management

One approach to the way in which cumulative land-use decisions can produce degradation has been outlined by Blaikie (1985a: 107). That model focuses on why soil erosion occurs rather than on the decision-making involved to stop it. Figure 4.1 focuses on a different set of decisions and indicates a simple decision-tree which traces through the stages in decision-making if and when the capability of the land declines. In box 1 are listed the social and environmental data at time *t* which form the initial *desiderata* for land-use and

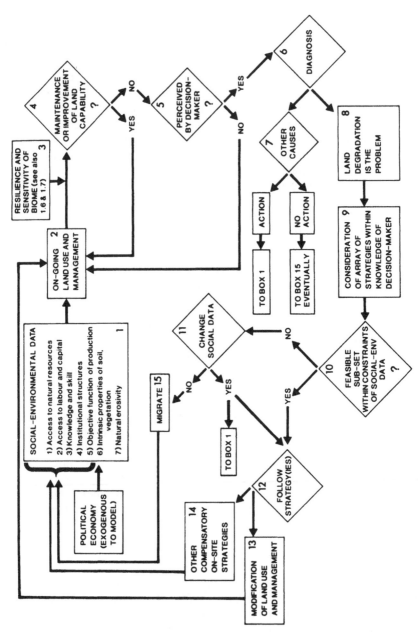

Figure 4.1 Decision-making in land management

management practice (box 2). The data consist of socioeconomic characteristics of the decision-maker(s) (box 1, points 1–5). Here the importance of access to resources by the decision-makers is central in determining land-use and management decisions. This has already been discussed earlier in chapter 2, section 3. Also the intrinsic properties of the land system (box 1, point 6) and the erosivity and erodibility of environment (box 1, point 7) are essential elements. The principles which govern decision-making over land use and management (the arrow between boxes 1 and 2) have been theorized mostly by economists. Basically, land use is conceived of as a matter of allocating resources of different relative scarcities to achieve or maintain an income stream from a piece of land. In each empirical case the objective function of the decision-maker will have to be specified. Typically one such is income maximization, perhaps modified or replaced by some considerations of the risk of income dropping below some critical level, or merely a target income along the lines originally suggested by Chayanov. Rather different are those models which indicate the circumstances under which land managers make decisions to *manage* the land, and to provide additional resources, or to forgo additional income to minimize or repair degradation, particularly those decisions which emphasize control rather than avoidance strategies. In these cases it is easier to distinguish land management from land use (see chapter 1, section 3). These models are concerned with present investments to maintain or enhance a future income stream. They typically involve a cost–benefit calculation. The decision-making tree in Figure 4.1 traces the outcomes of land-use and management practice as a function of the social and environmental data (box 1). The political economy is exogenous to this model, although extremely important, since it determines and provides the dynamic for changes in the agrarian structure, which is immediately reflected in changes in the circumstances of the land manager (listed in summary form in box 1). For example, access to natural resources may be restricted by a government forestry programme or by private encroachment. A Small Farmers Development Project may improve access to capital, and land reforms access to land, and so on. These change the *desiderata* and may change land use and management too, and they all derive from the political economy of agrarian change.

The pattern of on-going land use and management may or may not maintain or improve land capability. This will depend upon the resilience and sensitivity of the land system (box 3). If land capability is maintained, the cropping system and land management will remain unchanged (unless there is a prior change in the social-environmental data, as stated above). If capability declines, it may or may not be perceived by the decision-maker. Hudson (1981) says 'my experience is that the small farmer has little idea of how much soil he is losing', while many others report an intimate knowledge of what is happening. Richards (1985) for West Africa, Johnson *et al.* (1982) for Nepal, Berry and Townshend (1972) for Tanzania, all offer clear examples of the latter state of affairs.

If the decline in land capability *is* perceived, the next step is diagnosis (box 6). It can be attributed to causes quite other than land degradation. Homewood (1985) affords an excellent example in Baringo District, Kenya. The plateau and hill area surrounding the Baringo swamp have long been deteriorating due to overstocking and physical erosion, and this is clear for all to see. Advisers diagnosed the problem as pasture degradation caused by overstocking and have proposed destocking, group ranches and so on, but without much success. The Il Chamus pastoralists do *not* perceive the cause of their parlous state to be erosion, and a reduction in the capability of the dry land areas on the hills and plateau was, in their view, beside the point. Most of the Il Chamus cattle are based in the swamp area for most of the year and only move away on to the eroded uplands for very short periods. The swamp itself supports grasses (particularly *Cynodon* and *Echinochlon*) which are virtually impossible to graze out (i.e. highly resilient). Therefore the main land resource on which the Il Chamus depend has not declined in its capability. There is undoubted overstocking, but the Il Chamus identify the problem to lie in the differential ability of the various social groups using the swamp to rebuild their stocks after drought-induced losses. Absentee cattle-owners, often with office jobs and other sources of income, are able to purchase cattle quickly to replace losses while local people cannot, with the result that the Il Chamus actually own today far less cattle than twenty years ago – hence their poverty. Thus, theirs is quite a different perception of the problem and has not led to altered land use and management on their part.

If the diagnosis is that land degradation *is* the cause (box 8), an array of strategies which lies within the knowledge of the decision-maker is considered (box 9), and a feasible subset of strategies is then identified (box 10). This process will depend on a number of factors in the social data set. Many farmers will not have the labour or capital to invest in physical soil conservation works (as we have seen in Nepal, for example), but may be able to conserve moisture and soil by inter-cropping and other less demanding strategies. Others may not like to grow trees or maintain grass-covered contour strips because of a shortage of land. Also there may not be the institutional structures to ensure co-operation between farmers, without which a particular strategy carried out by an individual land manager will not be successful (e.g. the regulation of cattle on pasture, or desisting from the removal of trees from a communal forest). If no feasible subset can be identified, it may be possible to alter the social data (box 11) to enable a strategy to be adopted. The extent to which this is possible will depend on the economic and political power of the decision-makers. For example, landlords and, in the past, the landed aristocracy of many feudal societies compelled their serfs to construct terraces or plant trees. Modern farmers may seek advice from extension agents to enlarge their knowledge, or form co-operative work teams for conservation works, forest planting and so on. If none of this is possible, land capability continues to decline and the land managers, or some of them, will eventually be forced to leave (box 15).

If feasible strategies can be identified, then they are followed (box 12). Modifications of land-use practice and management take place (box 13), which are input into on-going land use and management (box 2) at time $t+1$. However, there also may be other compensating on-site strategies which do *not* involve modifications of land use and management, and the choice made will depend upon the access position of the decision-makers, and the perceived costs and benefits to them. For example, if land is available, extension of cultivation is another strategy. Alternatively, petty commodity production to provide cash to make good subsistence shortfalls is a common strategy for small peasant households faced with a reduction in the capability of their land. Wage labour may be another alternative. These compensatory strategies usually have implications for the social and environmental data set at time $t+1$. However, although they are responses to land degradation, they do not necessarily involve modifications of land-management practice.

This simple decision-making model allows us to draw a number of conclusions. The first is that uni-causal models of land degradation, intensification or de-intensification will founder, a point already made in chapter 2. Although a number of the more sophisticated models account for changes in land management in many important cases, there are always significant and widespread exceptions. For example, population pressure or decline can lead to all sorts of strategies other than intensification of land use or other land-management innovations, as we saw in chapter 2.

Secondly, and related to the first point, governments who seek to intervene to prevent land degradation may have to alter a great many of the parameters in the decision-making process in order to achieve their objectives. Of course, the major one is to try to ensure that the perceived private benefits outweigh private costs over a discount period appropriate to the decision-makers involved. However, extension work may also be necessary to enhance the knowledge and skills of land managers, and other institutional changes made. Governments' perceptions of land degradation may disagree with land users', as in the Baringo example above. Jamieson (1984) and Dove (1984) provide two excellent examples of 'government versus peasants' in perceiving environmental degradation in Indonesia. The history of colonial conservation in the British colonies of Africa is also full of such cases (Anderson and Millington 1985; Ranger 1971; Robinson 1978) and these are further discussed in chapter 6. Again, land users may come to other conclusions than that which holds that land degradation is the real problem. Political conclusions, not about land management at all, were reached by the Kikuyu in response to British colonial soil conservation efforts and there was a similar or less violent response throughout much of British East Africa (Fosbrooke and Young 1976). Also the land-management strategies have to be technically appropriate, and fit in with the patterns of access to resources and land use of the managers themselves.

Thirdly, it is important to identify both who the decision-makers are and their social and environmental data. This sounds commonplace, but it is often a considerable task. One of the most difficult elements is to identify precisely who it is that perceives, diagnoses and considers strategies. It may well be that both private individuals and local institutions of various combinations are the decision-makers. In addition, other decisions are made at a further remove which only *indirectly* affect decision-makers. Even if the state intervenes strongly in land management, provided that the land users have effective means of *choosing* to carry out alternative strategies, they are still the land managers. One essential element in the analysis is to identify the land managers and their 'social data' and this is pursued in the next section.

2 Land management: who is the manager?

2.1 Relations of production, scale and the decision-maker

A useful way of identifying the various decision-making environments in which land managers operate is to analyse the social relations of production in which they are involved. These relations are primarily defined by who has control of land, labour, implements, inputs and outputs; who decides upon cropping or grazing strategy, and upon investments, including, in this case, the creation of landesque capital, irrigation, tree planting and the like; and the type and rate of surplus creation and extraction through rents in labour services, cash or kind, usury, or through the employment of wage labour. These relations are often underpinned by particular property rights. In the rest of this chapter, the impact of changing relations of production upon the land manager is discussed, following a broadly historical route. First of all pre-capitalist relations of production are examined, with a case study from the Chimbu of Papua New Guinea. This is followed by a review of the transition from feudalism to capitalism in Europe and a discussion of the implications of an 'interdependent world' upon farmers in the developed world.

Many land-management decisions in developing countries are made in groups about resources which are held and managed in common; chapter 10B discusses an extended example in India. It remains here to provide a few briefer examples. In some areas of the world, particularly in many parts of Africa where 'customary tenure' exists, in pastoral areas there and in Asia, and in Melanesia, simple common property management without full private property rights applying to any major natural resource can be found. Elsewhere, as in most of South Asia and Latin America, private property often applies to arable land while forests and pastures remain as common property nominally owned by the state. In the following example, a decision-making arrangement prevails whereby most of the land-based resources are managed both communally and privately.

Among the Chimbu of the highlands of Papua New Guinea cultivation is an individual affair on land that is at least notionally inheritable and transferable between individuals. Yet land is also held in group territories and many aspects of management are organized by the group and between groups. Chimbu hold a large number of individually owned pigs which are fed from the owners' crops, but which are also allowed to browse and root in unenclosed and common land. The croplands and the common land need to be separated by strong fences. Croplands are tilled and ditched by hand, and some areas remain almost continuously within the fences. But the system depends in part on land rotation for reproduction of capability, aided by the planting of *Casuarina* which fixes nitrogen through root nodules, and permits a richer ground flora to flourish in its shade. A fence has a life of about four years before it begins to become rotten and require replacement, and often its position is moved when it is rebuilt by collective decision either to take in, or to fence out, large blocks of land (Brookfield and Brown 1963; Brookfield 1973).

Though pigs add manure, they also do much damage by rooting, and in landslip-prone areas they may also undermine fences. Within the fences in such areas, maintenance of deep field-drains is also necessary to prevent the saturation of the ground. A section of fence may collapse and the pigs invade the crops. The health of such a system of management, without either systematic crop rotation or manuring, depends on the achievement of co-operation between individuals, small groups and also larger groups. Where such co-operation fails, management is weakened, and evidence of degradation begins to emerge.

The role of co-operation in a rural community is a topic of great complexity. It enters areas of production and consumption, social control, religion and ceremony; land management is only one of its functions. But land management may be an important function, and has been neglected in the literature. Among the Chimbu, we could say that the land manager is sometimes the individual man or woman, sometimes the leading man in a small social group, sometimes a tribal leader, but sometimes none of these. Much of the 'land manager's' role is fulfilled by no one person, but by a process of collective decision-making that takes place around the fire at night, in casual meetings, and by the decision of others to follow the example of an individual. However, without the more than partial institution of private property, the technical demands upon the management of production tends to encourage common management in many instances. One Chimbu's failure to maintain his section of the fence, or to drain his own land adequately, may cause a breach or a landslip which affects many. Management responsibility is immediately shifted from the individual to the pressures of the group. But the very need for fencing, and for a division of land between the enclosed and unenclosed, arises from the large herds of pigs which are raised mainly for prestation purposes. Here, the responsibility is that of society as a whole. Both the problem and its management are diffuse, but they extend beyond the highest level of political organization. In order not to overstrain total resources, whole tribes need to co-ordinate their pig-raising, and hence their prestation cycle. Whereas the largest political groups in the pre-colonial highlands of Papua New Guinea numbered only 8000 to 12,000 people, and most were much smaller, such co-ordination extended across populations of 100,000.

In this case the co-ordinated management both of private and common property resources is involved. Clearly it is unwise to generalize about where the role of the decision-maker lies in the management of common property resources. Indeed, the problems of providing mutual assurance in management points to contradictions and tensions over *who* manages and decides – the individual or small group. Thus the decision-making model in the previous section can and should be extended to include possible conflicts in the perception, diagnosis, consideration of strategies and final choice of strategy(ies).

The production of surpluses in pre-capitalist societies has taken many forms. The system of prestation controlled by 'big men' has just been described, and is found to be basic to the explanation of degradation in chapter 8B. Few societies have ever been without a need for surplus production in order to sustain systems of prestation, exchange and trade, however rudimentary. Where prestation has been the object of production, it has been common for the goods to have high values in terms of the resources, especially labour, required for their preparation. The genesis of labour- and skill-intensive systems of land management may therefore, at least sometimes, be traced in the production of prestation commodities high in labour value for prestige reasons (Brookfield 1972). The motivation for trade and exchange is different; it is that of widening the available range of commodities, or of obtaining commodities at a lower cost than that of local production. The consequence may therefore be specialization, to the extent of throwing strains on the task of land management.

Pressures of a kind likely to affect management of land in both beneficial and detrimental ways can thus arise without private property and a land-owning class. Both kinds of pressures can occur in the same area if labour is massively diverted into intensive production of a prestation or trade commodity to the neglect of other land. Thus, again in the South Pacific, we find the physical remains of ancient terraces and massive yam-mounds in an environment widely degraded by the use of fire for more basic shifting cultivation. The 'management' which has produced this contradiction is that of a contradiction of social goals, with or without the emergence of a class of chiefs to control and benefit from surplus production.

The production of surpluses under conditions of private property in medieval Europe differed in their mode of resource management from the rural societies described above principally by a greater degree of formality and by a clearer emergence of a class of landholders. The responsibilities of individual members to the community, and the rights of such members, were adjudicated by a council, manorial court or parish vestry, bodies in which the larger landholders were the dominant voice. Some of these larger landholders owned their land, but others, and about four-fifths of all villagers in medieval England, were tenants or subtenants (Clay 1984). Some of the landowners also farmed part of their land and managed it directly, but other land belonged to the king, church or aristocracy, whose interests lay in the extraction of income to support their activities.

At the higher levels the principal intervention in land management was in the allocation of land, including the reservation of large forest areas for hunting of wildlife; the allocation of village lands into field and common property rested on custom which became enshrined in law. This was also so in Asia where, however, the state adopted a further role in large-scale water management, creating the so-called 'hydraulic societies' of Wittfogel (1957). In Asia as in Europe, however, detailed management of resources remained at village level. But the larger landowners and the state also acquired another role as market economies spread within and between organized states, and as towns and cities grew to serve trade, industry and government.

Almost all European and Asian farmers and farm-workers, however little they possessed, had some need of a form of currency with which to buy services and pay taxes. A minority of commercial farmers evolved wherever society was sufficiently organized to permit trade, division of labour and the emergence of non-agricultural populations in towns, temples and courts, and itinerant pedlars of goods and services. The commercial farmers found the protective restrictions of village production systems irksome and in some regions achieved a large measure of individualization of land and its management. Much greater pressures, however, arose from decisions by large landlords and rulers to embark on specialized production for trade. Pressures were then put on the peasantry themselves, by engrossment of their common property resources, increases in their rents and other payments, shortening of their leases and even the sale or resumption of their land, leading to their eviction. In eastern Europe, where the state added its legal force to the demands of the landlord class, peasants were forced back into serfdom in order to provide labour for their owners' enterprises as late as the eighteenth century. Almost everything that later happened to peasant societies under colonialism and the invasion of capitalism in the Third World had its precursor in some form in medieval and early modern Europe.

At its least, the landlord–tenant relationship was changed from paternalism, where it had that form, into exploitation. Scott describes it for a more recent Southeast Asia:

> In one area landlords might refuse customary pre-harvest loans, in another they might insist on the full rent in a bad year, in another they were no longer lenient if the tenant fell ill, in another tenants were replaced by more solvent competitors who required less assistance, in another share cropping gave way to fixed rent. The signs varied enormously to suit the peculiarities of each region, but they all pointed in the same direction. (1976: 67)

In most of the transitions from feudalism to capitalism in Europe and in rather different circumstances in Asia, the peasant remained the decision-maker for the land that the household cultivated. However, the social relations of production along with state interference involved multiple forms of surplus extraction. Over these the peasant had very little or no control. None the less, the choice of land use and management (or other compensating strategies) remained in the peasants' hands. It is also true that a rising agricultural bourgeoisie appropriated much of the land and became the individual decision-maker over that land instead, but with a less constraining set of production relations and therefore a wider range of options for land use and management.

The longer term effect in Europe was the introduction of improved husbandry, leading to greater yields and better land management, but the effect in contemporary Asia is very much more mixed. Pressures on the peasants in Europe led to the growth of a proletarian labour force for the better management of land by the remaining owners and tenants. But a period of change led to distress that is extensively reported, and this distress was probably also accompanied by damage to the land. Unfortunately, we have only one researched historical example of such a linkage, which is discussed later in chapter 7.

2.2 Management in an interdependent world

Regional interdependence on a large scale, leading to specialization and supported by trade, began a very long time ago, and was characteristic of all the ancient empires (Polanyi, Arensberg and Pearson 1957). There was substantial interdependence within late medieval Europe. However, it was only with the 'great age of discoveries' that interdependence on a world scale was initiated, and it was only in the nineteenth century that it began to affect land use in one way or another in almost all parts of the world. It is important to recognize that the transformation took effect at very different dates, and in different ways and at different speeds. Except by new settlers, it was resisted almost everywhere either actively or passively, and this record of resistance frequently involved, and had a deep impact upon, land-management decisions. The incorporation of most agricultural production into the world economy involves not only a global scale of analysis, but also a regional one as the pace and form of this incorporation varied across space and time.

Incorporation has been most drastic where plantation economies have been set up, and where specialization in export production has become the dominant activity of peasant farmers. This has happened most forcibly in areas newly settled on a large scale only during the colonial period, specifically to take advantage of new marketing opportunities. Two areas of this type are lower Burma and the Mekong delta, where Scott (1976) notes great and increasing inequality of landholding, a uniform cash-based tenancy, active money-lending, and exposure to the world market as mediated through a chain of middlemen. But while it is true to say that the cash nexus that enchained both peasants and landlords, and even the finances of colonial government, found its controlling points in the boardrooms of metropolitan countries, effective control over management was more local, residing in the hands of principal landlords and moneylenders, the Rangoon and Saigon offices of the principal banks, and the offices of colonial government. What is significant in these examples is exposure to the full fluctuations of world market prices without significant protection either from a surviving subsistence sector or the relative autonomy of traditional social relations of production.

The impact of the market on the rural producer is discussed in more detail in chapter 6 in the context of colonial rule. Obviously the 'objective function' of the production unit has changed, although this has varied according to the extent of entry into the market, size of farm and amount of assets and so on. Frequently this change has been in the direction of expanded production and profit. There are two aspects of the deepening of capitalist relations which deserve particular mention – agro-technology and farm organization.

Both the use of purchased fertilizer and the emergence of wholesalers in the marketing of agricultural produce can be traced back to the end of the Middle Ages. The emergence of corporate businesses in agriculture, however, came much later and their real origins seem to have been in colonial regions in the nineteenth century rather than in regions of older agricultural development. Plantations, initially run by their owners, later often by managers working for absentee owners, fell into the hands of trading companies through bad debts, and were operated as production units in a single corporation. The deliberate establishment of corporate plantations soon followed and became the dominant pattern in the great surge of plantation enterprise that took place after 1870.

The tree-crop or sugar, banana or pineapple plantation as it evolved in corporate hands was structurally different from the earlier slave plantations in a number of ways. Evolving in an industrial era, the modern plantation applied industrial forms of organization to all its operations, with vertical integration between production, processing, transport and marketing, and often also the supply of inputs including fertilizer and machinery. The empires of such corporations as Unilever, the United Fruit Company, Bookers, Guthries, Harrison and Crosfield, Dole and the like, as well as a host of smaller companies with only a few estates, managed their farming operations in a manner that had no real historical precedent. Their success, notwithstanding the changes in ownership that have taken place in recent times, is evidenced by the continuation of plantation management and the use of the plantation model in planned settlements and land-development areas in several parts of the world: the outstanding example is undoubtedly the Federal Land Development Authority (FELDA) in Malaysia.

The corporate mode of management has also been applied in other contexts. In pastoral regions, such as inland and northern Australia, chains of 'stations' are owned by companies – and now also by Asian governments – who manage the range by moving livestock between properties according to pasture conditions. Fruit and vegetable production in the southern United States is dominated by large companies which also maintain 'offshore' farms in Mexico and the Bahamas. Corporate management of specialized farms has also evolved within regions still dominated by freehold and tenant farms in Europe and other regions of old settlement; more generally, however, corporate control in such areas has evolved in a different way.

The pattern is well analysed by Smith (1984) in a study of marketing and control in the agriculture of southern Quebec. Farms are contracted to dealers who are in turn contracted to a small number of wholesalers, several of whom also operate major national and multinational supermarket chains. Price, quotas and delivery dates are laid down, and in effect determine all major production decisions taken by the farmer. This pattern is now widespread throughout the agricultural areas of such regions as North America, western Europe and southeastern Australia. Inputs to farming are only to a limited degree less centralized, and a high proportion even of land-management decisions is now partly out of the farmers' or farm managers' control. Parallel systems have been entered into voluntarily, as in the dairy co-operatives of western Europe, and have been supported by governments through marketing boards.

The effect of the growth of agribusiness on the family farm in North America has been analysed in some detail by Vogeler (1981), who argues that family farming is now a myth, preserved as such for the benefit of the larger interests which have taken effective control of American agriculture.

> Given the high cost of supplies, the low prices farmers receive, and the monopolization of the farmers' market, the choice is often between signing with a corporation or going out of business.... Once contracts are signed, processing companies usually make all the technical and market decisions about planting and harvesting and provide the necessary farm supplies – fertilizers, pesticides and harvest machinery. (Vogeler 1981: 141, 138)

Vogeler does not extend his analysis to the effect of these changes on the management of the land; some others have done so, as we shall see in chapter 12A, but not in a systematic way. Since the same trends are evident in Europe and many other regions, such an analysis would certainly be timely.

A different form of control over management has accompanied the 'Green Revolution' in Asia. Farmers participating in schemes such as the one in Indonesia are instructed what varieties to plant, and how much subsidized fertilizer to apply and when. While these instructions may not strictly be followed and have not always been successful, important elements of management are removed to a higher level. Growing in importance, too, is the partnership between agro-chemical multinational companies, large-scale finance companies and the governments of developing countries in the agricultural sector. Erstwhile peasants are involved in a variety of arrangements from the mere buying of new inputs with extended credit, to outgoing arrangements where the control of production decisions is in the hands of this partnership rather than that of the farmer. There are, furthermore, new settlement schemes where the small farmer is little more than a wage labourer and the control of the production process as well as the disposal of the surplus is determined by a local manager of a state or private corporation (for a summary, see Feder 1977; Frank 1980).

It is very difficult to assess the impact upon land management of the vertical integration of decision-making which corporate capital has tended to create. Certainly Dinham and Hines (1983) have documented some spectacular examples of land degradation as a result of speculative and short-term ventures by multinational companies in Africa. Much of the decision-making of large corporations with many estates will be determined by their discount rate for the benefits of investments in land management and this will depend upon a whole host of factors that only can be guessed at, since they will never be published. Political stability of the host governments where estates have been established, long-term prospects for the prices of the commodities produced, and the overall profitability of the corporation as a whole are three sources of uncertainty and can discourage investment in effective land management.

There are plenty of examples to show good and bad land management arising from capitalization of farming. For example, growing demand for quality wines in Alsace has led to recent capital intensification of vineyard methods, often removing old terraces in the interests of use of machinery, which is used more easily up and down slopes. The result is quite substantial erosion after only moderate summer rains, as in 1985. Across the Rhine, farmers in the Kaiserstuhl have created massive new terraces for their vineyards, with strong financial backing. The new terraces are of a size economic for the use of machinery. The consequence of somewhat imperfect terracing has been quite severe slumping, and it is clear that further investment in landesque capital is required in order to stabilize the new system. As the decision-making model has shown, it is not possible to make generalizations about the quality of land management in terms of one causal factor. In this case, the tendency for more centralized or non-local control of land use has been a major contributory factor, but this is not always the case. One may put forward the notion that the effective land-management decisions are rarely taken in government offices or company boardrooms. The decisions taken in these places are concerned more with production than with conservation of the means of production, without local knowledge and without the same imperative of long-term sustainability of production that a largely subsistence producer would keep in mind. However, the empirical verification of this generalization would be extremely difficult to sustain. Both the local decision-maker and the corporation have so many options in managing land or taking other strategies, that outcomes are extremely varied for both groups.

2.3 The state and land management

There is one further aspect of the scale and degree of centralization of land-management decisions that needs discussion, and it involves the state. The growth of government under capitalism is a remarkable phenomenon, the origins of which do not concern us here. We could say that it began with the earlier growth of absolutism, and more certainly that its massive role in economy and society came sooner and more pervasively in colonial territories than in the metropolitan countries, as chapter 6 will demonstrate.

In most countries the means available for intervention are limited, and are primarily technical. Soil conservation services were set up in many countries, in one guise or another, earlier this century, and in the United States particularly they have done major work, which we discuss in chapter 12A. Land managers have been advised, cajoled, assisted and sometimes compelled to follow technical advice. However, state bureaucracies lack the detailed information which is required as well as the authority on the social aspects of farming problems. We could say that one of the purposes of this book is to provide such information, but we have to note that it rarely exists except through localized research. Although of growing importance, therefore, the direct role of governments and their agencies in land management is still confined mainly to the technical problems of conservation and repair. There is, however, no doubt that it must grow in the future. But the challenge is immense since governments have to alter so many aspects of the social data of land managers and have to provide the means by which technically effective site-specific strategies are willingly carried out.

States which call themselves socialist contain about a third of the world's population, and claim a great deal more control over the production process than non-socialist states. While the degree of autonomy remaining in the hands of the communes, co-operatives and state farms, or the remaining individual farmers, differs greatly from country to country, and has also varied through time, management is in general much more centralized than in the so-called 'market economy' countries. Unfortunately, this centralization of management has not generally operated well from the point of view of management of the land. Whether the basic problem is bureaucratic bungling on a large scale or the habit of Marxian economics to consider land as a free input into production is open to question and is discussed in more detail in chapter 11. Certainly Komarov (1981) for the USSR and Smil (1984) (and chapter 11 in this book) indicate that the destruction of the capability of vast tracts of land occurs in the socialist countries as much as in the rest of the world. Along with the other problems of assessing the degree of land degradation, and its attribution to human agency, there is a strong element of ideological special pleading on both sides of the divide. A reading of Pryde (1972) or Tregubov (1981) about the situation in the USSR in simply not factually consistent with the presentations of Komarov (1981) or Brown (1982), the latter drawing particular attention to the impact of inflexible centralized decision-making upon land degradation.

A reader ideologically inclined to the left may put forward the notion that under 'real' socialism, even if that could be defined and agreed upon, the necessary co-operation between producers themselves, and between them and a democratic and representative state, would be easier to obtain. However, in all societies, there will continue to be conflicts between private and collective interests, between local and and national priorities in land use and management. Empirical proof would be useful here, but it is beset by so much that is still hypothetical. This is not to despair of social solutions, but to acknowledge that it is likely to be some time yet before any government can effectively operate as direct decision-maker in the management of land.

3 Conclusion

It is important to identify who makes the decision to manage land and how it is made. A decision-making model is useful in that it shows the large number of possible outcomes generated by an initial decline in the capability of the land. An analysis of the social relations of production of land users and the role of the state indicates that many *other* decisions are made which provide the context in which direct decision-makers operate. Sometimes this context is so constraining that it is a fine point to identify precisely who decides. The extension of the market economy and the penetration of centrally controlled institutions of private capital and the state, particularly but by no means only the centrally planned socialist state, makes this identification problem more severe. The issue of scale is also important because it helps to identify the many different levels from which relevant inputs into decision-making derive, and it helps too to focus upon the distribution of costs and benefits in land management amongst those who are involved. It is to the latter question that we turn in the following chapter.

5 Economic costs and benefits of degradation and its repair

A Issues in the economic evaluation of soil and water conservation programmes

David Seckler

1 Introduction: conceptual and empirical issues in soil and water conservation

The basic issues in the economics of soil and water conservation (SWC) are not so much of an economic as of a philosphical and empirical nature. Philosophically, there is a strong school of what may be called 'SWC fundamentalism' which contends that since all life depends on soil, conserving soil is worth whatever it costs. If this is true, then it is not necessary to estimate either the worth or the cost of SWC programmes. In this way SWC fundamentalism has led to serious neglect of the empirical base of this important field to the point that it is usually impossible to estimate what either the costs or the benefits of SWC programmes actually are.

The philosophy of SWC fundamentalism is implied, ironically enough, in the opening sentence of a report on 'Assessment and evaluation for soil conservation policy', 'The soil is a basic resource on which all life, including human survival and development depends. Unfortunately, it is also a resource that is being squandered by all nations of the earth' (Perrens and Trustrum 1983). When one begins with this premise the economics are lost.

SWC fundamentalism stems from a number of different modes of thought (O'Riordan 1981). Some of these are non-economic, involving bio-ethics and an eco-centric approach to the world, but a strong contributory factor in the social sciences is the logical fallacy which confuses the *total* value of something with its *marginal* value. This fallacy may be illustrated by the famous 'Diamonds and Water Paradox' that bedevilled economists until the latter part of the nineteenth century, after which it was laid safely to rest. 'How', it was asked, 'can diamonds be so expensive, when they are intrinsically so worthless; while water, which is vital to all human life, is so cheap?' The reason is because diamonds are scarce relative to demand, and therefore the value of one diamond more or less is very high. Water, on the other hand, is usually plentiful relative to demand, and therefore the value of a few gallons more or less is very low. If someone is dying of thirst he would, of course, gladly trade a diamond for a gallon of water, but most people are not dying of thirst.

So it is with soil. The value of soil depends on the supply and demand for a little more or less soil in specific locations, at specific times. A kilogram of soil for a potted house-plant may cost several dollars. Billions of tonnes of soil may wash out of the Amazon basin to the sea and not make any difference, except to fish. As this example indicates, erosion may in certain circumstances be beneficial. People who do not like the Aswan dam protest that it stops the flow of 'rich, sediment-laden, flood waters' on to agricultural land. In addition to the fact that the 'rich alluvial plains' that feed most of the world were created by geological erosion, a good part of the agricultural land in mountainous areas of Asia has been created by sediment traps and sediment deposition through irrigation, sometimes abetted by deliberate destruction of groundcover in the watershed to accelerate natural rates of erosion. As a good sceptic has observed:

> Soil erosion is a natural process which merely reflects the removal of water and soil from upland sites where both are, at best, indifferently used and their redeployment on lowland sites where both can be utilized more efficiently. (author unknown, cited in Soemitro, Anwar and Pawobo 1983)

In sum, it may be concluded that while soil and water are indeed vital to all life, this fact is irrelevant. Both soil and water are often in such plentiful supply relative to demand that the marginal value, while positive, is not very high. Thus conserving soil and water, far from being worth whatever it costs, may *not* be worth its costs. This is especially true if one considers the downstream benefits, as well as the downstream costs, of water runoff and soil erosion. On the other hand, there clearly are cases where investment in SWC is both necessary and desirable. It is the task of economists, working together with other physical, social and engineering sciences, to determine when the investments in SWC are likely to be desirable, and when not.

But even after the conceptual problem of SWC fundamentalism has been overcome, there remains the formidable problem created by its residue, as it were, of the neglect of the empirical basis of the field. In most of the developing countries actual rates of soil erosion and soil formation are not known; the effects of erosion on plant growth are even less well known; and the effectiveness of SWC techniques in reducing erosion and flooding are not known. As was rightly observed in a UN Conference on Desertification (1977: 177; cited in Blaikie 1985a: 15), 'Statistics [on soil erosion and deforestation] are seldom in the right form, are hard to come by and even harder to believe, let alone interpret.' (For a general discussion of this problem see chapter 2, and for a report and evaluation of efforts to improve the measurement of land degradation see chapter 3.)

Since very little is empirically known about SWC in developing countries, it is nearly impossible to devise rational, cost-effective SWC programmes in these countries. A concrete example of this problem was encountered by the United States Agency for International Development (USAID) Mission in Jakarta in planning for the Uplands Agricultural and Conservation Project (UACP, which is discussed further below) with a large SWC component. Naturally, if one is considering a project to reduce soil erosion it is desirable to know how much erosion there is, within some reasonable degree of error. The Mission searched the literature on erosion in Java and found estimates of erosion for the *same* areas differing by several orders of magnitude. The *locus classicus* is cited in Hamer (1980: 39) where two different groups studied the rate of erosion from cultivated land on the upper Solo basin of Central Java. One group estimated the rate of erosion at 50 t/ha/year; while the other group estimated it at between 1800 and 4800 t/ha/year! Similar discrepancies in estimates of rates of erosion have also occurred in Nepal, as chapter 2, section 4 has demonstrated, and some of the technical reasons have been explored already by chapter 3, section 2.

This example illustrates not only the problems created by estimates of erosion varying by factors of 36–96 in the same locations and conditions, but the utter nonsense sometimes created in the name of science. A depth of 1 cm/ha of soil weighs around 128 t. Thus, if the rate of erosion was in fact 4800 t/ha/year, then an average of *38 cm* depth of soil would be lost annually to erosion in this area! No crops could grow on such rapidly eroding soil. Further, these figures were evidently obtained by measuring sediment load in a river and then assuming that a high percentage of the sediment was deposited in the riverbed upstream of the measuring point. Assume for the purposes of illustration that the factor used was 10 per cent (i.e. 90 per cent upstream sediment deposition), and that the streambed above the measuring point amounts to 1 per cent of the cultivated land area being eroded in the watershed. Then the level of the streambed would be rising at a rate of [(38 cm × 0.9)/0.01], i.e. 34.2 m/year!

After contemplating such figures the Mission decided that the amount of erosion in the project area simply was not known, and because of this and other empirical problems discussed below, it was impossible to do a realistic appraisal of the SWC component of the project.

In this way SWC fundamentalism becomes its own worst enemy. Policy-makers are usually not fools, and they usually will not invest millions of dollars in projects justified in terms appropriate only for an evangelical preacher. Indeed, through guilt by association, SWC fundamentalism discredits the sound scientific work in this field, and has probably led to underinvestment in SWC in developing countries. Given the poor state of empirical knowledge, nobody can say how much investment is required, where, by which techniques, and to what effect.

The following statement from Hudson summarizes both the problems and the opportunities in SWC.

I believe that in the aid organizations too, there is increasing realization that agricultural development programs should always include a soil conservation component. A major difficulty is that both national governments and international organizations find it difficult to justify programs unless they can show a profit, or the right benefit–cost ratio, or the appropriate internal rate of return, or whatever measurement is used.

We should use a two-pronged strategy against this problem. First, we need to promote the idea that soil conservation should be placed in the same category as improving a nation's health or strengthening its education, desirable objectives that do not need to show an immediate cash profit.

Second, if the aid agencies require that the benefits of conservation be expressed in terms of cash values, then we must provide such information. A small army of graduate students and economists are working in this area at the moment. It is hoped that usable models on soil conservation economics will emerge (1983: 450).

Once the red herring implicit in the first prong of this strategy is eliminated, the second prong can develop a sound, scientific basis for rational SWC programmes in the future. The rest of this discussion in concerned with the means to that end.

2 Basic economic concepts

SWC programmes change the distribution of soil and water resources in space and time. In evaluating the benefits and costs of these programmes, there are four basic factors to consider. The first two refer to the spatial characteristics of costs and benefits – *on-site* and *external* or downstream; and the next two refer to the temporal characteristics – the *present value* of projects in which benefits and costs extend into a reasonably well-known future, and the *option value* of actions that extend into an essentially unknown future.

In principle, the benefits and costs of the spatial distribution of resources at a particular moment in time are rather easy to estimate, if the data are available, which they rarely are. These spatial benefits and costs are discussed in considerable detail in the following sections. In this section the focus is on the more conceptually difficult temporal dimensions of resource allocation, especially the problem of conserving resources in the present for future use. (See also the discussion in Barkley and Seckler 1972: part two.)

The first factor is estimating the present value of a future stream of costs and benefits through discounting. Discounting in effect allocates resources from the future to the present by making a unit of revenue (or cost) that accrues in the future worth less than if it accrued in the present. The formula is $A_1/(1+d)^t$ where A is the amount in time t; and d is the discount rate. Thus, for example, at a 12 per cent discount rate, a $100 net return fifty years from now is worth only $0.35. Such is the insidious power of compound interest.

Naturally, economists' use of discounting creates concern among people like foresters and SWC specialists who are involved in programmes with long-term payoffs. Even some economists believe that discounting should not be used (at least for their favoured programmes) on the grounds that discounting taxes future generations to subsidize present generations. On this point it should be noted that since discounting is a charge for capital, and capital is resource intensive, discounting favours the use of renewable resources, especially labour, over non-renewable resources. From this perspective, at least, discounting is the friend of conservation. This issue is returned to later, but for now there are two basic points to be made.

First, without discounting of *some* kind it is impossible to put programmes with different time horizons of costs and benefits on a comparable footing; and the ability to compare programmes that draw on common resources of time and money is essential.

Second, if in fact resources will be more valuable to future generations than to the present generations, then conventional economic analysis in effect adjusts the discount rate to include this fact. For example, if it is believed that scarcity of certain kinds of forestry products will cause their real value to increase by 12 per cent p.a. over the future, then these increasing values would be entered into the stream of future benefits of the economic calculation and wholly offset the effect of the discount rate, i.e. the present value of $100.00, twenty or fifty years from now, would still be $100.00. As this example shows, discounting forces one to be *explicit* about the projected value of resources in the future.

But it is clear that discounting does in a sense impose a tax on the future. If the choice is between using a tonne of coal, or some other non-renewable resource, now or in the future, it must be shown that its future use is sufficiently more valuable than its current value to justify sacrificing current for future consumption. Under discounting, the present in effect demands that the future pay interest on its savings (i.e. the difference between what the present could be consuming and what it does consume). What is wrong with that?

With this question one goes beyond economics into ethics. For example, one might accept the general ethical principle that present generations should take no actions which will make future generations *worse off* than present generations, while, on the other hand, there is no reason to take actions which will make them *better off*, if that entails making the present worse off. If this ethical rule is accepted, *and* if the future can be reasonably well forecasted, then the appropriateness of any given discount rate can be empirically evaluated. Further, over reasonably forecastable periods for market commodities the market rate of interest (minus a certain amount for the risk incurred by individuals that society as a whole does not bear, or the 'social rate of discount') probably provides a good approximation to the correct discount value. However, there are classes of goods that do require special treatment under the general ethical rule above. These are goods with 'option values'.

An option value is like an insurance policy, it is a *premium* one is willing to pay, over and above the normal cost of an operation, in order to be able to reduce risk and uncertainty. In the field of SWC, for example, it simply may not be known how much, if anything, a certain amount of soil at a particular site will be worth to future generations. But it may be known that without a SWC programme the soil will be eroded away and, thus, the *option* for future use will be lost. Certainly, if it does not cost anything to preserve this option it would be rational to do so; the option *itself* has an intrinsic value. But if it does cost something to preserve the option, then someone has to decide if the option is worth the cost. While economics can assist in making this decision (for example, by doing sensitivity analysis of the situation with and without soil under various technological and economic assumptions) it cannot, even in principle, estimate the *value* of the option, because the future is inherently unknown in the long term.

The short-run, long-run knowledge distinction is particularly important in the economics of SWC. In the short run, say over the next twenty years or so, conserving soil to provide livelihoods for the rural poor and for national food supply is clearly desirable in most of the developing countries. In the long run, say two or more generations from now, it is not at all clear that the need for soil in marginal areas will be nearly as acute as it is today. The food surplus problems of the developed nations are now also becoming the rule in some of the developing nations of Asia (especially in rice). Also, in most of the developing nations the percentage of the population that obtain their living from agriculture is decreasing, and in some countries the agricultural population is decreasing even in absolute numbers. The major exceptions to these generalizations are Nepal and Bangladesh in Asia, and many of the African nations. They are exceptions because of the failure of economic growth, largely because of political and social problems, not resource problems. By the end of the next generation population growth will have slowed considerably, and, quite possibly, there will be major breakthroughs in biotechnology to produce more food from much less land. After this transition period, the land may revert to natural pastures and forests, as it has already in much of the United States (especially in the northeast) and in much of Europe. At that time the problem of SWC in marginal areas will automatically be solved.

But again, we do not *know* that this long-run scenario is valid, or if so, when it might occur. If it is incorrect, or delayed for a considerable period of time, then there could be a severe shortage of soil resources, especially for providing the livelihoods for poor and otherwise unemployable rural people. Lack of present investment in SWC measures would then be an irreversible mistake that the future would regret. If it were possible, future generations, with *their* knowledge, would be willing to pay the present generation to invest more in SWC than we do on the basis of *our* knowledge. This is the essential meaning of an option value.

Thus, after the conventional economic evaluations of the benefits and costs of SWC programmes based on discounting are finished, there remains an option value that should be entered as a benefit. The amount of this option value cannot be estimated by economists. That is a task left to decision-makers who presumably reflect the ethical and other values of society as a whole. However, while economic analysis cannot define the *value* of the option it can define the *cost* of the option.

If a project has a favourable rate of return on the basis of *known* costs and benefits alone, then the cost of the option is zero. Nothing is given up by having the option, if the project that carries the option is desirable on other grounds. In this case, the option is 'frosting on the cake' of the project. On the other hand, if the project does not carry a favourable rate of return on the basis of known costs and benefits, the *cost* of the option is the amount of additional benefit necessary to make the project favourable. For example, if a benefit to cost (B/C) ratio of 1.2 is considered favourable, and the B/C ratio in the basis of known values alone is only 0.9, then the cost of the option is equal to 0.3 of the known costs. After this computation is finished decision-makers can decide if the project is worth doing or not, i.e. whether the value of the option is at least as high as 0.3 of the project costs, or not.

It is possible to carry this analysis one step further by way of a mental experiment that illustrates both the use of discounting and option values. It is conventionally assumed that soil is an irreplaceable asset which, once lost, cannot be produced or recovered except by the slow processes of natural soil formation. But this assumption is clearly incorrect: soil can be both recovered and produced if one is willing to pay the cost of doing so. Thus eroded soil can be gathered in sediment deposits and transported anywhere. Soil can be produced on particular sites by induced erosion from upstream sites and sedimentation. It is also conceivable that C and D strata could be converted into soil by on-site milling, with chemical nutrients and humus added as desired. Soil, in other words is *not* an irreplaceable resource like an animal species which, once it is extinct, is (so far as we know) lost for ever. Therefore, the *maximum* value of the option to conserve soil for the future is the present value of the cost of replacing/producing it at some future time when it might be needed.

Assume, for the purposes of illustration, that a particular site has 10 cm/ha of soil that probably will not be needed for at least twenty years but which is eroding at a rate such that it will be completely eroded away in twenty years. Further assume that the estimated future cost of replacing/producing this soil, together with nutrients and humus, is $10/t. Then, if no SWC techniques are used, about 1280 t/ha would have to be replaced/produced in twenty years at a cost of $12,800/ha. The present value of this amount at a 12 per cent discount is $1327. Under these assumptions, this would be the *maximum* amount of the option value to conserve this soil for twenty years, i.e. this would be the *most* that could be paid for SWC treatments. Of course, if the soil had a productive use over the twenty-year period, or if the erosion caused external damage, then the maximum investment would be correspondingly higher.

Thus it is possible to bracket the value of SWC programmes, including option values, between minimum and maximum values by such considerations. Once these brackets are established and the costs are specified it is up to the decision-makers to determine whether a particular SWC programme is socially desirable or not. This is true of all economic analysis, and has nothing to do with any peculiarities of SWC programmes *per se*.

With these basic concepts in mind, the rest of this discussion concentrates on some of the more important details of estimating the costs and benefits of SWC programmes. This discussion is largely based on material from a section of a Project Paper prepared by the writer for the USAID funded 'Uplands Agricultural and Conservation Project' (UACP) in East Java. While the details are specific to this project, the methodology and issues are relevant to all SWC projects.

3 The agricultural benefits of SWC programmes

The on-site agricultural benefits of SWC programmes (including trees, grasses, etc.) are of three basic kinds. First, the preservation of soil on-site as a medium for nutrients, water and plants; secondly, the prevention of loss of the natural nutrients in the soil; and thirdly, enabling better agricultural management practices through the above two factors, and through such items as cultivating flat areas on bench terraces rather than on steep slopes. All of these benefits depend, directly or indirectly, on the relationship between three basic parameters: (a) the net rate of erosion before and after a SWC programme; (b) the critical depth of the soil; and (c) agro-climatic and crop systems conditions, which in turn affect (a) and (b). Calculation of these parameters is beset by uncertainty and even orders of magnitude cannot always be treated with confidence (see chapter 3).

The net rate of erosion is the rate of loss of soil minus the natural rate of soil formation. In temperate climates, on non-agricultural land, the natural rate of soil formation is about 0.8 mm/year, or 10.0 t/ha/year. However, in the humid tropics the rate of soil formation may be three times higher, 2.4 mm/year or 30.0 t/ha/year (Hamer 1982: 5). It would perhaps also be higher in agricultural land because of greater mechanical and chemical activity. However, there are immense variations in the resilience of land systems throughout the world and in particular in both the impact of soil loss on yields and the natural regenerative powers of the soil, as chapter 3 particularly emphasizes. This is another example of an important lacuna in the present state of the art of measuring the impact of erosion. Be that as it may, these figures are the best available for the estimation of on-site effects. Transportation of soil from areas of low valued use to high valued use through erosion is also an important though neglected factor, but even more difficult to measure.

The critical depth of soil is the minimum depth at which crop yields are significantly reduced. This is mainly a function of the soil-moisture holding capacity of the soil in relation to agro-climatic conditions, mainly precipitation, evapo-transpiration and the root depth of plants. Adventitious roots, which provide most of the nutrient uptake, occupy only the upper 15 cm or so of the soil. Tap roots, which provide most of the water uptake, vary by crop but can extend as far down as 2 m or more under moisture stress conditions. Thus moisture stored in the C horizon can also be used by crops. How much soil depth is needed for these nutrient and moisture storage functions obviously depends on agro-climatic conditions and crop systems, and, as chapter 3 has indicated, this vital area of research still cannot give us reliable estimates. Hamer (1982: 4) has calculated some appropriate relationships between loss of soil depth and yields. Using these estimates, and assuming some amount of fertilizer application to compensate partly for loss of soil nutrients, and favourable agro-climatic conditions, something like 15 ± 5 cm soil depth may be taken as the point where yields begin to be reduced significantly due to erosion in much of Indonesia.

To estimate the rate of erosion, this analysis follows Hamer's (1982: 3) excellent discussion of the Universal Soil Loss Equation (USLE) under Indonesian conditions (see also Stocking's comments on USLE in chapter 3, 58–9). As Hamer and Stocking in this book both note, the USLE has only been verified for temperate climates and cropping system, medium textured soils, and slope gradients of 3–18 per cent.

The USLE may be written:[1]

(1) $E = f(C, S, T, L)$

where E is the average annual erosion (t/ha/year); C is the climatic factor, using Hamer's estimation from precipitation data; S is the soil factor: a classification based on soil texture, organic matter, structure and especially, permeability; T is the topography factor: slope gradient, in per cent, slope length and land form; and L is the land utilization factor: plant cover in relation to bare soil. The first two factors, C and S, are parameters that are not changeable by intervention. The last two factors, T and L, are the instrument variables that can be changed to control erosion. T involves mechanical control through changes in the slope and form of the soil surface. L involves biological control through changes in plant cover.

Addressing biological control first, various values for L have been calculated, giving a rating for each type of land use, from 1 for bare land, which is the worst condition for L, through about twenty other land uses (with ratings from 0.85 down to 0.05) to the best condition, which is irrigated

---

[1]It should be noted that the USLE usually contains a management factor (M) for conservation practices like terracing or grass stripping. However, since $M = f(T, L)$ this term is redundant in the USLE. In a regression analysis this would cause formidable problems of colinearity, since M and (T, L) are highly correlated. For regression analysis of E in relation to particular states of M, the function could be written $E = f(C, S, M)$; however, $E = f(C, S, T, L)$ is more precise for analytical purposes.

sawah with a rating of 0.01 (source: Soil Research Institute, Bogor, Indonesia). These L values must be interpreted under the condition that all other factors in the USLE are constant. However, the values would perhaps change, even in ranking, at different absolute values of the other factors, especially T.

Second, the primary function of mechanical control (T) is to decrease the slope of the land surface to lower the quantity and velocity of water runoff and hence soil erosion. Estimates for T are provided in table 5.1 (Hamer 1982: 1). As shown in the right-hand column of table 5.1, the rate of erosion is roughly *proportional* to the slope, over slope ranges greater than 15 per cent. Over undulating terrain, slope degree and length should be estimated as a weighted average of various segments of the land surface.

Table 5.2 (Hamer 1980: 21) shows values of different combinations of L and T or management (M) factors by different quality standards ($M = 1$ represents the worst value). It is interesting to note that the highest quality bench terraces have the same M value (0.4) as the best grass strips. Both are 2.5 times more effective than permanent ground cover with estate crops.

As noted before, neither the USLE nor empirical research in developing countries has given sufficiently reliable estimates of the actual magnitudes of erosion to be used in planning. This task of empirically estimating the parameters of the USLE in tropical areas (through multiple regression analysis) is both the single most important and interesting research area in SWC today. Various alternatives have been discussed already in chapter 3, and the reader will note a difference of view of the importance and usefulness of USLE in predicting soil loss in tropical soils. Stocking, in chapter 3, section 3, tends towards avoiding the problems associated with adapting USLE and to favour alternatives, while the author of this chapter broadly takes the view that the challenge should be taken up, and that the problems of data availability and calibration will eventually be solved in most circumstances.

4 The costs of SWC programmes: estimating the labour requirements of bench terracing

The choice of technique or package of techniques in a specific situation is a crucial aspect in the design of SWC programmes. Table 5.2 lists some of the choices available. In this instance bench terraces were selected although there must remain considerable uncertainties regarding their cost-effectiveness relative to other SWC techniques and the level of socio-economic acceptability to the farmers in question.

The design of bench (and other) terraces has been specified by Sheng (1981). This involves the calculation of the volume of earth required to be cut, mixed and compacted in bench terracing in relation to the slope of the land and the width of the terrace bench. Also the relationships between the

Table 5.1 Topographical factors (*T*) in erosion

SLOPE GRADIENT CLASS (%/M)		ASSUMED SLOPE LENGTH (M)	MEAN *T* RATING	RANGE IN *T* RATING	*T*/ MEASURE SLOPE
0–5	0.5	45	0.35	0.00–0.75	0.14
6–15	0.9	35	1.60	0.76–2.40	0.17
16–35	25.5	25	4.60	2.41–6.80	0.18
36–50	43.0	20	7.90	6.81–8.99	0.18
>50		20	9.00	8.99	0.18

Table 5.2 Conservation practices and erosion

CONSERVATION PRACTICES	RATING
Bench terraces	
high standard design/construction	0.04
medium standard design/construction	0.15
low standard design/construction	0.35
Traditional terraces	0.4
Colluvial terraces on grass strips or bamboo	0.5
Permanent grass strips, e.g. Bahlia grass	
high standard design and establishment	0.04
low standard design and establishment	0.4
Hillside trenches (silt pits)	0.3
Croatalaria sp. (legume) in rotation	0.6
Contour cropping, slope gradient 0–8%	0.5
9–20%	0.75
>20%	0.9
Surface mulch retention (litter or straw 6 t/ha/y)	0.3
(litter or straw 3 t/ha/y)	0.5
(litter or straw 1 t/ha/y)	0.8
Permanent ground cover with estate crops	
high density	0.1
medium density	0.5
Early reafforestation with cover crop	0.3

Source: Soil Research Institute, Bogor, Indonesia.

required number of cubic metres of earth per hectare of land area under bench terracing, the cubic metres of earth per hectare of bench surface of agricultural land (net of land used in the risers), and the numbers of man-days required per hectare of bench surface are all essential data and can be read off a graph prepared by Sheng (1981). The next task is to estimate the cost of earth cutting, moving and compaction per hectare of bench surface created. This was done bearing in mind that (a) the major part of the job would be done in the dry season when the cutting of the earth would be much tougher and therefore more expensive; and (b) that shadow labour prices should be used if no alternative opportunities for labour exist, and market prices if these opportunities exist. An example of this set of computations put the labour cost of preparing 1 ha of bench surface on a 17.6 per cent slope at $80 and for a maximum 44.5 per cent slope at $372/ha (both using shadow prices). Market wages would push these estimates up two- or threefold.

An important site selection rule in bench terracing is that the depth of top-soil should not be less than one-half the height of the riser (Sheng 1981). This rule assures that there will be sufficient top-soil in the bench area so that agricultural production will not *decrease* as a result of terracing – a very important rule indeed. With this rule, and the preceding defined parameters, the shape of the solution to the problem of site selection becomes clearer. It is essentially based upon three kinds of situation.

(1) Land so badly eroded it cannot support agricultural production. This land should be planted to grasses and trees where possible, or, if not, ignored, unless external costs of erosion, or aesthetic and other values indicate otherwise.
(2) Land that has soil depth and rates of erosion such that agricultural productivity will not be reduced for several years. Here, subject to the same *caveat* as above, this land should not *now* be treated.
(3) Land that is not beyond salvation but shows 'clear and present danger' of being in a state of reduced production through erosion. This is the proper target land for SWC programmes. Once this land is identified, SWC programmes can be fine-tuned in the light of the considerations described before.

5 Estimating external costs and benefits

The external or 'downstream' costs and benefits of SWC programmes can be estimated *if* the data are available. This section uses some very rough estimates to illustrate how this can be done. The Snowy Mountains Engineering Corporation (1982: 103) report on the Serang River project provides some figures on the impact of sedimentation on the Kedung Ombo Reservoir that may be used for illustration.

The reservoir has a dead storage capacity of 91 million m^3 (MCM) and a total storage capacity of 728 MCM. It is estimated that 2.4 MCM of sediment is trapped in the reservoir, displacing live storage. The estimate is a 1.4 MCM yearly loss of live storage. The economic analysis of the project indicates that the value of live storage is $0.065 m^3 resulting in an annual loss of live storage of $90,000 /year. The present value of this loss, at 12 per cent discount over 38 years, is $740,000. At this rate of sedimentation, total storage of the reservoir will be exhausted in 250 years.

A more significant loss, according to the report, is due to damage to canals and irrigation facilities below the dam from other areas in the catchment. A sediment trap on the canal might trap 1.25 MCM/year. The cost of mechanically removing sediment is $1.40 m^3 or $2.1 million /year. The present value of this cost over 38 years is $17.6 million.

It is assumed that the rate of sedimentation could be reduced by one-half by the conservation project (bench terraces as well as silvo-pasture techniques). Thus the total present value of reduced external costs would be about $9 million, or, say, $10 million with other benefits to fisheries, turbines, etc. included. The SWC part of the project costs $30 million. Thus the present value of reduced external costs is about one-third of the project cost on the assumption that present rates of erosion will not increase. If, without the project, erosion increased at a rate of 2 per cent/year, then this value would double to $20 million, or 60 per cent of project costs over the 38-year horizon. Longer horizons make no substantial difference to the present value.

6 Conclusion

The purpose of this discussion has been to concentrate on some of the major conceptual issues in the economics of SWC, rather than on the techniques of analysis themselves after these conceptual issues are settled. The major point to emphasize in conclusion is that there is nothing particularly unique about SWC from an economic (or any other) point of view. SWC is subject to the same kinds of economic analysis as anything else; and economics is as limited in the kinds of evaluations it can make in SWC as it is in anything else. As economists and SWC professionals realize the strengths and weaknesses of their respective fields, and the ways in which each can contribute to the other, a sound basis can be established for more rational and effective SWC programmes in the future.

B   On measurement and policy, scientific method and reality

Piers Blaikie and Harold Brookfield

This chapter has emphasized and expanded a number of themes which occur repeatedly in this book. First, it has taken a strong anti-fundamentalist stance in assessing the importance of land degradation, on the grounds that the value of soil is not absolute but depends upon the supply and demand for it in specific places at specific times. Elsewhere in the book other, but strongly related grounds, have also been suggested for opposing a fundamentalist line. Chapter 1 developed the idea that land degradation was a perceptual term. Different land users have different criteria for assessing degradation according to the demands they put on the land. Again, in chapter 7 a long-term historical perspective of land degradation in France shows how social and technical change in land use and management have adapted to land degradation. There is at least some supportive evidence from the past that the problems of marginal lands today could, as Seckler suggests, be solved in the future. However, chapter 6 argues that there can develop a number of mutually reinforcing symptoms of impoverishment of environment and people in marginal areas such that there is a tendency for the problem to reproduce itself and *not* be solved. Also, the great problems of empirical modelling of the physical aspects of erosion and declines in crop yields, outlined in chapter 3, section 3, are shown to be formidable in this worked-through example.

The grounds for debate over the importance of land degradation have clearly shifted from a universalist and fundamentalist viewpoint to one of assessing the likely course of the development of agriculture and land-management technology on the one hand, and the economic and political access to it of the people using the land on the other. Seckler maintains that we simply do not know about the longer term and therefore we are reduced to assuming the option value of the use of soil in the future to be the shortfall in the cost–benefit ratio between the calculated value and the target. Perhaps there is a middle ground here between professing not to know the future agrarian situation of a particular area at all, and making an informed guess over the next twenty to forty years. After all, much of the rest of the quantitative analysis (particularly relating to the USLE and the calculations of the topography and land-use factors) is also a matter of informed guesswork.

There are possibilities for assessing the long-term trends of the national or regional agrarian economy. National or sub-national food production trends and sample household surveys, along with data on natural population growth and migration, provide a basis for making estimates of the numbers of people who will need soil in a particular area in the future. Then it is perfectly feasible to ask on a country-by-country basis, what technical and societal changes have to be made so that presently eroding soil (of different broad categories) will *not* have a significant option value. Also, alternative options for the rural population have to be examined in broad terms – jobs in cities, or refugee camps, or new land for settlement? Likely scenarios can be constructed based upon broad statistical trends of population growth, crop yields, available land for cultivation or pastoralism and the likely course of political-economic development. This in turn leads to a conclusion that there is scope for assessing option values for usable soils and forests for the next generation or two. Maybe we can dispense with the second or even first decimal place, but we should try.

The second important contribution to the assessment of land degradation in this chapter is that it demonstrates the importance of carefully specifying the costs and benefits of degradation and conservation over space and time. This links with chapter 4 which distinguishes between the direct land manager and decision-maker and others who make decisions which indirectly affect the parameters of choice of the former. By putting together two considerations of specifying costs and benefits of degradation and conservation over space and time, and of direct and indirect decisions about land management, the underlying causes of land degradation become clearer. The overall calculus of costs and benefits of management and conservation including upstream–downstream effects usually can only be done effectively by a 'higher level' institution, and any action following this requires co-operation which often stretches beyond small groups and villages to whole watersheds or even nations. The institutions of state should, in theory, play the logical mediating and organizing role in achieving co-operation between different groups and ensuring a rational distribution of costs and benefits. Also the state, in theory, should intervene to apply intergenerational equity in conserving soil. Indeed, in the latter case the necessity of the state's intervention is even stronger since there are not future generations around to enter into the political process to press their case! In this way, the liberal theory of the state would assume that its role would be to intervene in a rational manner above competing interests (between upstream and downstream, and between present and unborn generations). However, there are two problems with this.

First, the management of land requires a detailed local knowledge based on experience, and it often requires on-the-spot and on-the-dot decision-making. A terrace wall collapses one night in the rains – is it worth repairing, and how? A field includes soils from a catena and shows evidence of rill action – what action is the farmer to take in different parts of the field? The state, or other centralized control outlined in chapter 4, naturally finds it difficult to assert higher level control and at the same time dispense extremely local expertise. This is a crucial contradiction to which much of the material of this book is directed. Secondly, the state is not a rational instrument of executive power for the greatest good of the greatest number. Nor is it in any reductionist interpretation of Marx's famous words 'but a committee to run the affairs of all the bourgeoisie'. The paradox is that states *do* get involved in soil conservation and initiate projects such as the one Seckler describes in Indonesia. There must be 'rational' (but not necessarily neutral) methods of decision-making, which are applied to the problem. The solution appears as a policy or project document, and then, according to conventional wisdom, gets implemented. However, reality is different, as Clay and Schaffer (1984) point out in general terms, and for soil erosion specifically by Blaikie (1985a: 61–4). Class interests, actual bureaucratic practice, different perceptions of the problem by various levels of the bureaucracy and the rural population itself, all serve to transform original intentions. That is the challenge and the problem.

The last important theme of this chapter concerns the problem of data and interpretation. This ties up with other contributions in this book. The problem of not having time-series of soil loss for most of the world inhibits a longer term perspective and evaluation of conservation programmes. The problem of not having enough detailed information on soil loss at *one* point in time together with not really knowing the relationship between soil loss and soil productivity is another (see Stocking in chapter 3). The implications are difficult for economists who rely upon fairly accurate measurements of scale to calculate benefits and costs (and to uphold the integrity of quantified social science!). Without those data, and without knowledge of the future to provide a rational basis for discounting, a curious dilemma arises, which is well illustrated in this chapter. A precise and internally logical methodology has to digest mere guesses about soil loss; its implications for productivity decline; actual yield data; responses to soil management in terms of a reversal of yield reductions; and the future of rural populations in order to calculate option values on soil. It seems churlish and Utopian to call for a 'new economics' to deal with this, and to abandon the basis of currently practised economics altogether. The technical aspects of improving data collection and relating soil loss and conservation to productivity are easier to tackle although, as chapter 3 indicates, there is still a very long way to go. The other 'non-economic' aspects, usually unquantifiable and hence dismissed as 'soft', continue to threaten the ability of an internally consistent scientific method to give rational answers to the question 'When is soil worth conserving?'

6 Colonialism, development and degradation

Piers Blaikie and Harold Brookfield

1 Introduction

We now move from a discussion of general principles, which has occupied most of the first five chapters, to the study of specific cases in which degradation may perhaps be explained. We have set out a framework of explanation in chapter 2 and illustrated it by a case study of Nepal. In that chapter, and again in chapter 4, we found it necessary to move not only 'upward' from the land manager to the social and political system, but also backward in time to understand the antecedents of modern conditions. This historical thrust is sustained through most of the chapters which follow. Were we writing in terms of systems theory, we might consider degradation to be a 'maladaptive' response to historical change, a failure to adapt to changing social facts. In our 'regional political ecology' approach we take the opposite tack and place our first emphasis on social and economic causation. In the two chapters which immediately follow we find this approach valuable even in studying problems that arose before the period of rapid modern change under capitalism, colonialism and socialism.

In this chapter we are concerned with the major change of modern times for much of the world – the temporary loss of independence under colonialism, both direct and indirect. Our object in focusing on this event is to develop an explanatory system of wide domain, for the forces that evolved under colonialism are but an exaggerated expression of forces encountered everywhere during the period of capitalist expansion. We begin with a brief discussion of colonialism itself, seeking to disaggregate elements which had the sharpest impact on land management. We then examine some aspects of this impact in the colonized regions of Africa, with only brief reference to the impact in Asia, and none to the Americas for want of space. Finally, we turn to those mainly upland areas in which direct colonialism was resisted more successfully than in the lowlands, but in which the new forces were none the less strong although experienced by people in those areas in a different form. Our discussion ranges over a long period of time, but its main focus is on the period since about, say, 1850, when for convenience an older mercantilist colonialism may be said to have given way to an era of more direct control (Cohen 1973).

Two hypotheses are advanced. The first is that certain relations of production and surplus extraction can put land managers under pressure, or deprive them of resources so that they skimp on management practices. The second is that, under certain conditions of accumulation, capitalist land users seek to employ the resources of the biome for short-term gain, so that they are transformed into profit and not replaced. Both hypotheses present different sides of the same coin. We shall find, as always, that they do not explain everything.

2   The contradictions of colonialism

2.1   Objectives of the colonizers

Colonialism, as Brookfield (1972: 1) argued earlier, was by intent a revolutionary process in the sense that it sought to terminate or divert former evolutionary trends in the interest of bringing certain people and their resources into a subordinate linked relationship with another. Rather more directly, Watts has written of this process as 'articulation with the colonial, and ultimately the global, economy . . . effected through the colonial triad of taxation, export commodity production and monetization' (1983: 249). Still more directly, Polanyi wrote how even in modern times 'white men may still occasionally practice . . . the smashing up of social structures in order to extract the element of labour from them' (1944: 164), and hence dissolve 'the body economic into its elements so that each element could fit that part of the system where it is most useful' (1944: 179).

Yet at the same time colonial administration in some of its phases was inspired by a civilizing, even missionary ideology. In regard to resources, there were considerable efforts by colonial authorities to conserve soils, forests and water supplies even in the teeth of opposition by their political masters. The civilizing ideology was often, even usually, imbued with a contempt for indigenous social institutions and methods of land management, and with racism, yet there was a real wish systematically to change the ways, the culture and (often) the religion of the colonized toward a rational, European model. Bohannan (1964) argued that it was an essential role of the missionary to hold together the pieces of society as it was smashed, and then to put them together in a new pattern. Never was this more relevant than to colonial conservation policy. This contradiction between exploitation and a benevolent purpose was sometimes present even in the minds and practices of the same people, though more often it is reflected in disagreements between different parties among the rulers. It runs right through the whole of colonial history from denunciations of the excesses of the Spanish Conquistadors by de las Casas at the start of the sixteenth century to the most modern times. Not always was the benevolence sincere, and as Furnivall remarks of the history of Dutch rule in Indonesia:

> The Liberals entered 'love for the Javanese' in the published accounts, but did not let it touch their pockets; and when the Ethical leaders hauled down the Jolly Roger and hoisted the Cross, they did not change the sailing orders ... any colonial policy in application is effective only in so far as economic circumstances are favourable. (1939: 392)

We shall find that land degradation is an imperfect stick with which to beat colonialism, although there have been implicit and explicit attempts to use it in this way. Blaikie (1986) has argued against this tendency as an anti-capitalist indulgence, one which does not spell out alternative paths to new and feasible social relations of production and land use. Marx stated 'all progress in capitalistic agriculture is a progress in the art of not only robbing the labourer, but of robbing the soil' (1887: 506). But capitalism is not *inherently* the cause of degradation, nor for that matter is socialism, feudalism or any other broad descriptive category of social process, including colonialism. A full record of land degradation and conservation in the colonized countries cannot be given here; that would take volumes. It is an area as beset by contradictions as almost any other aspect of colonialism. Policies conducive to exploitation and to the 'rape of the earth', and policies designed to restrict and repair the damage were both put in place and practised.

2.2   Acquisition of the land in Africa and Asia

The impact of colonialism on indigenous land users was extraordinarily diverse across space and time. In many areas the first impact was the loss of land by indigenous cultivators and pastoralists to settlers, ranchers and planters. In Africa this happened most heavily in the southern part of the continent but also in East and North Africa. In Algeria, for example, 20,000 Europeans took 2.5 million ha of the best land, much of it alluvial and irrigable, while some 630,000 Algerian peasants were left with 5 million ha of mainly dry, hilly and thin-soiled land (Stewart 1975). In Zimbabwe 6 million ha, a sixth of the entire country, passed nominally into European hands during the short 'age of fortune hunters' after 1890 (Palmer 1977: 227). In Kenya, after a confused period during which Europeans were allowed to hold large tracts in the highlands on 999-year leases, a Lands Commission in 1934 allocated 43,250 km^2 as a European preserve. By 1937, some 124,000 km^2 of mainly less fertile land had been gazetted for African occupation (Brett 1973: 172). In South Africa successive European invasions confined the rural African population into areas which, successively delimited by Acts between 1877 and 1913, then slightly enlarged in 1936, totalled only 12 per cent of the national area, 44 per cent of this in semi-arid parts of the country (South Africa 1955). Even in Swaziland, which retained its political identity, only 37 per cent of the land remained unalienated and Crush argues that:

> the Colonial State patently and repeatedly serviced the interests of capital
> accumulation ... in its development of a land policy.... This was
> exemplified in the alienation of a disproportionate quantity of land for
> settler-estate production ... the provision of accessibility to Swazi labour
> and the ensuring of its release from the indigenous mode of production (by
> taxation). (1980: 85)

In Asia, as well as in Africa, an important early step was the declaration of all
'uncultivated waste land' as the property of the state. In much of Asia
pre-capitalist states had claimed this title before colonial rule, but there was
an important difference in enforcement and purpose. Shifting cultivators
were, in general, not restricted in their access to such land before colonialism,
but subsequently they were and some of the 'uncultivated waste' land was
then granted to European planters. This happened quite early in Java and Sri
Lanka, and later in the century the process accelerated; we examine the
specific case of the modern province of North Sumatera in chapter 9. In India,
a major objective of colonial policy after the 1860s was to conserve the forests
in order to manage the supply of timber for the growing railway network, and
to this end the first of the colonial Forestry Departments was set up in 1860,
and a series of measures restricted both private felling and access by local
people. In many parts of India, cultivators used the forest for fuelwood,
construction timber, wild foods, spices and as a source of fodder for
livestock. Preservation thus adversely affected the self-provisioning ability of
local cultivators, and had far-reaching implications for the management also
of privately held land. We shall encounter such restrictions on access to the
'commons' again below, in chapters 7 (in France during the eighteenth
century) and 10 (in India in this century); however, in Asia their effect on
shifting cultivators has, where effective, been the most marked. They
required adoption of a different way of life. Even for settled agriculturalists,
however, Scott (1976: 64) calls attention to the need to purchase bamboo and
firewood that could no longer be collected, and to a loss of access to forest
food resources that made the poor more dependent on wage-work for the
wealthier farmers.

2.3 Taxation and the state as landlord

The role of the state as ultimate landlord was well established in much of Asia
in pre-colonial times. It was known, too, wherever states existed in both west
and southern Africa. In 1936 the Zulu king fined both the vendor and buyer
of a hut as they were trading his land on which it stood (Allan 1965: 364). In
Mughal and earlier times in India, taxes on land were the main means of
raising revenue and we have seen their impact in Nepal in chapter 2. The
taxes were high, commonly two-fifths or one-half of produce from the rice
fields and some other crops. Under colonialism, taxation was employed as a

major instrument of policy for the extraction of commercial produce from farmers, without direct acquisition of their land.

The classic case is that of Java, where a land revenue system formalized early in the nineteenth century was converted after 1830 into the 'Culture System' under which dues were payable in export crops, sold for the profit of government. The system was compulsory, and in growing to overshadow the limited amount of private production on alienated land turned Java into 'one large State business concern' (Furnivall 1939: 121). Where sugar was the crop, forced deliveries alone were insufficient for labour was required on an industrial basis and was organized through the village administrative system. Later in the century more direct industrial management came to be required, and corporate plantations were permitted to lease rights to Javanese land and labour on terms requiring a rotation of sugar with rice and other food crops, thus in effect 'privatizing' the state's revenue rights. As peasants were also excluded from the nationalized 'waste land', even the return to a cash-based land-rent system did not diminish the pressure to lease land to the 'estates' (Furnivall 1939; Geertz 1963). Elements of this system remain in operation even in modern independent Indonesia.

Land tax is in principle less regressive than a head tax or hut tax, such as was widely introduced in colonial Africa. However, if converted to a cash basis and rigidly administered, it provided the state with an income independent of fluctuations in crop yields and prices. In Burma (Scott 1976) and Sri Lanka (Bandarage 1983) the land of defaulters was forcibly sold during hard times, contributing to inequality and increasing the supply of wage labour to the emergent capitalist farmers. In Africa, the enforcement of a head tax or hut tax had different policy consequences according to whether or not indigenous farmers were confined into inadequate reserves. In the latter case the main product was labour for the mines and plantations, as throughout the southern part of the continent. In the former the product was new export crops grown initially to pay the tax and was particularly effective in this way in Uganda and much of West Africa. Among the Serer of Senegal, groundnuts were the tax crop, and cultivation expanded further as traders entered to mop up surplus money by supply of low-quality merchandise and alcohol, often advancing money at high interest rates against the next crop and holding cattle as security. Three-year and two-year crop rotations were adopted to accommodate groundnuts, mainly on hitherto lightly used land, with harmful consequences for capability since little manure was available for the new arable land. In one Serer village, only 20 per cent of the land was manured in 1966 (Lericollais 1970).

These instruments of land-use policy were successful in their object of developing commercial economies, and ultimately of creating the revenue basis for the independent states which succeeded colonial administration. Nor were the instruments themselves wholly new, for forced labour and deliveries – if not also taxation itself – were common in pre-colonial times. One instrument was abolished under later colonialism, this being slavery, once more effective methods had been devised for extraction of labour and produce. It is not therefore surprising that almost none of the independent states have abandoned the instruments developed under colonialism, or the elaborate district and local administrative systems devised to make them operative.

3 The impact of colonial rule on land management in Africa

3.1 The contradictions translated on to the land

The new forms of land use fell on a tremendous variety of lasting and usually successful indigenous land management practices in Africa. A recent report (IFAD 1985) lists about sixty farming systems with prominent land management and conservation practices. Some of these systems had evolved in refuge areas where people had concentrated to escape the wars and slave-raiding of the pre-colonial and mercantile colonial periods. For example, the Mandara and Mafa tribes of the Cameroon mountains sought escape to a sensitive environment, where they developed techniques that included extensive terracing, contour banks, trash lines and grid networks of mounds, ridging and deep furrows connected with peripheral drains, together with a range of biological conservation measures (Boutrais 1973; Boulet 1975). As people migrated, they were often quick to adapt to new environments. Allan (1965: 72) gives the example of the Mambwe whose northern population use a grass-mound system while the southern group uses the large-circle *citamene* variant of the shifting cultivation system, and suggests that the latter was adopted as people moved into a woodland region with poor, leached soils.

The adaptations amongst many others were made not only to safeguard subsistence but also to produce crops for sale. The Dagari rotational bush–fallow system in Burkina Faso was adapted to the tax-enforced introduction of groundnuts and cotton by the French, as the Serer also adapted their system. Richards (1985: 18) cites cases of management initiatives following expansion of groundnut and also tree crop cultivation in several parts of West Africa. Modern research suggests that these adaptations were more widespread than was commonly believed in the colonial period, when a blinkered arrogance could produce statements such as one cited by Clayton:

> The poor farming methods and soil depleting practices prevalent among [African] peasant cultivators stem from ignorance, custom and lethargy ... the main obstacle to overcome is the native's lack of understanding of the need for the prevention of soil erosion. (1964: 12)

The amount of degradation itself may indeed have been exaggerated by colonial observers who viewed land-use methods other than those preferred by themselves as necessarily degrading. For example, colonial forest authorities in Sierre Leone wanted only to preserve the oil palms for commercial proposes and decried their destruction in the cause of clearance for shifting cultivation (Millington 1985: 10). Foresters in India also took jaundiced views of indigenous practices (Tucker 1984). Conversion of primary into secondary forest may well increase the number and quantity of products available for local use, but it reduces the productivity of the forest as a source of commercial timber. In Tamil Nadu, south India, for example, invasive bushes make a reliable and quick growing fuel, but the Department of Forests regards them as harmful since they prevent regeneration of species useful to the timber industry (Blaikie, Harriss and Pain 1985). We see similar contrasts of perception in Indonesia in chapter 9B.

However, there is little doubt that there has been a large increase in degradation during and since the colonial period. The evidence, some of which we cite below, is massive and incontrovertible. The reasons are perhaps not too hard to seek.

3.2   The causes of degradation under colonialism

It seems fairly clear that the massive disruptions of society brought about under colonialism in Africa must bear the major share of any explanation of deteriorating quality of land management. The enforcement of rapid change led to substantial disintegration of social control (Lericollais 1970; Martin 1983) and moreover created a sort of malaise particularly well described by Allan:

> The African systems of land-use had their own protective devices against erosion.... When the necessity of cultivating steeply sloping land was forced upon a people in the past, as it was in the refuge areas, they evolved devices in no way inferior to our own anti-erosion practices.... That modern Africans do not react in this way to their more pressing land problems is symptomatic of a loss of group initiative and self-reliance, and of new attitudes, concepts and values developed under European tutelage and dominance. There was no refuge area safe from the alien conqueror with his incessant and insatiable demands for labour, tax money and economic crops. (1965: 386)

In parts of Africa there were additional reasons for massive disruption, in the form of irruptions of diseases which affected both humans and livestock (Vail 1983: 205). The spread of the tsetse fly, rinderpest, smallpox and other diseases, all of which occurred repeatedly towards the end of the nineteenth century, were powerfully aided as a disruptive force by the annexation of land from which people were forced to move and required to settle in formerly avoided areas (Kjekshus 1977; Vail 1983; McCracken 1985). Kjekshus, writing of Tanzania, succesfully links these ecological outcomes to colonial policy, but we find few other writers who do the same. Most general historical works (e.g. Brett 1973; Hopkins 1973; Palmer and Parsons 1977;

Kitching 1980; Birmingham and Martin 1983) concentrate on the socio-political consequences of colonialism and mention ecological effects, if at all, only in passing. Allan (1965), by contrast, deals only briefly with colonial rule in this wide-ranging discussion of land management and mismanagement under different environments. The 'radical' literature on the Sahel famine perhaps comes closest to our chain of explanation, but concentrates mainly on the induced breakdown of technical and social means of risk aversion and of the 'moral economy' that underpinned them. There is a wide gulf between this literature and the detailed physical studies of desertification. In seeking more general explanation we therefore have to rely mainly on case studies and inference. We concentrate on three aspects: the loss of land, commer-cialization and the abstraction of labour.

3.3   The loss of land

Throughout southern and much of eastern Africa enforced migration and the annexation of land by settlers created numerous cases of 'instant over-population' in relation to resources. The Swaka of Zambia were *citamene* farmers, gathering the biomass from large areas to add nutrients to small fields, and at a density of two persons per km² were able to sustain this system on land of very low capability. In 1929 their reserves were drastically reduced so that fallowing periods became insufficient and degradation set in (Ranger 1971; Robinson 1978). In the same country high densities created by land reservation led to erosion in the hilly areas of the east (Robinson 1978: 32). Wisner (1976) describes how the Kamba of Kenya were relocated to make way for settlers and suffered many setbacks; later commercialization improved opportunities for a minority, but the consequences of using the inferior land made available included overstocking, reduction of fallow periods, increased vulnerability to drought and reduced yields. By the 1920s there was extensive degradation (Silberfein 1984). Analysing the relationship of actual density to 'capacity' on the basis of measured food-crop yields in part of Kamba country in the 1960s, Porter (1970) however found evidence of subsequent adaptation, both in cash-crop farming and in agricultural methods, especially on individualized land, and of population redistribution so that some high-risk areas were in fact 'underpopulated' by the standards he employed.

3.4   The consequences of commercialization

The case of the Kamba, who adopted coffee as a cash crop, is not unusual. There have been many successful adaptations based on the introduction of commercial crops (Richards 1985; IFAD 1985). However, there are also contrary examples, where reduction in fallow periods, monoculture with soil-exhausting crops, overstocking, planting in rows and sometimes mechanization have had seriously degrading consequences. These consequences have arisen because compensating adjustments either were not or could not be made. Increased use of land through the addition of cash crops has sometimes led to a lack of land for cattle, while a widespread increase in stocking densities during and since the colonial period, a consequence of commercialization of the livestock economy as well as of population growth, has been reported as a cause of degradation in east Africa (FAO 1974), Botswana (Abel *et al.* 1985), Swaziland (Doran, Low and Kemp 1979) and throughout the Sahelian region, as discussed in a large literature since the 1970s (e.g. Franke and Chasin 1980).

The nature of the cash crop itself has been of major importance. Where grown in single stands, and not inter-cropped or relay-cropped as in traditional African gardens, groundnuts, tobacco, cotton and maize in particular tend to offer poor ground cover especially early in the growing season when erosivity is highest. Groundnuts are particularly demanding; in the Cassmance area of Senegal there was a loss of 30 per cent of soil organic matter and 60 per cent of colloidal humus after only two years under groundnuts (de Wilde 1967: 16). Adoption of these land-extensive field crops has not only absorbed fallow, while they themselves return little trash to the soil, but has also led to a reduction in the area of minor crops such as pigeon pea (*Cajanus cajan*) or cow pea (*Vigna unguiculata*) which have good soil-protection qualities (Juo and Lal 1977; Aina *et al.* 1976). Monocultural methods on European estates have sometimes led to even worse consequences; many tobacco farms in the Eastern Province of Zambia were abandoned because of degradation and low yields after the boom of the 1920s (Klepper 1980). Wherever monocropping has been adopted, with or without the plough and mechanization, the consequence has almost always been loss of nutrients and erosion. The introduction of the plough and mechanization have led to a deterioration in soil structure and sometimes pan-formation, and have been accompanied by greater erosion than hand tillage (Fournier 1967). Stocking (1983) compared the land management practices of commercial, semi-commercial and subsistence farmers in Zambia, and found that adoption of ploughing, row-cropping and other methods described above by the commercial farmers almost always led to greater erosion than that found on traditionally managed subsistence farming land. Moreover, clear felling and open-field farming after the European fashion robs the soil of many nutrients which it obtains from the vegetation itself under shifting cultivation. In the adoption of these practices European agricultural advisory services have played a critical role, but the abstraction of labour has also been an important factor in adoption, since the new methods are less labour intensive than the old.

Note should also be taken of the effect of the social changes produced by commercialization. Once money became available to all, extended family groups began to break up, thus atomizing land management. Among the Serer of Senegal, there were in one village thirty-two family households in the 1960s compared with six large extended-family clusters, each under tight control, around 1900 (Lericollais 1970). Inequalities grew, with the result that in times of famine poor farmers among the Hausa of Nigeria were obliged to sell first their labour power, then their livestock, then their land itself to the wealthy (Watts 1983). Sometimes the pattern of authority was directly disturbed by the appointment of village officials in place of traditional chiefs. While the innovation of new methods was facilitated by these changes, the breakdown of regular rotation cycles, atomization of livestock management and erosion of common property management systems all had serious consequences for the land.

3.5 The abstraction of labour

When 'native reserves' were first created in southern Africa the effect was to produce a labour shortage for European farmers (Hurwitz 1957: 27). However, the restriction of Africans to inadequate land and the enforcement of taxation soon made temporary labour migration to the mines and cities a necessary part of the strategy for survival. The loss of an important part of the male labour force not only placed heavier burdens on the women and the older men, but also led to an inevitable slackening of land management, the consequences of which may be seen throughout the 'reserves' of southern Africa. In some reserves near to centres of employment of male labour, all farm work is done by women, with no conservationist practices at all (Adler 1957). The specific denial to migrant Africans of an alternative security in the 'European' areas required the maintenance of farms for a growing *de jure* population, taking up more and more land without the innovation of management practices appropriate to the support of higher population densities. Even on the best land, the marginalization of people in this way can so lead to degradation that the land itself becomes ecologically marginal. Breakdown of the plant cover, the extension into humid regions of plant associations natural to the sub-humid land, and massive water and wind erosion over wide areas are consequences of the sort of land management to which most of southern Africa has been increasingly subjected under its maldistribution of land and labour in modern times (Acocks 1955).

Relative to its condition in pre-colonial times, the deterioration of vegetation and land in South Africa and adjacent countries, especially Lesotho, may well be the worst in the continent. However, the abstraction of male labour has had widespread consequences, for there are only limited parts of Africa in which labour is abundant. One such is Rwanda, where the Belgian authorities employed the forced labour of three million people to undertake extensive conservation works, including terracing, tree planting and a dictated land-use pattern. Since many Rwandan men also migrated to work in Uganda, this was difficult to organize and there was widespread resentment. The system broke down after independence (Allan 1965: 184).

The rather special case of the Lozi of western Zambia may be used to illustrate the significance of a declining labour force upon degradation. The Lozi live on patches of higher ground in an extensive plain annually flooded by the Zambezi, and in the past developed an intensive system of swamp cultivation, the main labour force for which was supplied by vassal peoples under conditions of slavery or corvée. This was abolished under British pressure in 1906, and while the swamp-cultivation system continued the lack of labour led to deterioration of the drains, so that after 1940 large areas went out of use and recurrent food shortages became established (Trapnell and Clothier 1937; Gluckman 1941; Hellen 1968). A critical new factor was labour migration, which by the late 1940s was absorbing half of the Lozi manpower, many more than before 1940. Dryland cultivation also deteriorated. Ploughing was introduced, leading to degradation of the light soils which in turn led to the ploughing of wider areas and a worsening of the problem. Woodland gardens were opened up in the sands on the edge of the plain, and an intensive system became extensive. Since that time modern resources have been applied to re-opening of the swamp gardens, bringing the food economy back into surplus but on a more capital-intensive basis (Allan 1965: 152).

4 The colonial state and the degradation problem

Colonial attitudes towards land degradation problems have varied considerably through time, being influenced as much by external circumstances as by what the officers themselves perceived to be happening. Before 1900 there was very little apparent concern, but then there ensued what Anderson and Millington (1986) describe as a 'forestry management phase', which owed its origins to the earlier experience of India. Erosion was also recognized as a 'settler' problem (Beinart 1984), principally in southern Africa, but no major state intervention took place until the 1940s (Adler 1957). After about 1920 and until the 1940s attention shifted to soil conservation. The first soil surveys were undertaken, and consciousness of erosion was raised by the early effects of commercialization and by the consequences of drought. The rising awareness of the soil erosion problem in the United States, and the action taken in the 1930s discussed in chapter 12A, also had their impact. A few professional critics of land-use policy, such as Stebbing (1935) who first warned of desert advance, also had some effect. Only limited work was however undertaken, and much of it took the form of prohibition or coercion (Berry and Townshend 1973). Most of the effort of the tiny agricultural services went into support of the settlers, and – as we have seen – a patronizing or contemptuous view was generally taken of indigenous agriculture (Stocking 1984c).

Remarkably little reliable information on the soils of Africa was available until after the Second World War, when a major drive to 'develop' Africa was initiated. Subsequent research has been rapid and comprehensive, but while conservation formed part of the development effort, especially after some disastrous attempts to introduce mechanized farming on a large scale as in Tanzania in the 1940s, the economic aims dominated. The paradox of colonialism remained, so that in a majority of the development schemes, plans and projects, the conservation element was subordinated to commercial goals (Stocking 1984c; Millington 1985; Anderson and Millington 1986).

What was done was important, none the less. It was unfortunate that early efforts were so often misguided, or carried out in a coercive manner. Unfortunate effects, such as the construction of terraces which were quickly washed away (Berry and Townshend 1973: 246) and the build-up of insect populations through prohibitions on burning, did not increase confidence, and many of the efforts aroused both resentment and suspicion. As African nationalist movements grew they not infrequently focused on coercive conservation as a political issue; additional heavy work thrown on Africans removed from their former land to make way for settlers make an obvious target for resentment, with particular force in East Africa (Young and Fosbrooke, 1960; Coulson, 1981: 56; Heyer, Roberts and Williams 1981). Soil conservation became such political dynamite that late-colonial officials were frequently exasperated by insistence on its importance from London and Paris, and by experts foisted on them by international bodies. Nor was the notion of superiority of European and American ways ever absent from these efforts, and indeed it still continues as the pages of the conservationist journals unfortunately demonstrate even today. What was actually done remained always congruent with the objectives of colonial capitalism, so that 'new' farming systems of the 1960s were always commercial systems, and designed for the incorporation of machinery and fertilizers. It is only after a long period since independence that the conservation efforts of the colonial period have again been taken up by the political authorities. We detail a parallel situation in the Pacific in chapter 9C below.

5 Some Asian contrasts

5.1 The plantation lands

Only a brief comment on Asia is possible, but is necessary in order to place the African experience into a wider context. We have seen what happened in Java above, and throughout Southeast Asia there was extensive development of plantations in the late nineteenth and early twentieth centuries, together with accelerated commercialization of indigenous agriculture. Except in Java, however, the basic food grain economy was far less severely impacted than in Africa, and it was also supported earlier through substantial irrigation and drainage works which began in the nineteenth century. The object of these was commercial production, but stabilization of water management had many secondary benefits. The pattern of labour recruitment was also different. Since the indigenous societies, though taxed and exploited in many onerous ways, were not subjected to the same deep disruption as occurred in Africa (at least after an initial military conquest), it was not at first possible to extract labour from them in large quantities. 'Surplus' labour from among the rural poor of India and China could readily be recruited in their stead, until the internal pressures of the crowded and exploited heartlands in Java, Tonkin and Luzon began to generate a large indigenous supply of workers for the plantations.

Though the same instruments of policy were used as in Africa, therefore, the effects on the people and on land management were not the same. Bandarage (1983) finds some parallels in Sri Lanka, and Scott (1976) gives a classic account of commercialization and its effects in Vietnam and Burma where, however, indigenous entrepreneurship quickly became a factor in change. In Malaysia, and to some extent also in Indonesia and Burma, 'dual economies' were created by selective use of the instruments of policy, to conserve as well as to dissolve. In Malaysia, colonial preoccupation with plantation development led to policies which stunted peasant economic change and sought to hold Malays within the subsistence sector (Lim 1977). Such efforts led to relative impoverishment rather than to satisfaction, the consequence of which is the modern drive of Malaysia's independent government to bring the Malay people into full participation in the new economy.

We tell some part of this complex story, and its consequences, for outer island Indonesia in chapter 9, but a review in parallel with that of Africa would demand a much larger treatment, since the complexity is great. All the people of these regions ultimately became enmeshed in a capitalist economy, but the manner in which they did so, and its consequences for the land, is a story that cannot be told here. One aspect, however, may usefully be told in more detail; this is the control of forests in British India where a seemingly small policy objective had large consequences.

5.2 Forest control in British India

British India had always known taxation and received only small and concentrated groups of European planters, but its labour was never recruited so massively – either for work at home or abroad – as to have more than localized impact on the agrarian economy. Only one major new cash crop was introduced by the British; this was jute, the particular ecological requirements of which meant that it did not contribute to degradation. Commoditization of production was already highly developed in precolonial times (Maddison 1971), and was merely accentuated by colonialism, as was the inequality of land-holding and incomes. However, the rural poor in many areas depended to an important degree on access to the forests, and here they came into conflict with the authorities as we saw in our introductory discussion of colonialism.

Timber was required for the railways and for export. Early British exploitation was haphazard and predatory (Schlich 1889/90: 198), and a Forestry Service was established in 1860 which managed the forests alongside the Revenue Department, but forests managed by the latter tended to be commercially exploited more ruthlessly (Tucker 1984) and such Revenue Forests as remain are severely degraded today. Forestry officials sought to conserve trees for later use, and conflict with the peasantry quickly developed. In the Himalayan foothills conflicts became so severe that large tracts were set on fire in 1921, with environmental consequences that went unrecorded. Attempts at compromise were made by giving control to village *Panchayats* in Madras between 1915 and 1945, but they were so devastated by local farmers and commercial loggers, that they had to be taken back by the Forestry Department again – a salutory lesson to those who would seek to give the forests back to local management without first establishing whether effective institutions existed (an issue further discussed in chapter 10A and B).

To a considerable degree the measures were successful. Since the forests earned revenue, conservationist management could pay for itself (Anderson 1985: 7). Colonial rule preserved the forests for controlled exploitation and for the future, but did so by denying access and with the peasantry bearing the social costs. Moreover, this colonial success was not sustained beyond independence, when corruption of politicians by businessmen quickly came to threaten the survival of the forests (Vohra 1984: 7), still without much benefit to the peasants. British conservation policy led to emigration from the hill districts, but it certainly helped preserve the land and its resources for the future. This 'success' of a colonial conservation policy is thus a useful counterweight to the litany of woe we have told above. It is also a useful corrective to review the course of events in regions which colonialism never fully penetrated, or penetrated only briefly and partially. This is discussed in the following section.

6   Colonialism, capitalism and the hills

6.1   The historical meaning of marginalization

We defined the margin and marginalization in chapter 1. Here we review the experience of a scattered set of lands united by being hill countries or regions, often marginal in ecological terms for agriculture, and by having been marginalized in the political economy sense by a capitalist colonialism which found the cost of establishing control to exploit their resources marginal in an economic sense. Substantial degradation has occurred in all of them, and it is important to consider why. The history we review is necessarily a little longer than that considered above, for most of these regions in addition had also been marginalized in pre-colonial times in relation to the pre-colonial states. Their history, like their ecology, is varied, but certain generalizations can be made.

The areas principally discussed include Afghanistan, the northwest frontier region of Pakistan, Bhutan, Ethiopia, Kabilya in Algeria, the Kurdish areas of Turkey, Iraq and Iran, Lesotho, the Rif mountains of Morocco, Nepal and northern Thailand. With some reservations, the same pattern also links elements of the history of parts of Burma, Andean Bolivia, Colombia and Ecuador, parts of Mexico, some princely states in India, Rwanda, Swaziland, and parts of southern Europe especially Albania, southern Yugoslavia, Sicily and Sardinia. In all these countries and areas serious land degradation has been reported. Brief country reviews can be found in Greenland and Lal (1977) (especially parts 5–7), USAID Environmental Reports from 1979 to 1982 in which many reports on developing countries were produced, and in United Nations publications such as FAO (1977).

Many, but by no means all, hill areas in developing countries had a tendency to find themselves on the edge of and in conflict with pre-capitalist states, and resisted wholesale conquest and incorporation into these states. They also resisted later imperialist expansion, although they were profoundly affected from without by the latter in spite of a degree of political independence which was frequently preserved. However, the subsequent course of history of these areas altered the course of people–land relationships as it did in the colonized lands, but here closing off certain productive responses to the changing demands and opportunities of the world economic system. In more recent post-colonial times political and economic marginalization continued, if not deepened, although it tended to take new forms. Today, as repositories of the last remnants of surviving pre-capitalist economies, they exhibit a syndrome of environmental degradation, out-migration, demographic pressure, dependency and neglect.

6.2   The historical development of hill areas

The historical development of hill areas can be divided into four phases: (a) pre-capitalist state formation; (b) relationships between pre-capitalist states and hill areas; (c) the north European imperialist period; and (d) the post-colonial period. The onset of these phases varies enormously between specific places. Hilly areas in the seasonally dry tropics and sub-tropics have historically been the locus of very ancient settlement and irrigated agriculture, which relied upon seasonal or perennial water sources. Usually, however, the amount of agricultural surplus was limited by the patchiness and small size of the areas which could be irrigated with hand-dug channels and simple lift mechanisms. The Zagros mountains and the irrigated valleys of the Rif and Atlas mountains are examples. As far as historical and archaeological accounts can show, however, land use in the larger tracts of hills and mountains tended to be extensive, involving the continuation of hunting and gathering, slash and burn cultivation, but dominantly pastoralism in the sub-humid regions. As surpluses from agriculture developed elsewhere, often in neighbouring plains, trading and raiding by hill inhabitants became common, providing additional support to their limited agricultural base.

The natural economy, which was originally defined by Luxemburg as 'the production of personal needs and the close connection between industry and agriculture' (1951: 402), tended to persist in hill areas, whereas expanded production controlled by and on behalf of the interests of a military, religious and/or landed aristocracy was technically more feasible elsewhere, in neighbouring plains or areas of flat and irrigable land. Except in the Americas, where different topographic and climatic conditions obtained, the earliest states grew up in the plains and the ruling classes of these plains states were in frequent conflict with the tribes and confederations of tribes from hill areas. The medieval period of Afghanistan, what is now Pakistan, India, northern Burma, and much of the Mediterranean littoral is characterized by these struggles.

Taxes, tribute, labour for public works, for building and to serve the army, and sometimes raw materials were frequently sought by rulers of lowland states from hilly areas and their inhabitants. Depending upon the relative military and political strength of the two sides – hill and plain – the states sometimes levied taxes, had access to raw materials from the hills (frequently forest products) and generally contained the military threat. Sometimes hill people who later formed states of their own were able to extend control into the plains for a while. Prithvi Narayan Shah achieved this after the unification of Nepal in the eighteenth century at the expense of the decaying Muslim kingdoms of the Ganges Plain, and the Berber lineages ruled much of what is present day Morocco during the seventeenth century.

A fundamental shift in the balance of forces occurred when the Europeans arrived. In the earliest conquests, in the Americas, the Spaniards took over the Aztec power base in the Valley of Mexico but added a new power base on the Atlantic coast. In Peru, by contrast, the Spaniards set up a wholly new base on the coast, and their conquest of the Andes relied mainly on a set of mining centres. Mariategui stresses that the Spaniards 'feared and distrusted the Andes, of which they never really felt themselves masters' (1971: 5).

In the Asian and African regions the major sources of surplus were already located in the plains. Here were found labour to work mines and plantations, land for plantations and estates, and the main sources of land revenue. For a long time the hill areas remained marginal to imperial economic interests. They frequently remained a troublesome frontier and many imperial wars concerned the pacification of hill peoples, although complete annexation of their territories was seldom considered worthwhile or was a downright military impossibility.

Many hill people were often skilled fighters, knew the terrain in which they were fighting, and could melt into the mountain vastness and forests as soon as the imperial column threatened. When combined with rugged terrain, forests have provided vital military advantages to guerillas defending their homeland. The Ottoman Empire destroyed the mountain forests of parts of

Greece, Yugoslavia and most of Albania for military and strategic reasons alone, and caused the stripping of most of the remaining topsoil from the limestone mountains around much of the eastern and northeastern Mediterranean littoral. In other cases, imperial powers were content to leave the hilly areas as buffers. While this outcome preserved political independence, there remained considerable political, economic and military pressure. Thus Abdur Rahman Khan, Amir of Afghanistan (1880–1901) wrote:

> my most powerful neighbours, England and Russia ... are the greatest absorbing nations upon the earth, and though the Eastern countries that they have already conquered are dying of perpetual famines they ... go on ... constantly crawling forward. My country is like a poor goat on whom both the lion and the bear have fixed their eyes. (Rahman 1900: II, 150)

Many bitter and long wars ensued between imperial powers and hill peoples. The Afghan wars of 1839–42 and 1878–80, the war between the East India Company and Nepal, concluded by the Treaty of Sugouli in 1816, and three wars against the hill tribes of what is now Burma before 1886 (Stewart 1972) are good examples. Another apposite case study is that of the Moroccan Rif, the extraordinary exploits of Abd Al Krim (Hart 1976), and the subsequent transformation of the peasantry after the pacification of the region (Seddon 1981: 141). All the attributes of the historical unfolding of hill areas are there: commerce and piracy with pastoralism and a fragile agriculture; a long history of war with more powerful lowland states; heroic resistance against two imperial powers (French and Spanish); eventual defeat; establishment of a pattern of outmigration to foreign armies and later to temporary employment in large-scale enterprises far away; differentiation of the peasantry at home; and serious environmental decline.

Varying degrees of political and legal independence were often retained, but with treaty clauses which allowed imperial economic interests to operate as best they could, given that occupation and administration was either an unreachable military objective or simply was not worthwhile. Imperial interests were varied. Some of them had little direct economic basis. In other places, commercial interests were uppermost. The case of the hilly tracts of upper Burma illustrates the combination of commercial and strategic interests. The Bombay Burmah Trading Corporation was interested in the exploitation of the teak of upper Burma, and the threat from the French far to the east was also given as a reason for the annexation (Keeton 1974; Nisbett 1901). Access to timber in northern Thailand was also part of the Bowering Treaty of 1850, where British and French imperial power secured the rights to teak and other valuable timbers. The deforestation of the Nepal *terai* was accelerated by demands for railway sleepers associated with railway construction in India, discussed above.

In some cases these imperial wars resulted in the ceding of valuable lowland territory to the victors. The case of Lesotho is apposite. The war of 1865–8 between the Sotho and the Boers resulted in the ceding of rich arable lands to the Orange Free State (Ellenberger 1969; Ström 1978), and the Sotho were pushed back to the less fertile highlands. This process of spatial marginalization was a common outcome of wars between imperial powers and indigenous hill peoples. Similar displacements occurred in the French possessions in the Maghreb and Sahel, although the mountainous areas remained as 'no-go' areas for colonial military forces and effective administration.

Labour was the third requirement of the imperial powers and their colonists. Labour was needed to work distant plantations and farms, and the mines where these existed. In Lesotho the majority of working males now work in the mines of South Africa (Mueller 1977; Murray 1981). Nepal found itself obliged under the terms of the 1816 Treaty to allow recruitment by the British East India Company, and later the British Crown, of hillmen into the army needed to police the British Indian Empire to the south. Likewise in Morocco and Algeria, Rifian peasants joined the Spanish and French armies in large numbers (and in the former case were decisive in helping Franco overthrow the Republic).

The same classic 'triad' of colonialism – taxation, export commodity production and monetization (Watts 1983: 249) – underlay all these pressures, whether or not enforced by treaty provision. Although direct taxation was not demanded of peoples not brought under political sway, the pressures placed on local rulers ensured that the latter would impose this burden also. What the marginalized areas lacked was the public investment in physical and governmental infrastructure which, however limited, provided some long-term return for the people of politically colonized areas. Nepal is an extreme case, since it provided labour for the British army and the tea plantations of Assam, as well as many 'forest-intensive' exports (timber, copper vessels, *ghee* and livestock to name a few). Other countries such as Afghanistan and Ethiopia provided little in the way of labour or commodities but remained politically ossified in a semi-feudal form, and the ruling classes continued to extract onerous taxes and corvée labour to finance the importation of foreign arms and luxuries.

6.3 Transformation and degradation

The brief gloss given above offers reason to revisit and expand the old notion of hills as 'cultural refuges' (Semple 1907, 1911). In all these examples, the relationship of hill states to European colonialism has in general tended to reproduce the marginal character of hill populations, and to bring about the social conditions conducive to accelerated degradation of the land. However, hills here are a sub-set of other areas which also could be termed 'marginal' in various senses. The discussion could be broadened to other non-hilly marginal areas (in various different senses) which might be merely inaccessible to colonial political influence on account of large tracts of forest, or unattractive to it on account of aridity or inhospitable climate to European settlement. However, the existence of hilly or mountainous areas within the

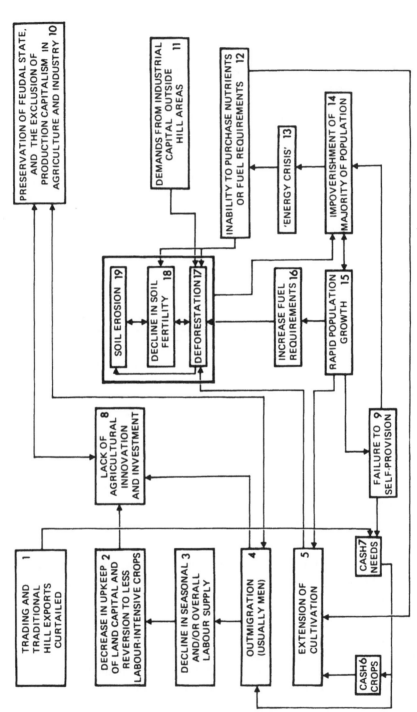

Figure 6.1 Historical development of marginalization and degradation in hill areas

reach of colonial expansion provided a general social and physical context within which different circumstances historically formed the basis of political, ecological and economic marginality in a particular historical fashion.

The pressures on farming systems can readily be characterized. The curtailment of ancient trading, partly by decree and partly by declining demand for traditional products, put increasing pressure on agriculture as the main source of income. The practice of migrant labour brought in money, but it meant that the reproduction of this labour was without cost to the employer, so that the peasant household cross-subsidized the commercial economy of the plains (Taussig 1978; Bernstein 1979). Reliance on migrant labour for income, now integral to the economy of many hill areas (Blaikie, Cameron and Seddon 1980; Murray 1981; Escobar and Beal 1982), can have the effect of reducing available labour for the maintenance of labour-intensive conservation works.

These and other forces are summarized in figure 6.1. Differentiation within the rural population is increased even though the pace proceeded more slowly than in areas where the direct effects of productive capitalism were experienced. Rudimentary health services and probably a greater variety of diet due to imports have contributed to declining mortality and increasing fertility, though in some areas the scale of population growth is not readily explained. In hill areas, however, the production base has remained largely stagnant since research has not yet bred improved strains of their crops and the ability to buy fertilizer and to capitalize agricultural production is very limited. All these factors accelerate the loss of ability to be self-provisioning, and contribute to degradation.

6.4 Changes in the post-colonial period

During the colonial phase described above, most of the lineaments of underdevelopment of hill areas had been drawn. Moving now into the post-colonial phase (say from 1950 onwards), a number of new political economic developments took place. Increased demographic pressure started to be felt in many places. The increasing importance of foreign aid and an altered allocation of resources within the post-colonial state also had its impact upon hilly areas.

Spatial differentiation of development did not, however, change to the advantage of the regions marginalized during the colonial period. New class alliances had their focus in the old capital cities and in the main ports, where growth in world trade had its major impact. Dominant classes, such as the urban bourgeoisie and large merchants, had their power reinforced by expansion in primary production for export. The transformation of agriculture, when it began, was largely confined to the plains. Allocation of funds for development and research have consistently shown a bias in favour of these areas, rather than the more 'difficult' hills. Posner and MacPherson

(1981) have outlined a clear example in South America, which has many parallels in Asia and Africa. First, the World Bank estimates that between 90 and 95 per cent of public investment in Mexico and Peru since the Second World War has been in irrigated agriculture, and practically none in watershed management. Second, the major impetus for colonization schemes in Colombia, Ecuador and Peru was not to solve the problems of the pressing needs of the hill areas at all, but to increase aggregate agricultural production and reduce rural–urban migration without causing political upheaval as a result of land reform. Third, a flat-land bias in research stations in tropical America is very marked. With the one signal exception of the Andean Potato Institute, research has focused almost exclusively upon export crops. Finally, most land classification systems, not only in Latin America but worldwide, put a ceiling of 10 to 15 per cent slope on annual cropping, thereby diverting researchers' attention away from the problems confronting hill farmers.

The central issue in the importance of hill areas to many post-colonial states is whether the environmental/social problems of hills threaten to undermine the prospects for accumulation on the part of the ruling and dominant classes. For a time, such a threat seemed to become a reality in Latin America (Stavenhagen 1970; Blanco 1972). However, even in the Andean regions, the threat of revolution seems now to have receded, and with it any impetus to allocate state funds to check the general decline that afflicts these regions. Once again, they can be safely ignored.

7   Conclusion

The syndrome of spatial marginality in hill regions on the periphery of the colonial world tends to make them an extreme case of many of the processes discussed in this chapter; even without direct colonial administration, many of the same instruments of policy were transmitted into the hills which were, moreover, neglected in the provision of infrastructure which colonial development did provide from its revenues.

The aftermath of the colonial episode has included accelerating land degradation, famine and drought on the one hand, especially in Africa, and a reduced state involvement in conservation until very recent times on the other. The further development of dependent capitalism in Africa has taken the colonial evolution yet further, while unresolved conflicts left over from the colonial period have precipitated an almost continuous series of wars and rebellions, in Asia as well as in Africa. Africa, though not Asia, has also suffered a prolonged series of famines in which the role of land degradation may never be capable of definition, but is certainly substantial.

The two hypotheses with which we began are certainly sustained in general terms, but to say this is only to simplify a much greater complexity. The same relations of production and accumulation have led to degradation in some cases, and to successful adaptation in others. Technical aspects are clearly of importance, including differences in sensitivity and resilience of the land, and in the nature of the actual crops introduced. The social relations of production do not offer a complete explanation of degradation, nor does the role of the state, but nor do the natural conditions of environment and environmental variability.

There are certain conclusions that may, however, be drawn from this analysis. The lack of either appreciation of or concern for indigenous institutions and technologies has had devastating consequences, and imported methods as well as imported crops have often proved quite unsuitable. It does seem that a greater emphasis on food crops, and a greater devolution of resource management to community level could be advantageous, though at risk of local corruption. The story of the Indian forests stands against this approach, yet in neighbouring Nepal quite the reverse has been true. Strong central management to allocate resources to people and areas, which, by definition, are weak politically and economically, is one route to effective escape from marginalization in all senses. This degree of technical ability and legitimation is something which many post-colonial states do not have.

However, as we shall see later in the book, there is not much evidence that central management applied to a whole economy and its production decisions can do any better. There is one basic problem, and that is the element of coercion. In colonial societies, as in others, state involvement in surplus extraction requires coercion. State involvement must be seen to benefit the majority of land users. If it does not, then conservation is rightly seen as an outcome of a surplus-extracting policy of whatever political hue. The success stories of conservation under colonialism were all too often the 'uncaptured peasantries' of Hyden (1980), which one tends not to hear about, or those farmers who *did* benefit from commercial opportunities.

We emphasized at the beginning that the colonial experience was an exaggeration of a much more common experience of disruptive change. Taxation, land grabbing or amassment, demands for cash-crop production and the extraction of labour and of surplus were not unique to colonialism; they happened everywhere. Examination of the colonial experience enables us to analyse their consequences when they struck with a new force and the backing of a conquering alien power. In the next two chapters we shall find some of the same elements present in two very different situations and times. Yet the same powerlessness in the face of change remains.

7 Questions from history in the Mediterranean and western Europe

Piers Blaikie and Harold Brookfield

1 The longer view and its implications for analysis

Discussion in this chapter switches from the historical overview of chapter 6 to a more detailed examination of a specific historical case in Europe, perhaps the only area in which research provides us with the means of studying a problem of degradation which was once at least locally serious, but which largely 'went away' at least until very modern times. Even though the main evidence is only 200–300 years old, this case does raise the question of the possible significance of climatic change, and for this reason is prefaced by a review of research on degradation in the Mediterranean region, where the human or natural question has been closely studied for some twenty years.

To some degree, therefore, this chapter constitutes a test of the weights to be put on the different parts of our 'equation' on page 7 in chapter 1. Although not in a quantitative manner, it becomes possible to test how far human interference and a weakening of management can be held responsible for degradation in a case where the natural processes have also been variable over time. We do not reach firm conclusions, indeed cannot, but are able to suggest that variability in the 'human' elements in the equation can be given some degree of precedence even though natural events outside the present range of normal experience may also have been significant.

We employ evidence from two regions contrasted in both climate and history. The sub-humid Mediterranean is characterized by the high erosivity of its seasonal rainfall, whereas in northwestern and west-central Europe erosivity is much less, and estimated modern soil-loss rates in this latter region are among the world's lowest. Both regions are largely deforested, but in the Mediterranean the 'maquis' is widely regarded as a degraded form, whereas in western Europe deforestation has led to the successful creation of agricultural and pastoral landscapes. There is, however, evidence of significant erosion in this latter region during the last few hundred years. Most of this chapter is devoted to the latter topic and especially to the evidence from France and to a lesser extent Germany. We begin, however, with the Mediterranean, where a brief overview of the natural/human controversy will usefully set the scene for a discussion of the west European evidence, and also has great value in its own right.

2 Erosion in the Mediterranean: human or natural?

2.1 The setting and the problem

The Mediterranean region is one in which major political, economic and social change has taken place during the past several thousand years, and in which there is also strong evidence of significant climatic fluctuation in historical times. The evidence of degradation is widespread and striking; whole regions, with all their human works, have gone out of cultivation. Little remains of the indigenous forest cover, and large montane areas, especially on limestone, are now almost bare of soil. Moreover, as Hutchinson (1969) noted, it has for long been accepted as established that the heavy decline in agricultural potential of the region has been due to millennia of exploitative agriculture. Reading some of the classical writers themselves supports this view.

Tomaselli (1977) has briefly summarized some of the evidence concerning degradation of vegetation, noting that it was already a problem in Roman times, involving the creation of large pastoral estates and a decline in the quality of management. On the southern coast, in North Africa, the breakdown of complex irrigation works all the way from Egypt to Algeria was reviewed by Murphey (1951) who accorded with later writers in firmly concluding that known causes of human origin were quite sufficient to account for all the evidence. The old suggestion of climatic desiccation was rejected, but it did not die. Hare, who has worked in North Africa, concluded cautiously that the expansion of desert surfaces is 'largely due to unwise response by human societies to the strains imposed by naturally recurring drought' (1977: 88).

But it is not only drought that is significant. Most erosion is produced by water, and it is water that has done most of the damage. In the European Mediterranean, the problem was first put in context by Braudel who presented a great deal of evidence on the woeful condition of the sixteenth-century environment, a picture of neglect and exploitation of land and people alike, attributed by Slicher van Bath (1963) to economic causes. In seeking explanation of the 'inelasticity of agricultural production' which led to frequent food shortages in the cities, Braudel relied on the onset of wetter and colder conditions in the nadir of the Little Ice Age (Braudel, 1972/73: 25–102, especially 94–6, 267–75, 420–27).

Disentanglement of the human from the natural is especially difficult in this region, both because of its violent human history and because of the environmental sensitivity of the semi-arid Mediterranean. The record of the eastern Mediterranean is clearly susceptible to alternative interpretations, in which the place of intensive land management also surfaces. A long-term view is provided by Naveh and Dan (1973), whose discussion concentrates on the area best termed Palestine – Israel, the West Bank and western Jordan. Intensive pastoral farming with localized agriculture on moist sites (Levy 1983) between 7500 and 5000 years before present (YBP) gave way to a phase of intensive management involving terracing of the hills. With several regressions, this management system attained peaks in the early Roman and Byzantine periods when Palestine supported between three and five million mainly rural people. Soil and water conservation, terracing, crop rotation and manuring sustained a varied production. The subsequent breakdown of this system was accompanied by war and depopulation. Terraces were broken down; there was severe erosion and valley and lowland swamps were formed.

There was a not dissimilar series of events in Lebanon, but after a long period of exploitative deforestation, the mountains then became intensively occupied by Maronite and Druze settlers who by AD 900 established an intensive management system apparently similar to that practised earlier in Palestine (Mikesell 1969). Naveh and Dan (1973) opt mainly for human causes of degradation, but note that damage was more far-reaching and recovery much slower on the more sensitive natural environments. On the more resilient areas, many of the degraded ecosystems have shown remarkable stability at their degraded level. Naveh and Dan identify several cycles of degradation.

2.2   Evaluating the climatic factor

There is, however, evidence of erosion sequences going far back into prehistory, first put in concrete terms for part of Greece by Higgs and Vita-Finzi (1966) and Higgs, Vita-Finzi, Harris and Fagg (1967). These erosion sequences can be traced back to before the beginning of agriculture, but have continued since Roman times with alternate phases of deposition and downcutting. Pastoralism in the hills with arable land in the lowlands is a very ancient combination in Greece, and the long-term effect has been a transfer of capability from one to the other, impoverishing the vegetation and soil of the hills while at the same time creating alluvial lowlands capable of intensive irrigated agriculture. But the successive alternation of cutting and filling is not easy to explain, and Vita-Finzi (1969) came down heavily on the side of a climatic interpretation.

The critical evidence was from Tripolitania, in a series of ephemeral wadis in which Carthaginian and Roman colonists had increased the soil and water available from eroded pre-Neolithic fill by constructing low dams which trapped silt and held runoff, increasing water available for agriculture by as much as eight times. This system broke down when social conditions deteriorated after the Roman period, and most of it was abandoned after warfare and repeated invasions. Vita-Finzi then, however, calls attention to a new period of deposition, smoothing and steepening the stepped courses of the wadis and burying many of the Roman works. Searching around the Mediterranean, he found a similar though not exactly synchronous period of deposition in every country, commencing between AD 400 and 900 and continuing in all areas until the sixteenth century and locally even into the eighteenth century. A new and very active phase of channel erosion and river-mouth aggradation was then initiated.

In a review of this evidence, Harris and Vita-Finzi (1968) attribute the transition primarily to climatic change with management breakdown playing only a subordinate role which differed in nature and period between areas. A regional problem, they assert, requires a regional explanation, and they seek it in the possible effect of the 'little optimum' between AD 900 and 1200 on summer rain, then of the generally wetter conditions of the Little Ice Age after AD 1300. Renewed modern downcutting is favoured by restoration of drier summers, reducing plant growth and increasing runoff; human activity is simply abetting this phase of downcutting (Vita-Finzi 1969: 110). Tomaselli (1977), on the other hand, calls attention to major deforestation after the sixteenth century, which he attributes to growth of population.

In regard to the woeful sixteenth century, Vita-Finzi (1969) argues that the widespread formation of valley swamps in southern Europe was the product of late medieval aggradation in the valleys, the consequence of wetter conditions in the Little Ice Age and not of a decline in farming practice. Moreover, this phase created the valley plains which, under a higher technology, are now the basis of arable farming in the countries of southern Europe so that, however created, it was a 'boon' to later generations.

It has since become clear however, that some of the most degraded landscapes in southern Europe have been so for a very long period of time. Examining the archaeological and geomorphological evidence of a region of 'badlands' in southern Spain, Wise, Thornes and Gilman (1982) show that the drainage pattern and general slope configuration have remained unchanged for some 4000 years, that subsequent erosion has been localized along drainage lines, and that 'modern' erosion is not catastrophic, otherwise the numerous archaeological sites in vulnerable positions would have been destroyed. Some high recent recordings of soil loss may have been affected by a decline in management standards, especially of terracing, during a modern period of rural population decline. Slope processes consist largely of slumping and solution-collapse, and the main work of high-volume, low-frequency rainfall events is to remove the product of this long-period trimming. Evidence of massive and seemingly active degradation is not always to be taken at face value. However, elsewhere in the Mediterranean, in Italy, for example, badland erosion is demonstrably active and very damaging to farmland (Alexander 1982).

On the basis of evidence such as that advanced by Vita-Finzi, together with results of other research in North America and New Zealand, Gregory and Walling concluded that 'whereas the effect of human activity is notable at the short term ... this has usually produced changes of degree rather than kind' (1973: 378). Since that time, at least in the Mediterranean context, there has been something of a swing of opinion back toward giving greater emphasis to the human factor. Davidson (1980), for example, disputes the synchronous nature of erosion and suggests that the question needs to be handled more in multivariate terms. Gross vegetation changes in the pre-classical period cannot be accounted for by climatic change, while the post-classical onset of aggradation also becomes more complex with further research.

In regard to the decline of agriculture in North Africa after Roman times, Shaw (1981) concludes from a thorough re-examination of the evidence that Roman climatic conditions were probably similar to those of today, and while coming down in favour of Vita-Finzi's medieval humid phase as a cause of general aggradation between about AD 1000/1100 and 1750/1800 offers a suggestion that is of value to our argument:

> Is it not possible that large-scale land use by Man in the circum-Mediterranean region may actually have served, on the whole, to impede massive erosion rather than to encourage it? And is it not possible that in the centuries after c.1000, in a period during which intensive agriculture was at its lowest ebb, 'natural' forces might have taken their toll of the environment when the artificial restraints of human cultivation were not at their most effective? (Shaw 1981: 395)

Naveh and Dan's (1973) material would fit this explanation quite well. Together with Davidson's (1980) and Shaw's (1981) work, it suggests that the human role may have been both undervalued and even misinterpreted in the argument.

3 Northern and eastern France in the eighteenth century
3.1 The problem and the evidence

During the period immediately after the Second World War, the effect of a large literature on erosion and conservation of the soil emanating from the United States, coupled with the disturbed and dislocated condition of agriculture in the war-ravaged regions, led to a short-lived surge of concern with the problems of soil erosion in France and Germany. A number of studies of contemporary problems appeared around that time (e.g. Furon 1947; Wolff 1950/51; Schultze 1952). While this brief concern faded as European agriculture so manifestly recovered, specific work on soil erosion has persisted, and has enjoyed something of a revival during the period of more anomalous weather that began in the 1970s. Although there has long been concern over land deterioration in the uplands (Mather 1983) it is only now suggested, for the first time in many years, that even the arable lowlands of Europe have undergone, and are undergoing degradation involving loss of fines at rates which threaten the structure of the soils (Morgan 1977, 1980a, 1980b, 1985; Kirkby 1980; Richter and Sperling 1976). There has also been a number of more specific inquires (e.g. Macar 1974; Gabriels, Pauwels and de Boodt 1977; Richter and Negendank 1977). One recent inquiry, designed simply to establish quasi-natural rates of erosion under forest in Luxembourg (Imeson, Kwaad and Mücher 1980) discovered that present rates have obtained for only some 200 years, before which there was much heavier erosion under a non-forest cover. References to older erosion also appear in other studies.

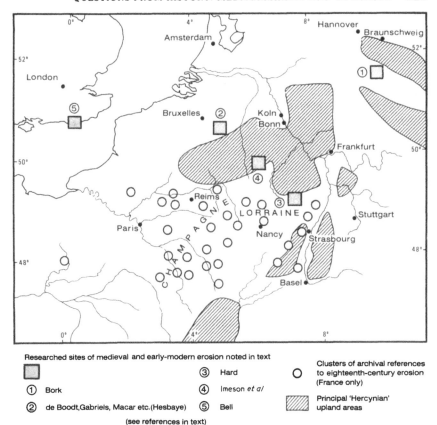

Researched sites of medieval and early-modern erosion noted in text

		Clusters of archival references
③ Hard	○	to eighteenth-century erosion (France only)
① Bork	④ Imeson *et al*	
② de Boodt,Gabriels, Macar etc.(Hesbaye)	⑤ Bell	Principal 'Hercynian' upland areas

(see references in text)

Figure 7.1 Part of northwestern Europe showing areas in which medieval and early modern erosion has been recorded

Up to the present time, however, the actual evidence of erosion in agricultural areas of Europe in early modern times has scarcely at all been incorporated into the literature. A small number of publications in major journals (e.g. Vogt 1953, 1958a, b; Hard 1970; Hempel 1968; Bork 1983) has presented clear evidence that such erosion occurred and was serious (see figure 7.1), but the research behind them in both Germany and France still suffers neglect. Discussion in this chapter concentrates mainly on the French material, being more accessible to the authors linguistically, and this research is almost entirely the work of Jean Vogt, a geographer trained first in agrarian historical geography, then as a geomorphologist, who became a professional geologist. What follows rests quite largely on his work, and on two intensive days of discussion between him and Brookfield in Strasbourg and in the field in 1985.

Most of the hard evidence is archival (Vogt 1977), although the traces of former erosion can be seen in the smoothed over gullies that notch many hillslopes, and discovered in the colluvium that buries medieval structures in valleys; it is argued that much of the Pleistocene 'limon' has been reworked in this way and in its present formation has a more modern interpretation (Hard 1970; Vogt 1970b; n.d.; personal communication). Birkenhauer (1980) reverses this relationship by arguing that the periglacial formations are particularly susceptible to erosion, a view strongly reinforced by the work of Bork (1983). The archival evidence rests on local inquiries and reports, some of them the precursors of modern 'disaster' reports and not unlike them in that the object was to establish a case for aid, in the form of tax- and tithe-remission. Some specific agricultural inquiries initiated by the *ancien régime* and under the Revolution are important, and so also are the *Cahiers de Doléances* prepared for the reconvened Third Estate in 1789. Partly because of the tight control established under the *ancien régime* and the heavy rural taxation that afflicted French agriculture (Abel 1980: 161), the record peaks in the eighteenth century and falls off badly in the nineteenth when many documents were destroyed. But it extends from the sixteenth century to the twentieth, and the fact that it peaks in Germany as well as in France in the eighteenth century lends support to Vogt's contention that this peak is not only an artefact of the data. More recent evidence from Germany points to the same conclusion, but by taking account also of stratigraphic evidence from the medieval period, it opens a larger question for research (Bork 1983).

3.2 Human or natural causation?

Given that the erosivity of rainfall in northern France is only a quarter of that in the Mediterranean and peaks only briefly in the early summer months (Pihan 1979) there would seem to be little reason to suppose natural causation of the events described in the archives, over so long a period of time. However, 'Little Ice Age' conditions did prevail during this period, and we shall see later that violent rainfall and snowmelt may have been more prevalent than they are today, though perhaps by no great degree. Moreover, a high proportion of the cases reported are concentrated on the leached soils developed on periglacial 'limon' (loess) which mantles quite large parts of the region. Within these Aqualf soils a large part of the clay fraction has been removed to depths greater than 40 cm, most probably under more humid conditions during the earlier Holocene; moreover, ploughing leads to the formation of a pan at the depth attained by the plough (Gras 1979). By its compactness this pan inhibits percolation and encourages runoff through and over the clay-poor surface horizon. As we shall see later, the depth of ploughing has been an important factor in the explanation of erosion.

The varying depth of limon on hilltop, hillside and valley is suggestive of substantial erosion in a probably ancient past, or during and since the 'great age of clearance' of the early medieval period. In the Hesbaye region of Belgium erosion has been mapped in relation to date of clearance from forest since the early 1800s, and there is a clear association between the distribution of truncated soils on valley sides and the period of time elapsed since clearance (Bolline 1979). Macar (1974) has calculated losses of between 3 and 9 t/ha/year on these slopes, while de Boodt and Gabriels (1979) measured losses up to 10.3 t/ha over a 16-week growing season, with from 70 to 90 per cent of the total loss occurring during the first week of heavy rain before a ground cover was established on the ploughed soil. Modern deep ploughing has increased percolation, and 'chisel-ploughing' has been employed to break up the pan; however, early land-consolidation schemes on limon areas of Lorraine and elsewhere, with removal of hedges and other erosion barriers, led to some alarming losses so that in 1975 the requirement of environmental impact assessment was incorporated in legislation for land-consolidation schemes in France.

The high erodibility of the loess regions in particular makes them particularly sensitive to unusual weather events as well as to the effects of inadequate management; the highest erosion rates in the world are on the loess lands of north China (chapter 11). It is this high sensitivity which makes the loess region critical for explanation, and recently Bork (1983, 1986) has established two periods of severe gully erosion and colluvium formation in an upland loess-mantled region between the Harz and Göttingerwald in Niedersachsen, West Germany. Using pottery and other buried artefacts to date carefully studied profiles, Bork has established a major period of erosion between about 1300 and 1400, some 200 years after the clearance of the region from forest, and lesser damage in the period between 1700 and 1800. Between these two periods, for the earlier of which a very high mean annual erosion rate of 48t/ha/y was established on the slopes, surface deposition of colluvial material has filled the gullies and smoothed the landscape. Figure 7.2, redrawn from Bork (1983: 43), summarizes the evidence. Especially in regard to the greater erosion of the fourteenth century, when gullies of 10 m depth were developed in formerly arable land, Bork relies on a climatic explanation, specifically heavy summer rains and hail associated with the unsettled climate of the early part of the Little Ice Age (Lamb 1977, 1984). This positive statement of a climatic explanation requires full consideration.

3.3 Erosion and agrarian pressures in the lowlands of northern and eastern France

Most of the erosion described in the eighteenth-century documents would seem to have been sheet and rill erosion, but there are also reports of substantial gullying, sometimes several metres in depth, and there are numerous reports of colluvial deposition in lowland fields that buried crops, blocked roads and sometimes invaded villages. Most reports relate to summer storms, but they extend through the year, and there is a further significant group of damage reports following rapid snowmelt, probably over frozen ground. A search through the records of northern and eastern *départements* at intervals over thirty years led Vogt to the discovery of such reports in all

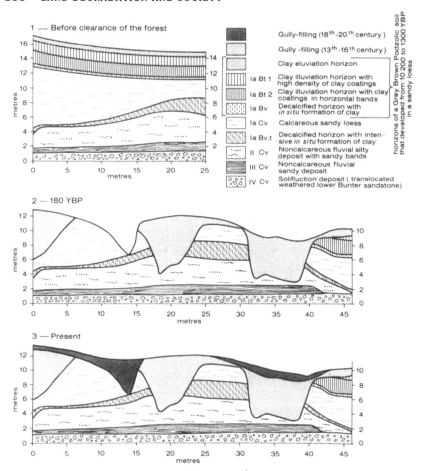

Figure 7.2 Profile section through a loess soil at Rüdershausen, Niedersachsen, West Germany, showing medieval and early modern gully erosion. *Redrawn from* Bork (1983) figure 15, *by permission of the author and Catena Verlag*

areas, but with a notable concentration in the eastern part of the Paris Basin, particularly in Champagne and Lorraine, largely, though by no means entirely, on the limon-mantled areas. Because of the spotty nature of the data, Vogt has not used cartographic methods of display and analysis (Vogt 1957a, b, 1968, 1970a, b, 1967–71, 1972a, b, c, 1966–74, 1975, 1982, n.d.).

Some reports are concerned with 'catastrophic' damage for its own sake, but many others are also concerned with the effect on production and livelihood. Serious decline in yields is often noted and some of these reports suggest agronomic explanations. Thus in Lorraine during the 1740s village priests suffered a severe diminution of their income from tithes and were asked to explain the reasons. They blamed widespread erosion, especially in the fallow fields, and in turn attributed this to shallow ploughing due to shortage of working livestock following epidemic mortality and heavy corvée demands for transport. Because fields were barely scratched percolation was limited and runoff excessive, carrying not only the soil but also the manure spread on it. A severe decline in yields was the result (Vogt 1957a). This sort of evidence is repeated widely.

A clear association with the practice of bare, ploughed fallow was noted by many contemporaries, and is particularly clearly demonstrated for western Germany by Hard (1970). Grain-crop or grass-protected land suffered less, and where crops such as hemp, barley vetch and other fodder mixtures and later clover were introduced in place of the bare fallow there was notable improvement; the introduction of root crops yielded intermediate results. This agricultural transformation was, however, only patchy before the nineteenth century, and in most of the eastern Paris Basin the three-field system persisted, though not without change. In the period of rising grain prices after 1740 there was growing pressure to enclose part of the common lands on which the poor depended to feed their livestock during most of the year. These livestock, principally sheep, provided the manure for the fallow field on which they were also pastured, and on the smaller area of spring-sown fields ('marsages') on to which they were turned in winter. Reduction of the common lands through creation of new commercial farms made the 'stinting' of common rights necessary so that livestock numbers were reduced, thus in turn reducing the supply of manure. To this must be added frequent epidemics which destroyed working livestock. It is this lack of manure which, added to shallow ploughing or insufficient ploughing, was probably at least as much responsible for reduction in yields as top-soil erosion itself. As Sutton cites from an eighteenth-century parish register: 'without wasteland there would be no sheep, without sheep there would be no manure for the farms, and without manure there would be no corn in the fields' (1977: 250).

Pressures on rural France were already more severe than in other parts of Europe even in the seventeenth and early eighteenth centuries due to heavy taxation added to the long-term decline in prices. On smallholdings in the Beauvais district taxes took as much as one-fifth of the whole yield (Abel 1980: 164). Thus, when yields were poor, farmers even on 10 ha might be unable to feed their families through the year. With declining yields due to erosion and lack of manure the situation grew worse. At Guindrecourt, Haute-Marne, more than half the smallholders were eliminated in a dozen years before 1788 (Vogt 1972a), and in Bourgogne in 1773:

For the past twenty years the land has been so barren and degraded by storms that in several hamlets they have combined two holdings into a single one, yet still insufficient to provide the cultivator with his share. (Archival statement quoted in Vogt 1970b: 34; our translation)

3.4 The final crisis of the ancien régime; *human and natural forces in conjunction?*

Pressures increased severely in the second half of the eighteenth century as grain prices rose, encouraging commercial farming while those earlier rendered landless could not buy their food, and at the same time rural population increased substantially. This was the period of which Tricart wrote in first introducing Vogt's work:

> The great plough-up of the eighteenth century, systematic organization of strip cultivation and the cultivation of waste land, tied to the growth of population and deep penetration of bourgeois capital, all provoked a violent crisis of erosion. (1953: 155; our translation)

The squeeze was twofold. On the one hand the urban demand raised the price of grain, encouraged consolidation of farms and clearing of the commons, and led to land speculation. Much land was bought by bourgeois entrepreneurs and other absentees (Hoffman 1982), yet few large landowners invested in agricultural improvements (Forster 1970). In consequence, and right down to the 1840s, increases of grain production were obtained by extension of area, and not by any significant improvement in yields (Morineau 1970). On the other hand the growth in rural population strengthened the tenacity with which the smallholders defended their established system with its fixed rotations and common land. Grantham (1980) argues that this latter was the more powerful force. Lefebvre's argument still has strength:

> the rural masses considered these collective rights as a property that was as sacred as any ... the very existence of most peasants depended on them. Those who worked only a small plot of land, and even those who did not have any land, were able to raise a cow, a pig or a few sheep thanks to the communal pastures. Thus progress in agricultural techniques could only be achieved at the expense of the poor (1929/1977: 36)

This is not an uncommon situation. As we shall see in chapter 10 it also holds good today. But in France new elements were added to the situation in the last decades of the *ancien régime*. The shortage of cattle became worse as numbers declined during a series of epidemics; they may have diminished by as much as a third between 1760 and 1787 (Festy 1947: 23). Moreover, lack of wood led to a growing use of straw for fuel, rather than for bedding with manure. The land was increasingly deprived of the means for reproduction of its capability. By the 1780s not only had rural poverty increased dramatically, but it seems likely that the land had also approached limits of resilience. In Vogt's (1972a) paper on Haute-Marne, for example, fifty of ninety reports of damage relate only to the years 1783–8.

From the 1770s onward, meteorological information begins to become available for substantial parts of western Europe. During several periods about this time a winter and early spring pattern of blocking high pressure over Europe was abruptly replaced by what has been called the 'European monsoon', giving rise to violent storms in late spring and summer; in other years a blocking high pattern persisted through the summer, and was the cause of severe droughts. The climatic pattern was more 'continental' than it became in the nineteenth and twentieth centuries (Lamb 1982), and the mean erosivity of rainfall was probably greater in consequence. As we shall see later, Bork (1983) argues that such conditions also existed in the fourteenth century. It is important to use this information to establish whether a climatic causation for the apparent peak of erosion in France and Germany in the late eighteenth century can be regarded as a plausible hypothesis.

Based on his reconstruction of daily weather maps for 1781–9, the years immediately before the French Revolution, Kington (1980) concludes that the decade was one of unusually high climatic variability in western Europe, even if conditions were not as severe as those of 1812–17 which gave rise to the last major subsistence crisis in continental Europe (Post 1977; Pfister 1981). A good period for agriculture from 1779 to 1781 gave way to a wet spring and summer in 1782, to a summer of heavy thunderstorms in 1783, then to an exceptionally cold winter in early 1784, followed by a rapid melt and spring hailstorms. This latter event seems to have been responsible for some of the worst erosion damage recorded in northern France. This was followed by a long drought extending into 1785, comparable in length and severity with the recent 1975–6 drought. In 1785–6 there was a further severe winter prolonged into the spring, and followed by a further drought; subsequent heavy rain and hail falling on dry soil with thin ground cover could certainly have triggered the substantial damage experienced in Haute-Marne in 1785 and 1786. There were then further major hailstorms in 1788, which did severe damage in Paris as well as to the land, while generally wet conditions gave rise to a very poor harvest and hence to the politically critical food shortage of 1789 (Neumann 1977). Was the damage simply the result of this unusual weather?

The detailed research that has gone into the social and economic conditions of this decade allows us to put this evidence on erosion and climate into a meaningful context. It seems clear that the most crippling climatic event was the drought of 1784–5, coming in the middle of a set of poor growing seasons. The drought led to great mortality of cattle, and to forced sale at low prices so that the effects of farmers' efforts to rebuild their stock were still reflected in high prices in Britanny as late as 1789 (Sutherland 1981). Loss of livestock would have had further serious consequences for soil fertility. In addition there was extensive infant and child mortality, and in northern France there was an increase in vagabondage and pauperization. 'Subsistence crises' occurred in the north in 1784, in the west in 1786, and throughout most of France in 1788–9, when the price of grain in northern France doubled (Labrousse et al. 1970: 551) and precipitated the revolution of 1789.

It seems clear that the production system as well as the social system was severely stressed by the cumulative effect of multiple bad seasons in the pre-revolutionary decade. The climatic events that precipitated distress were not only those violent events which triggered erosion. We cannot really say that climatic events 'caused' either however, since a long period of increasing distress was more fundamentally involved, accompanied by deterioration of the whole system of production. But we can at least identify the immediate causes of both reduced production and damage to the land in exceptional weather.

Most contemporary reports attribute low yields directly to erosion damage. For example, at Chevillon in Haute Marne in 1787, gully erosion had taken away the 'goodness of the land' (les sucs des terres labourables) so that yields were miserable (Vogt 1972a: 78), while at Petites-Armoises in Champagne the least storm took away soil, so that for many years the produce had been insufficient to feed the inhabitants (Vogt 1972c). However, the loss of fertility and the loss of soil, especially through gullying, were interrelated rather than forming part of one and the same process. A deeper sickness of the rural economy underlay both, depriving the hard-driven land of its capability as well as rendering it more sensitive to erosion.

This conclusion has to take account of some differences among modern historians concerning conditions in rural France around the time of the 1789 revolution. Le Roy Ladurie (1975) stresses that the eighteenth century was marked by progress from the 'final great crisis' in the 1690s toward steady growth, with commercialization responding to improved communications. The increase in the agricultural population to 18 million (of a total national population of 27 million) was, after the mid-century, increasingly due to a decline in the death rate, indicative of rising living standards, however slight. Tax, tithe and rent incomes increased, but so did incomes even among the peasants among whom a new class structure began to emerge, leaving a proportion – small farmers and the landless – increasingly apart from the more prosperous. To Ladurie, the essential fact was the greater awareness of the peasantry, more mobile and often with migrant relatives in the cities, that created the dissatisfaction and hostility which caused a large part of rural France to join the initially bourgeois revolution. The 'double modernization' of the landowners, who became more innovative and more commercial, and of the peasantry, who gained only a little but came to expect much more, and the contradictions thus created were no more than exacerbated by the economic problems of the 1780s. The revolution arose on a rising, not on a falling tide.

Agulhon and Désert, more concerned to explain the conflicts that quickly developed between the peasantry and the revolution, stress two fundamental considerations:

> to be peasants, that is to say to have their fate bound up with a fragile rural economy, and dependent upon climatic accident and upon the inertia of the commercial system; to be subjects of the King of France, and, through him, bound to the vicissitudes of what was already a global political economy. In the two years before 1789, it was the simultaneous crises of

the economy and the government under the monarch which...propelled the peasants into the struggle.' (1976: 20; our translation)

Great numbers of peasants without land, without reserves and without resources, found themselves reduced to begging and vagabondage by the bad seasons of the 1780s. Already marginalized (PE), and driven by desperation and fear to revolution, they were hastily appeased by the abolition of feudal obligations and tithes but not relieved of other charges; hence their rapid disillusionment.

Both these interpretations clearly have truth, and the difference is one of emphasis. The 1780s were a period of crisis in rural France, and erosion and land degradation were more symptoms of crisis than the cause. For the peasantry, even harder times were to follow in the 1790s, and conditions remained hard well into the nineteenth century. So far as there was land redistribution, the main benefit was to the larger commercial farmers. Whatever the reality of the 'agricultural revolution' in France, it was slow (Clout 1977). It had already begun in Alsace and some other areas in the 1750s (Vogt 1983) and it was not everywhere complete early in this century (Lamartine-Yates 1940). It was only gradually that the 'fluid' landless population first recognized by Meuvret (1946) as a distinct class disappeared through migration or absorption into a new *mini-fundista* class who acquired pieces from larger landholders, and tended to farm them more carefully (Fel 1977).

Catastrophic erosion continued to be reported in the 1790s, and again early in the nineteenth century, but then only occasionally in subsequent years. Some diminution in the erosivity of storm rainfall as climatic conditions became more maritime may perhaps be partly responsible, but it is more likely that improved agricultural practices provide the principal reason. There was already substantial interest in contour ploughing and other measures in the eighteenth century (Vogt 1970b, n.d.), but more important was gradual economic improvement which facilitated greater fertilization, elimination of bare fallow and especially deeper ploughing which led to greater percolation and reduced runoff (Vogt personal communication). Most of these slow changes went unrecorded except at random, but they led Festy (1947) to conclude that what had happened was a technical evolution in agriculture, simply facilitated by social and political revolution. Today, deep ploughing permits farmers in Champagne to plough their fields up and down slope without suffering severe erosion on land where extensive gullying was reported in the eighteenth century (Vogt 1972b, c). It is only on light soils with intertilled crops, as in the vineyards of Alsace, that erosion is today a major hazard. Except in the most sensitive areas, the health of farming and the health of the land seem to be closely related, and this is the relationship that seems to be the dominant explanation. This is perhaps the most important lesson to be derived from this long-term view of the experience of one region in western Europe.

The case of the upheavals in the Soviet Union from the 1920s and their impact on land management is outlined in chapter 11A, and bears comparison with this account of eighteenth-century France. In both cases the soil concerned was loess which, although potentially very fertile, needs careful management and is highly sensitive to human interference and climatic events. In both cases profound and rapid change altered the economic and social conditions of those who worked and managed the soil. There were important elements of pressure too, from the state to produce surpluses without the resources to do it in a sustainable manner. At this point the two 'chains of explanation', while broadly similar in some important respects, clearly part company when the social explanation of these changes is attempted. In the French case it was a pre-revolutionary change in response to capitalism, whilst in the Soviet case it was a set of post-revolutionary changes which set the scene for serious land degradation.

3.5 The situation in the uplands

The record in the uplands of western Europe tells a different story, for erosion appears to have been of a more massive nature and was associated with a cultivation system that in many areas was basically shifting cultivation. The consequence has not been repair so much as emigration and abandonment of farming. This was the situation discovered in Luxembourg by Imeson, Kwaad and Mücher (1980), and it is described for different localities by Vogt (1953, 1957b, 1960, 1972d). Erosion seems to have become severe at an early date, for Vogt (personal communication) found a seventeenth-century archival reference from the Massif Central in which it was recorded that the landholders were returning their land to the king, rather than continue to pay taxes on it, because it was so eroded as to be no longer worth their while to cultivate. The early commencement of massive emigration from the uplands of France to the cities may reflect such degradation on land of high sensitivity and low resilience (Labrousse *et al.* 1970).

The problem became more severe in the eighteenth century as settlement by refugees from war-torn areas was followed by demographic expansion, and closer settlement was facilitated by the introduction of the potato as a food crop. Fallow periods of from ten to twenty years were shortened in an all too familiar manner. By 1790 some sloping land was no longer able to produce crops, and became deeply gullied. Debris cones formed in valleys where they can still be recognized; ponds became silted, and a parlous economic condition took the place of expansion fifty years earlier (Vogt 1953). A particularly severe storm in the early summer of 1774 created deep gullies and extensive debris cones in the Vosges, especially on land used to grow potatoes (Vogt 1972d). In time, controls led to better management of mountain land, but the real control lay in the abandonment of large areas to recover under the forest that still mantles much of the uplands today.

4 Questions over a longer history in western Europe

4.1 Scale and the visibility of a problem

Despite the severity of the erosion in contemporary documents, it does not seem to have attained great visibility at national level so that historians of agrarian France have remained largely unaware of it. Nor did that competent observer Arthur Young (1892: 31) encounter more than a single case in his travels in France in 1787–9, and that case was in the south. This fact alone suggests that erosion itself was localized, the 'tip of a larger iceberg' of degradation, static or declining yields, and rural distress. There are other reasons for supposing that this was so, for the accounts cited by Vogt describe only few cases of floodplain aggradation or deposition in the major rivers, and many more of colluvial deposition in the fields, though sometimes to depths of a metre. For central Germany, Bork (1983) finds that 87.5 per cent of eroded material was redeposited within his research area, with only 12.5 per cent carried by water beyond its boundary. Discussing the relationship of colluvial and alluvial sequences in Britain, Bell (1981: 87) stresses that 'colluvium basically reflects localized erosion whereas alluvium reflects the larger scale changes affecting a significant portion of the catchment'. Bell, who found medieval pottery sealed under colluvium on the Sussex chalklands, provides one of the very few tangible reports of post-medieval erosion in southern Britain where Burrin and Scaife (1984) find that there has been little floodplain aggradation since the end of the Bronze Age.

The inference, already foreshadowed above, that erosion was only a small part of total degradation in eighteenth-century lowland France, has implications for our argument. It suggests that the presence of erosion in unusual or unexpected places and times may be an indicator of much deeper problems. If we are to draw the full value from the examples discussed in this chapter and the preceding chapter, it is important to explore this question.

4.2 Degradation and stress in the agrarian history of Europe

The economic history of Europe is now being established in terms of long cycles of population growth and decline, and of rising and falling prices for grain and other commodities. The seminal work of Slicher van Bath (1963) and of Abel (1980) has brought these trends together on a continent-wide scale, and similar trends have been established in Britain (Wrigley and Schofield 1981; Clay 1984). Without seeking to go into the detail of a large literature, it is clear that a period of growth in the early medieval period saw great expansion of cultivation, followed by decline and relative depression in the later middle ages; a new surge in prices and population in the sixteenth century was followed by stagnation and decline between about 1650 and 1750 after which a further surge was only broken for short periods into modern times. Underlying this, however, was an almost continuous decline in real wages after about 1300 so that thirteenth-century standards of living were not restored until around the turn of the present century.

The strength of the economic forces at work is such that environmental aspects have been discounted to the margin in the historical literature, except by Braudel and Le Roy Ladurie, both with reservations. Climatic historians have now shown that climatic trends partly paralleled the economic trends (Lamb 1977, 1982, 1984) but this evidence has been put aside or ignored in a kind of non-debate which demonstrates yet again the ability of different groups of scholars to talk past one another. Slicher van Bath (1963) alone showed how degradation of some marginal land surfaced during the fourteenth and fifteenth centuries and again in the eighteenth; his examples are of light, sandy soils where wind erosion followed neglect of good management. Abel (1980: 80) called attention to the geographical distribution of 'lost villages' in late medieval Europe, and suggested that marginal areas occupied only late during the preceding 'great age of clearance' lost most heavily in the subsequent decline, a point also made for Britain by Postan (1959). We cannot enter this debate, but we can raise certain questions of significance to our own argument.

If we were to accept the hypothesis that most modern land degradation is due to increasing population pressure on resources, we would note the late eighteenth century peak as confirmation, and would look also at the sixteenth century and especially the growth period between the eleventh and thirteenth centuries when, around 1300, it has been argued that western and central Europe were overpopulated or at least crowded in relation to technically possible production (Postan 1959; Abel 1980: 40). The rapid extension of settlement, some of it on to land never subsequently cultivated, and the marked subdivision of farms suggest the presence of agrarian pressure, and peasant yields may have been declining.

4.3 The German evidence of medieval erosion

The incontrovertible evidence of massive erosion in Niedersachsen presented by Bork (1983, 1986) belongs, however, to the succeeding phase of rapid population decline, and in his later paper Bork indexes it more precisely to the first half of the fourteenth century, the period of the great famine of 1315–17, the Black Death and subsequent disasters. Moreover, Bork (personal communication) suggests that this was a period of widespread erosion in central Europe. The erosion continued into the fifteenth century after which it was replaced by colluvial gully filling until the eighteenth century. Bork's explanation is climatic, relying on 'catastrophic rainfall events' associated with the stormy period in the onset of the Little Ice Age. However, inferred peaks of late summer rainfall intensity are identified at several periods between the fourteenth and eighteenth centuries (Flohn 1949/50; Lamb 1984: 40). Is it not perhaps also relevant that this was a time of great agrarian distress (Slicher van Bath 1963)? Kriedte (1983:8) argues that the distress was aggravated by the imposition of greater burdens on the surviving peasantry by an aristocracy deprived of rent-paying tenants and seeking to resolve its own problems by military adventure. This much is, at least as of now, more firmly established than a sufficient increase in erosivity which could account for the damage and devastation of the order clearly identified.

The German evidence is none the less startling, and it is perhaps more startling that it is not paralleled elsewhere. Bayliss-Smith (1979) has examined fifteen researched cases of alluvial aggradation in Britain and finds only two in which the inferred cause is arable cultivation during the medieval period. In the British literature the emphasis is on pre-Roman degradation and erosion (e.g. Smith 1975), and notwithstanding all the unusual features of the loess regions it seems remarkable that events of such magnitude as those identified by Bork (1983) seem to have been absent from other west European environments.

What we do find in Europe is evidence of fairly massive erosion in the eighteenth century and locally of greater damage in the fourteenth century, periods respectively of rising population and economic growth and of falling population and economic decline. In so far as we are seeking social causes, this is by itself unhelpful. Perhaps we need also to look for some other common complex of causes.

4.4 Towards a resolution of the problem

May we hypothesize that such a complex of causes may be found in rural distress on the one hand, coupled with greater environmental sensitivity induced by climatic variability on the other? The pressures which built up toward the end of the 'great age of clearance' were compensated for by a large population with which to manage the land. The removal of much of this labour, coupled with distress and severe pressures, exposed other weaknesses. In particular, animal husbandry had been weakened because grain farming yielded more food than livestock, and the subsequent return to animal husbandry was more for commercial than subsistence purposes, accompanying therefore an increase in wage employment, and exposing more of the agrarian population to the decline in real wages that set in before 1300. These trends continued right through to the eighteenth century and contributed to successive agrarian crises from which only one country – Britain – succeeded in escaping by the end of the period. In Britain alone, by 1750, the poor were no longer the managers of much of the land.

The Little Ice Age itself may have affected yields, but hard winters were followed by what were generally regarded as good harvests (Titow 1960). When there were poor harvests, the rural poor suffered because of their dependence on purchased food. Although more substantial farmers often made efforts to combat erosion problems, some of them using methods of surprising 'modernity' (Hard 1970), the increase in rural poverty had as its consequence a deterioration of the quality of land management which became critical in times of severe pressure such as the later eighteenth century.

With Vogt and Hard, therefore, we prefer an agrarian to a climatic explanation for the incidence of land degradation in western and central Europe, except in ecologically marginal areas, though we must recognize the probable incidence of rainfall of higher erosivity at periods during the Little Ice Age as an *immediate* rather than as a basic cause. It seems more productive to focus on declining quality of land management due to the increasing poverty and distress of those who worked most of the land, both in the later middle ages under heavy feudal exactions and in the eighteenth century under growing commercialization and attendant differentiation and hardship for the disadvantaged. It is the social condition of those who work most of the land that is relevant. The climatic element must be separated into long-term trends, which have an effect on welfare and hence on the social facts of the time, and 'accidental' events which have their worst consequences for land that is ill-managed. This hypothetical system of explanation can, to some degree, be supported from evidence in the late eighteenth century, but for the longer historical period it can remain only a hypothesis until work such as that of Bork (1983) in Germany and of Bell (1981) in southern England are more widely replicated, and can tell us more of what happened to the land during the long period between the Dark Ages and the early modern period. Then it will be possible to relate the history of the land to that of its managers, but until now this question seems not seriously to have been asked.

As a final footnote to this discussion of west European evidence, it is therefore perhaps worth remarking that the British case remains puzzling. Perhaps a closer look at the British colluvium might yield more evidence than now exists of localized erosion in the past? On the Wiltshire chalklands, and elsewhere in Britain, one can see the same sort of grassed-over gullies that one sees in France, and the micro-relief of many tracts of East Anglian waste shows the same indications of former erosion that Hard (1970) illustrated in such land in Germany. It would surely be too much to suggest that the experience of the land of Britain was altogether different from that of its neighbours, for the same price and population swings and the same climatic events occurred on both sides of that narrow strip of stormy water that provides such a barrier to mutual understanding and exchange of ideas and information.

5 Conclusion

The two examples discussed in this chapter are both inconclusive, although in both some reconciliation of competing explanations seems possible. In the west European example, the greater problem seems to be lack of any full evaluation of the evidence, as well as lack of agreement among different groups of scholars. While land degradation in western Europe has not been such a problem as it has been in the Mediterranean, it has nonetheless occurred, and while fewer areas have been degraded so much that their whole landscape has been modified there are some such, for example, in the highlands of Scotland (Watson 1939; McVean and Lockie 1969) as well as in the Hercynian uplands of continental Europe. But while in the Mediterranean case there remains some substantial room for doubt concerning the relative roles of people and of natural change, there is less room for doubt in western Europe, unless it can be established that frequent heavy storms so augmented erosivity during the Little Ice Age as to change the whole environment of farming. Present evidence is insufficient to sustain this alternative.

Three types of conclusions of relevance to our problem can be drawn from this analysis. First is the importance of the quality of land management in relation to crops and land-use system. Variability in depth of ploughing and application of manure emerge as central issues, the understanding of which involves consideration of the whole agrarian system and its changes through time. The constraints of a rigid land-allocation system reduced adaptability. The introduction of new crops, particularly the potato in uplands, required adaptations which were not easily made, given the social constraints. These are conclusions applicable in other regions and other times, and they illustrate the need to focus on those 'social facts' which are of greatest relevance to agronomic practice and its rigidity or flexibility.

Second is the complexity of the explanation, especially where it becomes necessary to take full account of environmental variability and popular change. We might refer back to our discussion of degradation under colonialism (chapter 6). Different though the two sets of circumstances are in almost all respects, there are elements in common. Exploitation of the peasantry and state and private extraction of surplus to such an extent as to constrain severely the ability to manage the land emerge in both cases, and the 'chain of explanation' – from the farmer to the conditions of the state and its economy – has parallels.

In eighteenth-century France, modern India (chapter 10) and both early modern and modern Nepal, essentially feudal systems retaining communal elements are seen under the pressure of externally driven transformation, supplemented by rising pressure of population on resources (PPR). Because they lie in different historical periods comparison has to be handled with caution, but it would be folly on that account to reject the lessons of a case in which degradation did *not* lead ultimately to disaster. The social transformations which later took place in France had the effect of facilitating better management of the land. Certainly, we cannot say that the huge social and economic transformations now taking place in Asia – harsh in their immediate impact on sectors of the population, as they were in Europe – will lead to any similar result. Equally, however, we cannot say that they will *not* have such a beneficial long-term effect. The lessons to be learned from France concern the 'how' of the 'why' this occurred.

Third, the importance of the occasional extreme event is exhibited in this case, more clearly than in any other discussed in this book. The point that emerges is not that a single disaster creates degradation, but that a succession of disasters has a particularly damaging effect when ongoing social and economic conditions are such as to expose the production system and the land to abnormal harm from such events. This is a conclusion that is certainly not bounded in time or place, and it emphasizes the vulnerability of the several systems under stress which we describe elsewhere in these pages. This conclusion also points up the potential linkage between the study of land degradation and that of natural hazards and their impact, a question which we have already developed in chapter 1, section 8.

8 Degradation under pre-capitalist social systems

A Degradation and adaptive land management in the ancient Pacific

Piers Blaikie and Harold Brookfield,
with William Clarke

The Pacific islands, and the great island of New Guinea/Irian, have been the focus of a remarkable amount of research into the history of people and their land. There follows a detailed examination of the unique situation in the central highlands of Papua New Guinea where sustained-yield wetland farming evolved some 9000 years ago – as long ago as in the Middle East – but where society remained isolated and evolved its production system without any direct contact with the outer world. In the New Guinea highlands, one crop introduction is shown to have had a dramatic effect during the last 300 years, but except in the swamps the earlier history of land use and its consequences remains imperfectly known. We begin, therefore, with a discussion of an older degradation in some small islands further east in the Pacific, where highly significant changes took place between 2000 and 1000 years ago.

While the history of people in New Guinea/Irian goes back at least 40,000 years (Golson personal communication), long before the evolution of agriculture, the islands to the east were not occupied until suitable sea-going vessels were developed. This happened only about 3000 years ago, and the people who came to these islands were marine gatherers with some agriculture. During this 3000 years, and especially during the past 2000, we find some remarkable examples of the interplay between a land management that degrades, particularly by causing severe soil erosion, and adaptive land management, actions that adjust land use toward viable sustained-yield agro-ecosystems. The evolutionary term 'adaptive' is used to call attention to the uncertain, halting and sometimes unexpectedly advantageous process that enables human and other organisms to move toward their future. Sometimes benign, sometimes destructive, the dynamic interplay of people with island land has not ceased since the first human tenancy.

Evidence from three Pacific islands, Lakeba in Fiji, Tikopia in the Solomons, and Anatom (formerly Aneityum) in Vanuatu tells a somewhat similar story, despite the strong ecological contrasts between the three (Hughes, Hope, Latham and Brookfield 1979; Spriggs, 1981; Kirch and Yen, 1982; Latham, 1983). In each case there was a period of human-induced erosion and vegetation impoverishment on the slopes, leading to the development of low-lying colluvial and alluvial deposits which in turn became the islands' most productive agricultural areas. The erosion was substantial: on Lakeba it lasted several hundred years with a mean denudation rate in one catchment of 1.7 mm/year, and was accompanied by the enlargement of a degraded soil–vegetation complex the origin of which seems to be pre-human, arising from climatic changes during the Pleistocene period. Fire was the main agent of this enlargement, leaving abundant charcoal in the record of the swamps, and it continued for several hundred more years after erosion itself abated, perhaps because the more easily removed weathered horizons had all gone. Yet in the last 200 years even the most heavily eroded catchment has become reforested. On wetter Anatom the story is similar, though the deforestation is less, while on young, volcanic Tikopia the more resilient soils of the uplands have been reoccupied and cultivated in a form of arboriculture in which almost every plant has use value.

The lowlands and swamps created by this process have, however, created an ideal environment for wetland taro cultivation, an intensive system with potential for sustained-yield stability (Brookfield 1979). Spriggs (1981), writing of Anatom, called this process 'landscape enhancement' and noted its similar occurrence in New Caledonia, Futuna, Hawaii and Rarotonga (Cook Islands) as well as in Tikopia and Lakeba. Spriggs raises another interesting question; all these islands now have, or had before conquest, a village-based society with a chiefly, redistributive system of organization under which chiefs have great privilege. Spriggs (1986), following Sahlins (1974) and others, interrelated the development of intensive farming with exploitation of the labour power of the commoners. However, we do not know what was the social organization of the period during which the heavy damage was done to the uplands, although from Lakeba it would appear that both settlement and power were less concentrated in the remote past (Best 1985).

These examples reveal the dynamism of people–land interaction not only in pre-colonial times, but in times before the hierarchically structured pre-colonial societies came into being. More difficult to determine is the long-term balance between gain and loss. Even some of the degraded land left behind on the slopes could be, and widely was, made productive by the intensive labour of terracing, mulching, composting and arboriculture (Barrau 1956; Ward 1965; Bellwood 1978; Watling 1984). Where this was not possible, as on the unproductive pyrophytic fernlands of Lakeba, and on some other Pacific high islands, it seems likely that the origin of these formations and of the ferralitic soils beneath them is at least partly climatic, although the area of degradation has certainly been enlarged since human occupation. In the eastern South Pacific on remote Easter Island, Flenley and

King (1984) suggest that severe deforestation over the past 1000 years could have led to the decline of that island's famed megalithic culture. At this point in our argument, it is worthwhile recording that all this happened in the total absence of outside conquest and exploitation or acquisitive traders, unless the ancient Polynesian voyagers, whose travels probably flourished much more in the 'little climatic optimum' of 1000 years ago than in the stormier conditions of the Little Ice Age (Bridgman 1983), acted in such a role.

B Degradation and a pre-capitalist political economy: the case of the New Guinea highlands

Bryant Allen and Robert Crittenden

1 Argument and background

1.1 A new crop, population growth and social exchange

In the highlands of New Guinea/Irian there are large horticultural societies which have only recently become integrated into a colonial or national state, and whose direct links with the global economy are of less than 50 years duration. These groups, their systems of production and their social economies have attracted a great deal of ethnographic and ecological interest since the 1950s. They provide an opportunity to explore the relationship of land degradation to a political economy that arose in only the most indirect connection with the global economy. Its development followed the introduction of a new crop – the sweet potato – from the Philippines about a hundred years after the Spaniards carried it thither from South America in the sixteenth century. The fact of this introduction indicates that the central highlanders were not out of contact with the world, but this contact was mediated only through tenuous trading chains that linked highlands to the coast, and so remained right up to the time of the irruption of Europeans into the region in the 1930s. Although elements of this contact are crucial in understanding the new social order that developed in the highlands after the seventeenth century, the evolution of that order was essentially indigenous, and we argue that it is critical in understanding the patchy incidence of land degradation that has occurred during the same period.

The discussion that follows is based mainly on data from only a part of the highlands, the Enga, Western and Southern Highland Provinces of Papua New Guinea. Trends elsewhere were similar to the pattern described here, but there were some differences referred to only in passing. The experience of this region suggests a number of conclusions of significance. Degradation is shown to occur when the growth of population is powerfully augmented by a competitive system of social exchange in which the items involved make heavy demands on production from the land. It is also shown that differences in the resilience and sensitivity of the land, exposed by the general spread of a new economy and social order, emerged in a contrast between degradation in some areas and the successful establishment of a sustained-yield production system in others. Changes in the pattern of labour are also shown to be important. The degradation described in this chapter is not dramatic: it includes little erosion and mainly takes the form of impoverishment of land capability – real or only perceived in relation to the demands of production. The processes described are slow in operation, so that a long time-span is necessary for them to become exhibited as patterns on the land.

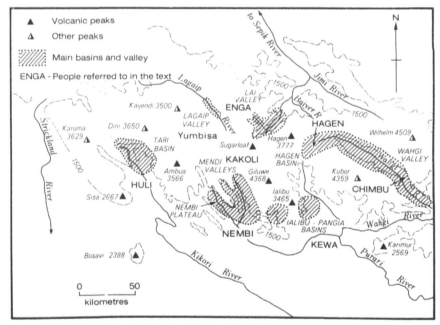

Figure 8.1 Part of the highlands of Papua New Guinea, showing the localities and people referred to in the text

1.2 The highland environment

The major structural feature of the island of New Guinea/Irian is a 'spine' of mountains with high narrow-crested ridges and steep-sided V-shaped valleys, but in the centre are a series of wide intermontane valleys, separated by high mountains and a number of large, extinct stratovolcanoes (figure 8.1); they have flat to rolling valley floors between 1500 and 1600 m above sea level; volcanism has resulted in the deposition of large amounts of laharic material in the form of mudflows and has covered much of the area in air-fall tephra and ash showers from which the majority of soils in the Papua New Guinea section of the highlands are derived. Volcanism has also disturbed drainage patterns creating large swamps. Limestone plateaux, lake-bed sediments, outwash fans, colluvium and narrow alluvial flood plains complete the ensemble of landscape features (Löffler 1977).

Mean maximum and minimum temperatures in the highland valleys range between 24° and 13° with very little seasonal variation in either temperature or humidity. Diurnal variation is considerable. Mean annual rainfall is between 1500 and 2000 mm in the east and 2000 and 3000 mm in the west. Periods of up to three and four months of significantly lower rainfall and increased risk of frosts occur on average every ten years in association with a reversal of the trans-Pacific circulation, or Walker circulation, the 'El Nino' event. Rainfall intensity is low, with few falls of over 100 mm in one day, although shorter very intense falls do occur (McAlpine, Keig and Falls 1983).

The most common soils in the highlands are the hydrandepts or humic brown clay soils. They are distributed widely over a range of landforms, an outcome of their derivation from air-fall andesitic volcanic ash (Bleeker 1983). The humic brown soils are characterized by a deep, well-developed, black surface horizon rich in organic matter; the A horizons have a strong, stable granular structure, are porous, friable and very resistant to erosion (Wood 1984). Alluvial soils occur on active floodplains and colluvial soils occur on recent mudflows and are highly fertile. The swamp soils are dark brown to black, composed of decomposing peat and are very high in organic matter. On steep slopes where the ash mantle has been lost, other soils based on underlying parent materials become important. Soil erosion in its more spectacular forms of gullying or mass movement is not, however, widespread and even 'soil losses by water erosion are less serious than expected' for a tropical environment (Bleeker 1983: 198).

1.3 Agriculture in the highlands

When foreigners first entered the highland valleys in the 1930s they encountered almost one million people – more if the highland valleys of Irian Jaya are also included – engaged throughout in an intensive system of horticulture based on the sweet potato (*Ipomoea batatas*) with only minor cultivation of other crops, and in rearing large numbers of pigs. These intensive systems differed greatly in nature from the swidden systems of the surrounding lowlands and slopes, yet in so far as they were based on the sweet potato and adapted to this crop they cannot have had an evolution time much longer than 300 years since sweet potatoes were introduced into these valleys (Yen 1974: 317). The discovery and excavation of drainage systems in the Kuk swamp near Mount Hagen have however established the long duration of intensive cultivation in the region, for these systems date back to 9000 years before present (YBP) (Golson 1977, 1981). The discovery has not resolved what happened in the drylands around this and other swamps, in which similar evidence of ancient drainage has since been found.

Golson (1982) has recently reappraised the evidence. With regard to the drylands around the site, an extensive deposit of grey clay in the swamp is correlated with catchment erosion between about 9000 and 6000 YBP, probably indicating that shifting cultivation of some form was practised around the swamp. The archaeological and sedimentological evidence only partly correlates with the palaeobotanic evidence, for at least at higher altitudes evidence for disturbance begins only between 5100 and 4000 YBP, and is well established by 2300 YBP (Walker and Flenley 1979). The early swidden cultivation was probably for taro and other crops, associated with more intensive cultivation in drained swamps (Bayliss-Smith, 1985). Within the last 2500 years, however, dryland soil tillage seems to have become established as a cultivation method, allowing continuous use of grassland which was previously not usable in this way, once a deflected succession had become established. Pigs were almost certainly an important component in the system, because forest clearing would have reduced access to feral marsupials and required the cultivation of fodder (Morren 1977).

Around 400 YBP a sharp reduction of forest pollens at high altitude and an increase of secondary-growth species is interpreted as expansion of grassland and controlled tree-fallow (Walker and Flenley 1979). This is presumably associated with the introduction of the sweet potato (Powell 1982: 225). The record is also strengthened by sedimentation chronologies for small lakes. In two, now surrounded by cultivation, sedimentation rates increased suddenly between 300 and 150 YBP, but there is no such increase in a lake still surrounded by forest (Oldfield, Appleby, Brown and Thompson 1980). Around the time at which sweet potato was introduced, therefore, there began a large expansion of cultivation.

1.4   The sweet potato

The effect of the introduction of the sweet potato was twofold. By yielding much better at high altitudes than taro, it allowed cultivation to expand above 2300 m and to become more secure at middle altitudes (Clarke 1977a). Because it is more tolerant of poor soil it enabled cultivation on one site to continue for longer. But in addition to permitting spatial expansion, the introduction of the sweet potato also triggered both intensification and innovation. The critical question for our argument is why this occurred.

The answer is not at first sight obvious. The new crop would have allowed an increase of production on the same land area, with less labour. The key possibly lies in the concept of 'social production' over and above subsistence production, introduced by Brookfield (1972) in a critique of Boserup (1965). Social production, meaning production for social goals, may lead to intensification in the total absence of pressure of population on resources (PPR). Taking up where this argument left off, Modjeska (1982) has combined 'ecological causation' with 'social causation' in an elegant exploration of production and the creation of inequality in Papua New Guinea. He argues that increased domestic pig production can provide higher levels of protein than hunting, and would facilitate population increase through improved fertility and decreased mortality. But a cycle of social production was also under way. The pig was employed as an indemnifying item in social relations, and became increasingly important as population expanded, social relations became more complex and the possibility of conflict increased. Modjeska sees 'pig production and exchange creatively develop a new and more "economic" version of social control and social order' (1982: 55). It is on this major innovation that we now focus attention.

1.5 Competitive exchange and prestation

Between the seventeenth and twentieth centuries highlanders evolved a new social order, one based on the competitive exchange of pigs, the use of pigs in marriage, child and mortuary payments, and their use in compensation payments for injuries or deaths caused in warfare. It is the competitive exchanges which are of greatest interest in examining pressures leading to land degradation. These ceremonies involved individuals and groups in the exchange of pigs and other valuables mainly gained through trade. Marine shells were particularly important, and were commonly acquired through trading chains linking highlands with the coast, in exchange for pigs. Two main types of exchange evolved (Strathern 1969): in one, massive periodic prestations combined the efforts of thousands of individuals, while in the other long chains of exchange relations (*te*, *tee* or *moka*) were formed in which live pigs, meat and shells passed along the chain. These chains indirectly linked groups more than 150 km apart, and at their extremities were in turn linked to other groups and individuals through informal trading connections that were 'important in siphoning certain valuables out of the *te* network and maintaining the volume of goods moving in the *te* system' (Healey 1978: 203).

There was a marked tendency for these ceremonial exchanges to become inflationary, through a perceived need to give more than was received on a former occasion. This inflationary trend is seen as the outcome of competition between ambitious men who could gain power by manipulating wealth in large public prestations (Strathern 1982). All members of a group were put under pressure to participate and those who would or could not were derided: they were 'rubbish men' or among one people 'stink-bug men' (Modjeska 1982). The power of a 'big man' rested on his ability to marshal the efforts of the reluctant (Strathern 1971).

1.6   Innovation in farming

While the sweet potato, readily consumed by pigs, facilitated these changes, a number of agronomic innovations was also adopted. Cultivation and waste were separated by fences, and the pigs foraged in the latter while being also fed from the former. New techniques of soil tillage evolved. East of Mount Hagen a checker-board or 'grid-iron' system of tillage through close-spaced ditches was elaborated, also facilitating soil drainage (Brookfield and Brown 1963). To the west, among the Enga and Huli, large plano-convex mounds composted with grass and old vines were used (Waddell 1972). Production from such fields continued for long periods with only short fallow intervals. Production became continuous, with plants at all stages of maturity within any large field. There is some evidence that mounding was spreading from west to east at the time of contact in the 1930s (Bowers 1968; R.M. Bourke personal communication), perhaps as a response to the need for increased production on the ash soils.

The swamp garden was a variation on the open sweet potato field. Large drains were constructed through suitable swamps, subsidiary drains were dug and beds formed between them above the water table. This system seems to have been adapted from the older system of swamp gardening, but throughout the highlands swamp gardens were converted to sweet potato. Some, including that at Kuk, were abandoned for reasons that are unclear, but elsewhere they flourished, attaining an apogee of complexity in the wide floor of the Baliem valley of Irian Jaya, where large areas of dryland gardens have been so severely eroded that, especially on limestone, only traces of an old system of stone-walled fields still survive (H.C. Brookfield personal communication).

1.7   Gender and class

If we are correct in believing that this system, with its range of agronomic methods specific to the needs of the sweet potato, has evolved in less than 300 years, it has large implications for patterns of labour. Men clear the trees, fence the land, and plant certain crops; women do all weeding. In the open fields dominated by sweet potato men do the basic work of land preparation, but in the mounded areas women then break down old mounds and reform them, collect compost and planting material, plant, weed and harvest. They are also either wholly or mainly responsible for raising pigs. Men's labour thus creates capital improvements in the land; women's labour, on the other hand, is repetitious and accrues little benefit from cycle to cycle.

Modjeska considers that men have exploited the advantages of technological innovation to further their dominance within the social formation. However, it is not only women who are exploited in this way. The more developed the system of exchange, and the more intensive the agriculture associated with it, the greater is the tendency for male society itself to become stratified. Early visitors to Mount Hagen were struck by the degree of stratification they encountered in the 1930s: 'big men' had more pigs, shells and wives than ordinary men, while poor men were often unmarried and worked for big men in return for food, shelter and security (Strathern 1966). Waddell suggests how this situation evolved among the Raiapu Enga:

> wealth is founded primarily in agricultural production, and power is achieved largely through the manipulation of the flow of this wealth between individuals and groups. Hence 'big-men' achieve and strengthen their position through careful agricultural planning, in which culturally determined limitations on both the availability of and access to open fields lead to major inequalities in levels of production of the staple food within the local group. These in turn, result in interhousehold variations in the numbers of pigs supported, and therefore in the opportunities of adult males for marriage. (1972: 216)

This statement can be extended to inequalities between groups and locales.

2   The conditions of degradation

Few writers on highland societies have attempted to address political, economic and ecological problems together within a single framework. If we are to undertake this task we must isolate those considerations which seem most relevant. We focus first on prestation and trade: individuals and groups are linked across long distances sometimes directly, sometimes in a series of chained exchanges, sometimes through trade especially in pigs and shells. Both the exchanges and the prestations have a tendency to grow incrementally through a constant process of 'bidding up'. We suggest that, over time, agricultural intensification has occurred, and innovation has been adopted, in order to meet the demands of this economy. The innovations have been developed primarily to allow sustained cultivation on volcanic ash, colluvial and alluvial soils, on steep slopes, and at high altitudes.

This 'model' needs, however, to take account of the differing capability of different environments, and of the wide variations in resilience and sensitivity of the land. It cannot therefore be expected that, even with a high rate of innovation, intensification can everywhere take place without leading to land degradation. It seems highly probable that groups on sensitive land, some of it marginal in terms of capability, have caused degradation by attempting to maintain their position in the exchange and prestation systems by producing at levels which cannot be sustained over long periods of time. In more favoured areas a remarkable degree of intensification has been possible.

This bald statement cannot encompass the extremely complex relationships between population, production, environment and social organization discussed, for example, by Brookfield and Brown (1963) for

Chimbu, by Meggitt (1965) for the Mae Enga, and by Wohlt (1978) for the fringe Enga at Yumbisa. These studies examine the relationships between agnation, patrilocal residence, population density and the available land resources. While these aspects are important, however, we suggest that land degradation can be brought about simply by the mechanism described above, even while populations are in the process of making cultural and agro-technical adaptations which would have the effect of improving the distribution of population against resources. We suggest that this occurs because, in at least most of the New Guinea highlands, degradation takes place at slow rates over a relatively long time, and because the growing complexity of group social organization has complicated farmers' responses to such degradation as they perceive to be taking place.

Successful management is possible on most areas of volcanic ash and alluvial soils and some colluvial soils; on most slopes less than 15° and some steeper slopes, though not in areas liable to severe mass wasting; between 1500 and 1800 m, the altitude of the lower parts of most of the highland valleys. The success of the system in these areas has, however, caused it to spread on to steeper land, inferior soils and to higher altitudes. Under these circumstances, degradation occurs in the form of destruction of capability for further cultivation, though the land may still be usable for pig foraging. Such a process is described in detail for the Kakoli area of the Kaugel valley by Bowers (1968), and we now examine her findings.

3   The Kakoli case and some other evidence

3.1   The Kakoli case

Most of the Kakoli people live between 2100 and 2200 m, and occupy and cultivate land up to 2600 m. Some groups live at this higher altitude. The Kakoli participate in a chained prestation which links the Mount Hagen *moka* chains with the Enga *te*. They use land in three ways: permanent open fields of sweet potato on the valley floor and flood plains; swiddens with a 10- to 15-year fallow on small areas of steep slopes such as gully sides; and a 'crop-and-abandon' method on the upper slopes, which results in:

> the failure of reafforestation to occur. It is this latter cycle which results in land degradation. Results can be erosion, laterization, or the regrowth of certain grassland vegetation which is difficult for gardeners to deal with. (Bowers 1968: 65)

On the forest–grassland edge, gardens are cultivated for up to two years and then fallowed. During this fallow, weeds and woody regrowth develop. Between three and ten years from the initial clearing, the land is again cleared, and turned into a permanently cultivated, tilled, mounded and composted sweet potato garden. Cultivation will continue until yields fall to the point where it is considered not worth replanting. The land is then abandoned and is colonized by a thick growth of *Miscanthus* cane grass. Such areas may be used for pig foraging, but they are considered as no longer suitable for cultivation. Forest further upslope is then cleared and the cycle begins again. Climax vegetation is continually replaced upslope by grassland, and natural successions by deflected successions.

3.2 Other evidence

This process is not restricted to the Kaugel Valley, nor to high altitude zones. In the central Lai Valley at around 1900 m in the territory of a clan deeply involved in the *te*, 44 per cent of the total land area of the Lyupini clan is classified by clansmen as '*kaka*', or unsuitable for cultivation. This land has slopes from 25° to 35°. It is covered with thick *Miscanthus* and a short stunted tree *Vaccinium albicans*, which is said to be indicative of very poor soils. Soils in these areas do not have the black to dark brown A-horizon which is found on similar but less steep land nearby. Air photographs clearly show old field marks in the *kaka* areas, although local men say it has not been cultivated in their lifetimes. In the same locality land with an even steeper slope has lost another soil horizon to expose a blocky, bright red clay (Allen 1982: 98).

In the Tari area, an upland basin at an altitude of about 1600 m, swamps, floodplains and rolling volcanic ash plains are intensively cultivated with mounded sweet potato gardens. The Huli people who occupy this area are not involved in chained prestations, but were linked to Enga groups to their north and to lower altitude groups to the south by trading. Huli kinship organization causes social group members to be distributed widely over the occupied area, effectively spreading the demands for pigs for prestation over the whole of the occupied area. Huli farmers use the composted mound method of cultivation across a range of land types, once the initial opening up of the land is completed.

On the upper slopes around the basin, forest clearing continues and large areas of *Miscanthus* grassland have been created, said by their owners to be unsuitable for cultivation (Wood 1984). The sequence of cultivation which has formed these areas is very similar to that described by Bowers, but in addition there are widespread fires during dry periods. Fires and foraging pigs will successfully eliminate small secondary forest trees which have attempted to colonize the grasslands. Wood (1984) used genealogies to give him a time depth to the cultivation of Tari gardens on different land types and found significant differences in yields over time between ash soils at different altitudes, and on different slopes, and between ash soils and alluvial, colluvial and swamp soils (figure 8.2). Above 1800 m on steeper slopes, yields fell from 8 t/ha to less than 2 t/ha in under an estimated forty years. Yields on ash soils declined more slowly, but still fell significantly faster than yields on alluvial, colluvial and swamp soils. The exchange relationships between Huli groups is not well known. It is clear, however, that an individual farmer on the higher ash soils can maintain his position

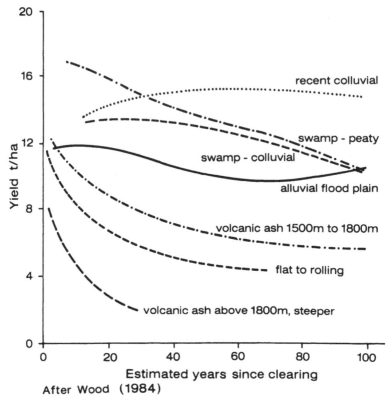

Figure 8.2   Yields and time since clearing in the Tari basin, Papua New Guinea. *Reproduced by permission of A. Wood*

relative to that of farmers on lower ash or alluvial and swamp soils only by clearing forest regularly and abandoning older grassland gardens.

3.3   The Nembi plateau: absolute or relative degradation?

Although highlands farmers say the *Miscanthus* grasslands are not suitable for cultivation, it is likely that this is a relative statement and that they are comparing these soils with other more fertile soils on other land types. In Enga, Scott and Pain (1982) found no chemical reasons why some *Miscanthus* soils should be able to be cultivated, and noted that, in some areas where there are no longer reserves of forested land to be cleared, the grasslands can be successfully cultivated.

But while a group recultivating grassland soils could probably support themselves satisfactorily, it is unlikely they could maintain or improve their standing in the regional economy, and it is likely they would be unable to support a population large enough to ensure their security from attack by neighbours. In the long term such a group would probably decline in size and might disappear, absorbed by surrounding groups (Meggitt 1977). If they try to maintain their position by cultivating grassland, they may begin to experience some of the symptoms of 'population pressure', but which may better be termed 'social production pressure'. This is probably the situation on the Nembi Plateau in the Southern Highlands.

The Nembi Plateau is a limestone area, with soils derived from air-fall ash. The plateau was an important trade route from the lowlands into the highlands and it is likely that the population increased on the plateau in conjunction with trading (Crittenden 1982; Strathern 1969). Shells and tree oil from the lowlands which passed across the Nembi Plateau entered the *te* and *moka* chains, 50 km to the northeast.

Almost no forested land suitable for cultivation now remains. Large areas of *Miscanthus* grassland surround open-field sweet potato gardens. Mixed crops are grown in swiddens in limestone dolines. Sweet potato yields from the open fields are very low for the highlands and periodic shortages of food occur. Despite this men maintain their involvement in local and regional prestations. Meanwhile their wives, and many men are polygamists, work hard to maintain production on land from which yields are declining. The Nembi do very little composting, although they know of the technique (Crittenden 1982; Allen 1984). The outcome is a high level of child malnutrition which is almost certainly associated with womens' work patterns (Allen *et al.* 1980; Baines 1983; Crittenden 1984).

4 Conclusion

The Papua New Guinea case contributes importantly to the discussion on the social and economic causes of land degradation. First, it provides one of the longest time perspectives available, and in so doing demonstrates that forest clearance, the intensification of agriculture and accompanying population increase over a long period of time, do not of themselves necessarily cause lasting degradation of land.

Second, it provides an example of the ability of social and economic pressures to intensify agriculture in the absence of a colonial or nation state, or of the global economy. In doing this, however, it emphasizes that the individual 'land manager' cannot be viewed in isolation from the social relations of production, or the access to and control over the means of production and the allocation of the product among various groups (see also chapter 4, section 3).

Third, the highlands case suggests that land degradation is associated with the exploitation of weaker groups within a society. It also suggests that where the weaker group are women, then indirectly the health and well-being of children will be affected, and will be seriously threatened if land degradation begins to apply pressure on this group to maintain production for social purposes in the face of declining yields.

Fourth, the importance of the time dimension, and of the type of land degradation, are illustrated. In the absence of the more spectacular degradation demonstrated by widespread soil erosion, the decline in yields and the creation of 'unusable' or 'unrewarding' grasslands did not create a problem so long as further forest reserves remained. This may have been the condition in the highlands for as much as 150 years, and it came to be accepted as 'normal'. Hence when forest reserves become depleted and people are forced to occupy grasslands their adjustment in either agro-technology or economy may be retarded, though this is not true everywhere. Retarded adjustment is further probable, however, where the hardships which accrue from land degradation fall first on the weaker sections of the population, be they small population groups or the women, since these latter have little say in the major social decisions affecting land use and exercise little political influence.

9 Management, enterprise and politics in the development of the tropical rain forest lands

A From forest and grass into cropland
Piers Blaikie and Harold Brookfield

1 Relevant questions

In this chapter we return to the colonial and post-colonial transformations discussed in chapter 6, but around a narrower focus. The area of inquiry is the major transformation of lowland tropical landscapes into productive, or supposedly productive extensions of the agricultural and pastoral oecumene. Together with the demands of forestry, this involves the clearance of large areas of tropical rain forest at a rate which is currently causing widespread concern and is described by some as the 'tragedy of our tropical rain forests' (Jackson 1983). Consequences of the disappearance of large tracts of evergreen forest are described in many sources, with emphasis on the breaking of delicate nutrient recycling mechanisms which leave the underlying soil, often poor, under grassland ecosystems of much lower primary productivity. This new 'great age of clearance' has a qualitatively different impact from earlier great clearances of temperate forest which occurred in Europe during the medieval period and in North America in the nineteenth century. The resulting landscape is not only of lower primary productivity but the soils suffer a loss of capability and resilience, and increase of sensitivity which can be permanent when forest is wholly replaced by short grassland (Nye and Greenland 1960: 134). Yet the grassland ecosystem is not useless to field-crop farmers provided suitable management methods are adopted, while under other forms of agricultural use the former forest areas can have sustained productivity of high value. Both the case studies which follow this introduction are concerned with the management of already deforested areas, and they present somewhat contrasting conclusions.

One major cause of deforestation in the humid tropics, and of the increased use of the deforested regions, is certainly increasing pressure of population on resources (PPR) which causes shifting cultivation systems, which are well adapted to forest management under low population densities, to press more heavily on the resource and, with the aid of fire, to eliminate the forest and then further degrade the replacement ecosystem. However, it would be wrong to generalize and to follow the still widespread belief that shifting cultivation leads almost automatically to degradation. In Papua New Guinea, for example, about a quarter of the country's forested area is well-developed secondary forest created and maintained by viable systems of shifting cultivation (Allen 1985). Shifting cultivation can be a permanent sustainable system (Clarke 1977b; Grandstaff 1978). In Kalimantan, Kartawinata and Vayda (1984) have shown, however, that where shifting cultivators follow loggers into an area, the consequence is often an increase in grassland and hence in erosion. It is in situations of this kind, where new forces intrude into the old adaptive system, that some of the most dynamic vegetation and soil changes are now taking place.

One sequence, common in Latin America and encountered also in other parts of the world, sees shifting cultivators following loggers and oil prospectors along their new roads, then seeding the land with grass for cattle after cropping is finished. Commercial cattle producers then buy the land for ranches, and with the support of well-funded development infrastructures produce beef for the North American and European markets. Heavily grazed, the land quickly declines in carrying capacity so that it is then abandoned and the ranchers move to new areas (Nations and Komer 1983). A similar pattern is reported along the new trans-Amazon highways in Brazil, where settlers with little capital either revert to shifting cultivation methods or convert their land into pasture (Smith 1978). Large ranches are also created, but soil compaction and nutrient loss quickly reduce productivity, and in the absence of methods to sustain capability much of the settled land is abandoned (Hiraoka and Yamamoto 1980; Moran 1982). Yet the acid, infertile soils of the Amazon basin are dotted with patches of more productive soil, rich in organic matter, and now believed to be the product of more effective systems of utilization under higher population densities in the past (Smith 1980).

The problem is one of management, and the conversion of lowland tropical rain forest into sustained human-use systems, or into a 'red desert' (Goodland and Irwin 1975), depends more on management than on the inherent qualities of the land. It seems clear, both from the evidence of past occupation in parts of the tropical rain forest and from modern research (Sanchez, Bandy, Villachica and Nicholaides 1982), that appropriate forms of management exist for much of it. The increasing use of mechanical methods of forest clearance, involving the use of bulldozers and root-rakes, often leads to major soil disturbance, compaction, erosion and erosion-induced degradation. However, it is not necessarily true that mechanical methods should be more damaging than manual clearance, and the management of the land after clearance can be of greater significance (IBSRAM 1985).

2   Peasants, planters and settlers

2.1   Tropical grasslands and land management

In the main, this chapter is concerned with management of land that has already been cleared from forest and is now under grassland or shrub. The fertility status of such land is often low, with low levels of available nutrients, high acidity, low cation exchange capacity, poor structural conditions and high susceptibility to erosion (Nye and Greenland 1960; Gillman 1984; IBSRAM 1985). Toxic trace elements are sometimes a problem. In the past it has commonly been said that such land should be taken out of use, but this is less and less possible and in the case studies which follow, as in many other areas, the need is to enhance production from such land. Modern research has suggested that very old practices such as deep tillage and green manuring may often be the best in terms of results (IBSRAM 1985). Although the loss of original fertility may be permanent, forest land converted into short grassland is not, as was once rather widely thought, degraded beyond effective sustained use.

The process by which invasive grasses replace forest is closely linked to the employment of fire, which kills young shrubs and tree seedlings but permits grass to survive. Among the most fire-resistant of all tropical plants is the Asian variety of the pan-tropical *Imperata* grass; *Imperata cylindrica* has replaced forest over as much as a fifth of the Philippines, and in Indonesia its area is estimated at 10 per cent of the former extent of forest in the outer islands (Kartawinata 1979). Chapter 9B is concerned with the management of an area largely converted into *Imperata* grassland on the island of Kalimantan (Borneo). In this introductory section we briefly review some aspects of its ecology and social setting.

A certain amount of *Imperata* grassland is useful. When young it is palatable and attracts wild animals which can be hunted; by the same token it can feed livestock. It is used for thatch. Southeast Asian shifting cultivators created fields of *Imperata* and sustained them by burning. However, fire used in clearing forest also encourages the invasion of *Imperata*, and escaping fires, especially in areas already grassland, extend the area of this invasive and fire-resistant plant. In seasonally dry areas the process of replacement can be very rapid, and in the Cagayan valley of Luzon, Philippines, great areas of former forest not much cleared before the present century are now under grass, and are heavily eroded.

Imperata land is difficult to manage. The grass can be replaced by others under heavy grazing, but since *Imperata* develops extensive rhizomes it can be eradicated for cultivation only by heavy hoe-work, or by ploughing, as Potter shows later (chapter 9B), unless it can be shaded out. Hence while forest remains, little use is made of the grassland for agriculture. Shifting cultivators leave it behind, and work a hollow frontier of surviving forest. Much of the criticism of the shifting cultivation system in the Southeast Asian literature rests on this pattern.

To no small degree, the initial implantation of commercial agriculture followed a similar pattern. Jackson (1968) describes how nineteenth-century Chinese tapioca growers around Malacca in West Malaysia created a hollow frontier of farmland on the low hills between the rice-growing Malay villages, and how this moved inland leaving behind it extensive areas of *Imperata*, most of it taken up early in this century by rubber planters whose undemanding tree-crop could use this land. In the same way, the first Dutch tobacco planters in North Sumatera in the 1860s and 1870s believed that only a single crop of fine-quality wrapper tobacco could be obtained from the cleared forests, and so constantly moved on leaving an *Imperata* 'wasteland' behind them. Only towards the end of the nineteenth century was it discovered that after a fallow period of seven to ten years a second crop could be obtained, but that this was better if secondary forest had grown up on the site; immediate steps were then taken to control the use of fire in order to encourage a forest fallow (Pelzer 1978).

2.2 Conflict and land management in North Sumatera

North Sumatera provides a prime example of the conflicting management goals and practices of planters and peasants on the same land (see figure 9.1 showing the location of the case studies in this chapter). When the first Dutch planters arrived and obtained land the region was only sparsely populated by shifting cultivators. In obtaining the land the planters were obliged to permit the peasants to take a food crop after the tobacco crop. Once a managed fallow was introduced, however, conflicts arose. It became necessary to regulate access to land, and also to drain it and to limit activities which, on friable volcanic soils, would lead to silting of the drains. Small areas allocated to villagers were inadequate, and were quickly invaded by *Imperata* and eroded. As more labour was introduced from Java, the region became a major food-deficit area, so that when imports became impossible during the Second World War the plantations were largely taken over for shifting cultivation, and major conflicts arose when the plantation companies sought to resume their lands after the war. Under what was by this time a high density of population, little *Imperata* survived, and more intensive management systems, including irrigated rice, took over more and more of the land. By 1954

> Land occupied during the war and postwar periods had often been abandoned by the squatters because the soil had become completely exhausted. Only heavy applications of fertilizer and repeated ploughing under of green manure crops could restore these soils to use for the raising of tobacco. . . . Furthermore, the continual cultivation of maize had caused heavy erosion, with resultant damage to the drainage ditches. . . . In other instances the construction of improperly designed irrigation had caused

serious damage to standing crops.... The local villagers were also guilty (*sic*) of abandoning their own exhausted soil for estate lands, repeating there the process of soil exploitation. (Pelzer 1982: 91–2)

The conflicts between commercial planters, the resident population and the food needs of labourers and other immigrants were never resolved in North Sumatera. The tobacco industry is now almost dead, and tobacco is a smallholder crop. The estates are largely converted to oil palm, and irrigated rice is the main food crop, but:

The principal production centres of *sawah* [irrigated rice] are in the same coastal plains as the main estate crops and compete for the same land.... The plantation drainage systems also affect rice land, and flooding is common both from drainage and from silting up of rivers due to almost unlimited forest cutting in the foothills and mountains. Finally, diseases and pests have been a problem.... Land problems arising out of the early concession agreements still plague the government today. (Ginting and Daroesman 1982: 59–62)

2.3 Land development: land in the service of politics

The burst of commercial plantation enterprise in the second half of the nineteenth century provided invaluable information on the capability of tropical lowland soils outside the existing cultivated areas. The value of tree-crops was quickly perceived and substantial further development by commercial interests took place in the first quarter of this century. By the 1920s the tin and rubber of British Malaya, the latter replacing both forest and grassland, yielded exports worth more than those of all other British dependencies combined – including both India and Egypt at that time – so that 'the export value per head of population in British Malaya in 1925/26 was the highest in the world' (Ormsby-Gore 1928: 21). While about a third of this income went to form foreign profits (Khor 1983: 56), the potential of the tropical lowlands under a tree-crop plantation economy was strikingly revealed.

The formation of pioneer settlements in these same environments began in the same period and had, from the first, mixed objectives. The contrast between crowding in the old settled areas and the emptiness of the forests and grasslands prompted a desire to redistribute population more 'effectively'. Thus, while the Dutch continued to encourage their planters in central Java to lease indigenous lands, and the services of the labour living on them on an alternate-year basis (Geertz 1963; Pelzer 1978: 18), they simultaneously initiated an outer-islands settlement scheme as part of a 'debt of honour' to the Javanese peasants (Pelzer 1945: 191).

The schemes expanded greatly in the 1930s, when low export-crop prices created considerable distress among peasantries who had come to depend on export-crop income and wages for part of their livelihood. The Dutch then sought to settle Javanese peasants on one-hectare blocks in outer island locations, where a 'Javanese way of life' could be reproduced. It was not until the 1970s that the land allocated to settlers in a much enlarged transmigration scheme was usually made sufficient to permit a significant cash-crop element in production (Arndt 1983). The aim remained the relief of poverty on Java, and the creation of concentrated Javanese populations among the heterogeneous but sparser populations of the outer islands (Guinness 1977; Hardjono 1977). The Madurese settlement in Kalimantan discussed in chapter 9B was unusual in this respect. By the late 1970s over 75,000 people per year were being resettled. There has also been substantial spontaneous migration out of Java, though not all of this has been to rural areas. There was, and continues to be, substantial loss of settlers from some transmigration schemes, partly because of land degradation, but more because of the hard life, the difficulties of adaptation of Javanese cultivation practices to much poorer soils, and the very low income levels that have often been experienced (Arndt 1983).

Meantime, however, a wholly different concept of land settlement had been initiated in Malaysia in 1955, where the object was to develop land resources for cash crops by settlement of landless and near-landless Malay peasants on blocks of adequate size, organized in an industrial manner on settlements modelled on plantations, except that the land was subdivided. Planned land development had multiple objectives: to increase and diversify production; to create a population of cash-crop farmers, supporting new regional economies; to relieve Malay rural poverty; and to expand the national role in agricultural development. Land is cleared by machine methods and planted under contract before the settlers arrive, and production and marketing are managed by Federal and State organizations on an agribusiness basis (Bahrin and Perera 1977). Over 6000 km^2 had been developed in this way by the mid-1980s. Almost all is under tree crops (Goh 1982).

The establishment of tree crops certainly reduces erosion and soil degradation by comparison with an *Imperata* cover, but not by comparison with the former forest. But while the degradation that takes place is insufficient to damage the capability of the land for tree crops themselves there are important downstream consequences. Long memory reveals the changes that have taken place. In a region of low hills in West Malaysia, where forest has largely been replaced by rubber since early in the present century, Yusoff thus describes the deterioration since his boyhood:

> The destruction of the natural vegetation cover around the kampung had caused much silting in the valleys and the yield per crop of padi was becoming less and less. People continued to plant except in areas where the level of the land had become too high to be irrigated. (1983: 375)

Today, the bulldozer has become a principal instrument of land clearance and levelling in Malaysia, not only on new settlement schemes and for urban development, but even to create larger fields in place of former hand-built terraces and field beds in so sensitive an area as the Cameron Highlands, where the former careful system of management is described by Clarkson (1968). The consequence is severe erosion and landslipping, so that in periods of heavy rain silt is deposited in the streets and even houses of small towns (*New Straits Times*, Kuala Lumpur, 24 November 1985). Current proposals to replace unprofitable rubber in the hilly country around Malay villages by cocoa, oil palm and vegetable crops create the prospect of much more widespread degradation if capital-intensive methods of land preparation are employed. However, some more conservationist methods are now sometimes incorporated, including the use of a biodegradable latex emulsion to hold the soil until a plant cover is re-established (IBSRAM 1985).

Land development is initiated primarily for political purposes, with an economic role fully elaborated only in the Malaysian case. Its objective is to develop national resources, and to relieve poverty and overcrowding in a demonstrable manner. Sometimes it is a substitute for more difficult land reform, or is used in association with partial land reform to relieve pressure for more drastic change. Sometimes it has an internal political objective, as in the settlement of nationally dominant Javanese in the outer islands of heterogeneous Indonesia, or providing Malays with a larger stake in the multi-ethnic Malaysian economy. The land is seen as a resource to be used, but the manner in which it is used has its origins in the politics of the country.

3 Conclusion

Management of the tropical lowlands is too often discussed only in terms of the forests. At the same time, however, the tropical lowlands are also the site of the most intensive and conservationist systems of land management known, especially the irrigated rice-terrace systems, while between these and the forests there is a large area that is 'degraded' from the point of view of shifting cultivation and forest use, but much of which is capable of productive use.

This 'intermediate' area is far from uniform in terms of its environmental characteristics, but is particularly characterized by the dominance of invasive grasses especially where fire is employed and breaks loose. Part of the present 'problem' of the lowland tropics is that the 'intermediate' area is being extended more rapidly by forest depletion than it is being reduced by the introduction of successful systems of management. In some areas, especially the Amazon, the 'intermediate' area is being extended alarmingly by the use of inappropriate methods of colonization and exploitation on land newly cleared from forest and severe degradation results. The fear of creation of a 'red desert' in such areas has a real basis. On the other hand, this 'intermediate' land can also be managed in ways that offer sustained productivity, though with potentially severe downstream consequences. The Malaysian case illustrated this. The 'intermediate' area is also seen as a major field for land colonization, but, where the motivation is not backed up by sound planning and steps to ensure appropriate management, results are at best disappointing, and at worst disastrous.

The chapter immediately following (9B) describes and analyses a case in which adaptation is taking place painfully. Official and local perceptions of the region and its future are sharply at variance, and their conflict underlines the essentially political nature of the modern development problem. This is even more sharply demonstrated in chapter 9C, where a rigid inequality of land distribution creates severe problems in implementing a resource–development policy, leading to a sharp contradiction between perception of land as a valued resource, and its non-perception as a resource that needs adaptive management.

Development of the tropical lowlands once reflected cultural differences between sedentary and shifting cultivators, leading to the slow replacement of the latter by the former. Colonialism and capitalism created new and destructive forces which did, however, lead to successful and highly profitable management in some areas. Capitalism also created conflicts, and they have been exacerbated by the political conflicts inherited by and developed within the post-colonial states. Since forms of state capitalism are now important, even dominant, in many of these countries, the conflicts of the colonial period have often been sustained and augmented. Politics are now perhaps the major controlling force in the management of these regions, and politics are only rarely in harmony with the needs of the land. It is for this reason that the rhetoric of conservationists concerning destruction of the tropical rainforest should not be decried, despite its hyperbole, for rhetoric is the language of politics.

B Degradation, innovation and social welfare in the Riam Kiwa valley, Kalimantan, Indonesia
Lesley Potter

1 Introduction
1.1 Three deforested areas

Large areas in the outer islands of Indonesia have been converted from forest into *Imperata cylindrica* (*alang-alang*) grassland, perhaps 14 million ha in Kalimantan alone (Suryatna and McIntosh 1980). The idea that the *alang-alang* country is a 'green desert' (Geertz 1963) still persists in some quarters (e.g. Soerjani, Eussen and Titrosudirdjo 1983) but its extent is prompting reappraisal as a site for future transmigration schemes, rather than the presently preferred forest or tidal swamp (e.g. Burbridge, Dixon and Soewardi 1980; Soewardi, Burbridge and Djokosudardjo 1980).

This chapter studies three areas in the upper Riam Kiwa valley of South Kalimantan, among the most heavily deforested areas of Indonesia. A small population, composed a century ago only of indigenous Dayaks and Malay migrants (Banjarese) of several hundred years' standing, made the 'empty' area of South Kalimantan just across the sea from crowded Java attractive for colonization schemes in Dutch times (Pelzer 1945). While most early efforts were directed at wet-rice cultivation, a part of the Riam Kiwa valley was selected in 1938 for an experiment in dry upland agriculture by Madurese colonists. These grassland farmers have more recently been joined by 'fugitives' from official transmigrant schemes of the post-Independence period in the difficult tidal swamps to the west. Javanese and Madurese now intermingle with the Malay (Banjarese) people in the lower and middle parts of the valley, leaving only the remote areas, still mainly in forest, to the Dayaks. The Javanese and Madurese are adapting their agriculture to the

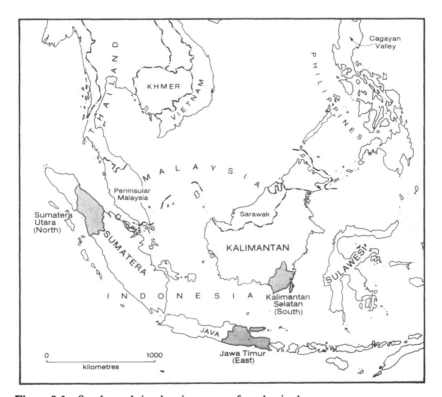

Figure 9.1 Southeast Asia, showing areas referred to in the text

grasslands, using cattle for the heavy ploughing required. The success of this system is by no means assured, and we have in this valley both deforestation and the reclamation of deforested land for agriculture; both the seemingly successful establishment of sustained production, and evident failure of management leading to degradation. Added to the two forms of degradation

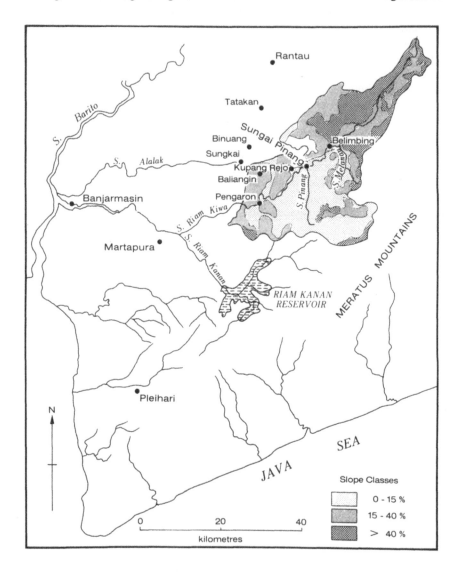

Figure 9.2 Part of South Kalimantan, Indonesia, showing slope categories in the Riam Kiwa basin
Source: Kabupaten Daerah Tingkat-tingkat Banjar, Sheet 3

– creation of *alang-alang*, and degradation of soils under attempted management – we have four different groups of people with different agricultural traditions. Moreover, there is a set of government policies concerning the preservation of forest and reclamation of the grasslands, and certain inputs are made by the authorities to supplement somewhat spasmodic guidance and regulation.

The experience of people and of the land in these three villages recalls the discussion of the impact of colonialism in chapter 6; there are many features in common. The present situation in the Riam Kiwa certainly had its origins under colonialism, but the continuing forces of commercialization, population growth and the introduction of new methods and ideas can still be identified forty years beyond the end of the colonial period.

1.2   The upper Riam Kiwa basin

The upper Riam Kiwa basin occupies an area of 1420 km². The land is hilly, about 60 per cent of the basin having slopes above 15 per cent, and half of this area with slopes of 40 per cent or steeper (figure 9.2). The current soil map places the whole upper basin under the rubric 'complex red-yellow podsolics, latosols, lithosols'. Yet in the earlier Dutch reports the Riam Kiwa was well known for its fertility (Schreuder 1923). A detailed survey was carried out for Madurejo, the Madurese colonization area, the soils of which were described in 1938 as 'red-brown clays of andesitic origin, of satisfactory depth and good structure' (*Kolonisatie Bulletin* 1938, 2: 21, translated).

Average annual rainfall over the basin has been estimated at 2500 mm (Japan International Co-operation Agency 1982). Rainfall is markedly seasonal. In drought years the dry season is more pronounced and the rains are sharply truncated. Shifting cultivators in this district seldom experience a problem in finding a season dry enough for the burn (Dove 1980). It is the contrast between the seasons, with the likelihood of high totals during the rainy months and the ease of burning in the dry, which makes this part of Kalimantan prone both to erosion and to the development of *alang-alang*. Climatic extremes may also be expected, with droughts one year followed by excessive rain in the next. It is hardly surprising that Dutch administrators commented on the frequency of famine (MvO 1909).

1.3   The creation and management of grassland

Active creation of further grassland is now confined to the upper part of the basin, the lower part having made this transition long ago. The maps in figure 9.3 indicate as accurately as possible the extent of the grassland from maps of 1926 and 1972. If the earlier map is correct, large areas of secondary forest still existed in 1926. However, the 1972 map shows as 'primary forest' some areas designated 'secondary forest' in 1926. Dutch sources show concern about the spread of *alang-alang* in the 1890s, with the mining engineer Hooze

(1893: 22) saying that the lower district near Pengaron had been completely deforested for shifting cultivation. Hooze described enormous grass and bush fires. In 1902 a prohibition was imposed on burning grass and secondary

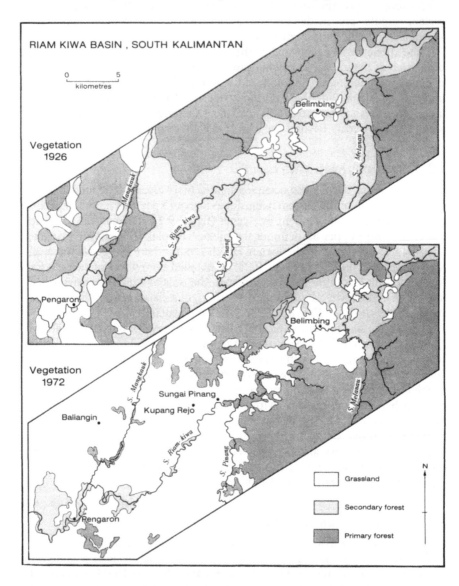

Figure 9.3 Distribution of vegetation types in the Riam Kiwa basin, South Kalimantan, Indonesia, 1926 and 1972 *Sources:* 1926–Reproductiebedrijf Topografische Dienst, Weltevreden (Java), 1926: Zuid-en-oost Borneo, 1: 100,000. 1972–P.N. Aerial Survey: Topographic Map Series Kalimantan Selatan (1977): 1: 50,000

bush but this failed because the *alang-alang* areas were too vast for all fires to be controlled (MvO 1938); in 1909 it was said that 'the main part of the Riams' was covered in continuous *alang-alang* (MvO 1909). The blame for this situation is laid squarely on the Banjarese. Wentholt (MvO 1938) went so far as to say that Banjarese movement into Dayak districts was equivalent to an invasion of *alang-alang*. The Banjarese are continually described as caring little for the heavy labour which accompanies 'decent' agriculture. In this they are contrasted with the Dayaks, who never burn young secondary scrub and do not like to see *alang-alang*.

Although the soil under *alang-alang* was described as in general useless for agriculture, it was admitted that an 'andesite strip' near Pengaron was favourable for permanent farming, some *cangkul ladangs* (spaded fields) having been laid out there (MvO 1929). Gerlach (1938: 449) describes these fields as the work of Javanese immigrants, sometimes hired by the Banjarese, who dug the soil twice to clear out the rhizomes. Banjarese also worked the land by shallower digging, then burned the residue before planting cassava and dry rice. The *cangkul ladangs*, however, gave better yields and could be occupied for five or six years after complete clearance of the grass.

Policies favouring control over forest use, particularly burning, with prohibitions on shifting cultivation in some villages, date back to the early 1900s. The low Dutch opinion of the Banjarese and their farming follows logically from general attitudes during the nineteenth century and earlier (Matheson 1981), but it has continued to the present, now being believed not only by government officers (often from Java) but even by the Banjarese themselves. They will often comment that Javanese, Madurese or Balinese transmigrants are much better farmers than they are.

2 Three communities compared

2.1 Belimbing

The village of Belimbing is an old Banjarese settlement, the furthest up-river before the Dayak lands are reached. Its total population of 1029 includes outlying riverine settlements as well as some Dayak farmers living in the forest. The population of Belimbing was greater a hundred years ago than it is now. The 1886 population was given as 1126 (Hooze 1893: 424) and in 1938 the village was listed as having 608 adult men; the present total is probably not more than half the 1938 figure. Some Javanese families have settled, but Belimbing has lost importance, remaining isolated along a difficult dry-weather track. Currently, four environments may be distinguished here: along the river bank, in and beyond the house yards, are groves of productive trees – coffee, kemiri (candlenut), cloves, kapok and a variety of fruit trees. In Dutch times, rubber was also grown close to the stream. Beyond this *kebun* area the nearer and lower hillsides are covered with *alang-alang* and the tall grass *Saccharum spontaneum*, known locally as *hariung* or *tampukas*. Moderate slopes within 4–6 km of the village are clothed in *Eupatorium*[1] scrub (*kumpae jepang*) with some secondary forest, *belukar*. These form the main cropping areas for rice, peanuts and bananas. Distant hilltops carry primary forest. Village shifting cultivation plots have been restricted to the scrub zone since the government ban on the cutting of primary forest swiddens. Available land for farming in this zone is gradually becoming difficult to find and is further from the village: already most farmers live in their field houses during the week. About half, the poorer farmers, have no houses in the village.

Land managers at Belimbing are faced with three choices: to work their present lands more frequently; to move to the *alang-alang* land; or to go even further from the village to seek new secondary forest. Present practice is to use a piece of land for a year and a half, taking off one crop of rice, perhaps two of peanuts and one year's bananas. The land is then returned to scrub for five years. It is known that any increase in the frequency of cropping will result in the incursion of *alang-alang*. There are some signs of erosion, but cultivation with digging sticks creates minimal disturbance and crop growth is generally vigorous enough to hold the soil.

The second option is to work the *alang-alang*. The grass is more difficult to farm without access to cattle for ploughing. Most labour effort comes in clearing and land preparation: initial opening is estimated to take 30 man-days/ha using a hoe but only 6 man-days by ploughing, activities which must then be repeated twice with breaks in between to dry out and break up the grass sod, and destroy the rhizomes. In contrast, clearing secondary forest is claimed to take about 32 man-days/ha and clearing *Eupatorium* 8 man-days; in each case it is a once-only operation. Although teams may be hired, ploughing is an expensive operation, costing about $US100/ha. Without schemes of government assistance and credit for livestock and fertilizers, grassland farming is beyond the capacity of the poor farmer.

Rapid changes have been occurring in Belimbing since 1980 as the possibilities of working the *alang-alang* have been assessed by local entrepreneurs, both Banjarese and Javanese. Increased numbers of plough cattle available for hire have induced those with capital to open large fields in the best locations. Members of the village élite, who formerly concentrated on merchant activities and the ownership of fruit and candlenut trees, are now increasingly taking possession of extensive areas of grassland for groundnut and rice cultivation. Two bad seasons on the slopes, beginning with the drought of 1982/3, and low rice yields have left many poorer farmers in a position of serious indebtedness. They gladly spend part of their time

---

[1] Species of the woody shrub *Eupatorium*, a native of tropical America, have become widespread in Africa and Asia since the Second World War. Banjarese call it 'Japanese weed', and believe it was air-sown by the occupying Japanese as a form of *Imperata* control (see Dove 1984, for discussion of *Eupatorium* origin-myths in Indonesia). The shrub competes well with *Imperata* when the latter has been overgrazed or the frequency of burning reduced. Although in Nepal the name of the upland species *E. adenophorium*, means 'forest killer' (Mahat 1985), the more common *E. odoratum* is not seen as impeding regeneration in forest-fringe areas of South Kalimantan. Rather it is regarded favourably as indicating better soil than under grassland with greater ease of clearing.

working as labourers on these new farms. The position in Belimbing seems to be that the poorer farmers, lacking any form of government assistance, are confined to the slopes which will continue to degrade, while those who are in a position to do so are reclaiming the older *alang-alang* close to the village.

2.2 Kupang Rejo

Located about 15 km downstream of Belimbing and near the administrative centre of Sungai Pinang, Kupang Rejo has a largely Javanese population who moved to the area from a tidal-swamp transmigrant settlement in the early 1960s. In this village a further stage in the pattern of landscape change and social differentiation may be perceived. Kupang Rejo has basically two environments: gently sloping to rolling land near the village centre and much steeper country beyond. Both areas are largely under *alang-alang*. Valleys close to the village have been converted to valley-bottom *sawahs*, but there is no terracing. The flatlands have been worked since 1967 and fertility has declined sharply: crops of rice and peanuts now depend on chemical fertilizer. The hillsides are worked on a rotation of one to three years' farming with one year fallow and regular burning of the *alang-alang*, with some cattle manure also being applied, though mainly to tree crops such as bananas. Possession of cattle for ploughing is now almost universal, Kupang Rejo being one of the major recipients under an Asian Development Bank (ADB) funded scheme. The cow-and-plough technology, while making it possible to work the grassland with relatively little effort, leaves large stretches of ground bare for months at a time at the beginning of the rainy season. As in Belimbing, most farmers grow upland rice, groundnuts and bananas, but it is the former which are the most important source of cash income, necessitating wide areas devoted to this field crop. The poorest farmers are relegated to 2 ha of the steepest land, one of which they must rest each year. They are generally young and sometimes recent arrivals: Kupang Rejo has experienced a rapid increase in population over the last nine years as a result of spontaneous in-migration from Java. Numbers have risen from 347 in 1976 to 598 in January 1985.

Like Belimbing, Kupang Rejo has suffered from economic stress during the past two or three seasons. During 1982 the BIMAS programme of credit, which assists farmers to purchase inputs such as fertilizer and pesticide, was introduced to the village; general household credit was also available at an interest rate of 12 per cent p.a. The programme was aimed at increasing farming intensity, to reduce the need for fallow and to permit dry-season crops to be produced. About one-third of households now participate but the drought of 1982/3, followed by excess rain which badly affected groundnut crops in 1983/4, has meant that most have been unable to repay their loans. Fertilizer use has dropped as people are unwilling to incur further debts. Fortunately, the ADB cattle scheme will enable more land to be worked per farmer. The enterprising, most of whom are already in the 'better off' category, are beginning to hire out ploughing teams to cattle-poor villages, even as far as Belimbing.

Farmers say they realize that erosion is occurring and some efforts are made to curtail it. On the steepest slopes, vertical cuts enable rapid escape of runoff and sometimes trees are planted along the contour, but farmers say they are 'too busy' to make terraces. This village is now part-way along the track toward the kind of higher technology farming system approved by the Indonesian government. Extensive areas of *alang-alang* are regularly ploughed and cropped, with varying amounts of chemical inputs aided by government credit. Those who possess *sawahs* are able to reap much higher yields than from the hill rice, ensuring their self-sufficiency in the staple as long as irrigation water is available.

2.3   Baliangin

Kupang Rejo is in the throes of greater intensification, including an increase in land farmed. In Baliangin land has now become a scarce resource, leading both to movement out of agriculture and some outmigration. This third village, twice the size of Kupang Rejo, is occupied by Madurese trans-migrants, many of whom are descendants of the Madurejo colonization settlement of 1938. The location of Madurejo was carefully selected on what were perceived as the best soils in the district. Even then, certain precautions were incorporated: terracing was seen to be necessary; *Crotalaria* species were to be planted along the ridge tops 'to hold the soil' and used as green manure; the Madurese, used to dry farming and stock raising in their home island, were expected to make use of animal manure (*Kolonisatie Bulletin* 1939).

The district consists of moderately steep hills about 100 m in elevation, which by 1938 were already covered in *alang-alang* and *hariung* (*Saccharum spontaneum*), with some scrub remaining in the valleys. On these, the Madurese established their typically dispersed farming system, each household being surrounded by its lands with no nucleated village as such. From the original core of about 400 families they spread out through the district, founding a number of daughter settlements close by. Baliangin is one such. Total numbers are now about 1150. All the available territory is occupied, so that young men in particular are forced to move out to seek new lands.

Although the Dutch were enthusiastic about early rice yields in Madurejo – the second crop produced eighty times the amount of seed planted – fertility has declined dramatically throughout the district so that present yields may be as low as ten to fifteen times the seed (about 360–540 kg/ha). Very few now aim to be self-sufficient in rice. The most assured crops and the best yields come, as in Kupang Rejo, from the small *sawahs* which now occupy almost every valley. Social differentiation in Baliangin exists on many levels but one of the most basic is between those who possess *sawah* and those who do not. *Sawah* lands are terraced and pampered; hill lands are ploughed repeatedly, no longer to remove the *alang-alang*, which has almost disappeared, but to eliminate the weed seedlings which are very prominent, particularly following increased fertilization. The planned *Crotalaria* is not favoured for cows will not eat it, and it has not been incorporated into cropping systems. Terracing is not much in evidence away from the lower valleys and the steep slopes generally remain untouched. As in Kupang Rejo, bananas are sometimes tried as a means of erosion control. Cattle have been incorporated into this system from the beginning, so that Baliangin's lands have been ploughed at all times.

Although the Madurese have persisted with a mixed farming system and an emphasis on tree crops which is more protective of the soil than monoculture, nevertheless, the intensification of their system has led to degradation. The land shortage which forces many young men either to seek new hill country, or to try their luck in the city, has also meant that less land is now left fallow as farmers can no longer afford to rest it. They operate on the philosophy *Tanah mati, pupuk hidup* (the soil is dead, fertilizer is life) and say that, although fallowing the soil improves yields to a certain extent, they can no longer expect a 'reasonable' crop without heavy fertilizer input.

3 Discussion

3.1 Pressure and perception

In a restricted Boserup-type interpretation, the three villages described above could be seen as three stages in a continuum of pressure on land resources. However, the adoption of radically different farming systems, including the 'capital' of working livestock and the 'landesque capital' of terraced valley *sawah* destroys the simple continuum since different responses to labour input are involved, and the adoption of these innovations seems to owe as much to opportunity as to pressure.

Moreover, the adoption of different farming systems, with their different demands on resources, creates important differences in the evaluation of these resources. The shifting cultivators of Belimbing regard the growth of scrub and *Eupatorium* on the hills as favourable indicators of the return of forest, whereas *alang-alang* is a pest to be eradicated, or a sign that the land must soon be abandoned. In the cattle-owning villages, on the other hand, *alang-alang* country is seen both as potential arable land that can be opened up with the plough, and also as essential grazing for livestock. *Eupatorium*, which cattle will not eat, is here seen as a useless weed. *Alang-alang* is also used for thatch, and where it has been almost wholly ploughed out, as at Baliangin, it becomes a scarce and valuable resource, one that must be fetched from a distance.

These differences in perception are widespread in Indonesia (Sherman 1980a, b; Dove 1984, 1985). They reflect the difficulty of comparing degradation through vegetation. Degradation of the soil is another matter, and we therefore turn to some data on the soils, and on yields derived from the land, which are a common concern to all farmers wherever they are located.

3.2 Soils

In the absence of any complete survey of soils, the author obtained some top-soil samples and had these analysed by the Provincial Department of Agriculture. While top-soil samples cannot provide full information it is noteworthy that the pH, organic carbon, nitrogen, phosphorus and total exchangeable bases were all much higher on the forest- and scrub-covered soils around Belimbing and in the hills at Kupang Rejo than on the grasslands, and even than in the valley-bottom *sawahs* further down the valley. For example, pH ranged from 7.1 to 6.3 in the former, but from 5.1 to 4.5 in the latter; organic carbon from 11.1 to 4.8 per cent and 5.7 to 3.2 per cent, and nitrogen from 0.6 to 0.3 per cent and 0.3 to 0.1 per cent in the forest-scrub and grassland areas, respectively. More remarkably, available phosphorus ranged from 23.2 to 11.4 ppm in the forest-scrub soils and only from 0.9 to 0.1 ppm in the grasslands. Using ranges developed by Bleeker (1983) for New Guinea, potassium and nitrogen levels are 'high' to 'moderate' in the first group, but 'low' in the second. The *sawah* soils do not stand out, but there is nutrient uptake from the circulating water while nitrogen is fixed from the atmosphere by *Azotabacter* in association with blue-green algae. Certainly *sawah* give much better rice yields than do the dry fields.

Charley has commented on the changes in soil nutrient levels under forest and under grassland such as *Imperata*, particularly relating to the quality of soil organic matter and the release of nitrogen and other nutrients: 'Degradation of the quality of the organic soil reserve ... could be a significant factor in fertility rundown and it could well affect supply of nitrogen, phosphorus and sulphur' (1983: 393).

3.3 Crop yields

Hill rice is common to all three villages. Yields over the past two years and expected yields in 'normal' and 'good' years were obtained and compared. Although Belimbing's expected yields in 'normal' years, according to farmers' perceptions, are the highest claimed (1.8 t/ha), Kupang Rejo's hill lands are also reasonably productive (1.6 t/ha); the flat at Kupang Rejo and Baliangin's hills are much less so (0.6–0.7 t/ha). Yields experienced in 1982–4 were all below these levels. Especially in Baliangin, the *sawah* lands assume much greater importance in helping to make up deficits from the hills. In 1983/4 the ratio of seed sown to crop harvested on Baliangin's hill lands was 16:1; the *sawah* returned 78:1. The average hill rice plot size was only 0.3 ha, while *sawah* plots owned by half the farmers averaged 0.15 ha.

The *sawah* and hill lands together yielded only one-third of the average family's needs, even taking into account the usual Madurese custom of mixing rice with maize and cassava to reduce consumption. Among the farmers sampled in Baliangin in 1983/4 only 8 per cent of the rice growers were self-sufficient.

In Kupang Rejo only a quarter of the farmers had *sawah* but hill plots were larger and more productive than at Baliangin, the ratio of seed sown to crop harvested being 28:1 in the hills and 110:1 in the *sawah*. With regard to the level of self-sufficiency, Kupang Rejo farmers were divided into two main groups: 33 per cent had to buy for several months while 41 per cent were not only self-sufficient but had surplus for sale. In evaluating this evidence, however, it is important to note that 1983/4 was a below-average season. Even in Belimbing, where yields from hill plots are usually ample, 54 per cent of farmers experienced shortfalls and had to purchase supplies for at least one or two months.

4 Does degradation matter? And for whom?

Within each village, it is clear that degradation matters most to the poor who are shortest of available resources. If fertility declines on his small plot, the farmer must either acquire fertilizer, take more of his land out of production for a longer period, or, as in the case of Belimbing, be faced with changing his entire farming system. Yet it is precisely on the small farmer that most pressures will come to work the land harder and longer and increase the risk of degradation. To the larger farmer, land degradation matters less, at least in the still fluid, 'frontier' situation of the Riam Kiwa, where new opportunities may still be seized and innovations tested. To the possessor of *sawah*, erosion of the uplands may have no direct impact other than to provide new soil for his *sawah* land. It is possible that the slopes will eventually become so infertile that deposition of this soil in the *sawah* will cease to be favourable. The overall impact of increased fertilizer use on all levels of the system remains at present an unknown factor in shaping the future.

Beyond the sphere of the village, degradation matters for officials of the regional and provincial government who are concerned with the implementation of national policies at the local level. Such policies are not always co-ordinated and objectives may even conflict, but all would argue that land degradation should be kept to a minimum. The forestry department is particularly concerned at the rate of spread of *alang-alang* at the expense of forest and has a major re-afforestation programme. All the grassland areas of the Riam Kiwa are designated 'productive forest' on departmental planning maps. To the foresters, the population of the valley is a hindrance, especially through its burning activities, to the proper implementation of re-afforestation plans.

The electricity authority which supplies power to the cities of Banjarmasin and Martapura would like to construct a large reservoir and hydropower plant in the Riam Kiwa, to supplement the inadequate existing scheme in the Riam Kanan basin to the south. The hydropower authority is also concerned with the re-afforestation of the catchment, arguing that the presence of *alang-alang* around the Riam Kanan reservoir is one reason for its inadequacy. If the Riam Kiwa dam project were implemented, the catchment would be designated 'protected forest' from which population would probably be excluded.

The agriculture department, on the other hand, is concerned to promote farming in grassland rather than forest, to supply credit for fertilizer, and to expand both the range of crops and the use of techniques such as terracing to reduce erosion, while the livestock division will assist with plough animals and give some attention to pasture improvement within the possibilities of village economies. The transmigration department, while concentrating its present activities on the south and east coasts, has identified an area near Pengaron for a possible new scheme.

Some of these conflicting plans for the district are people-oriented and designed to assist the small farmer, while in others people are simply a nuisance or irrelevant. While the farmers themselves perceive the constraints on their actions largely at village levels, the final responsibility may be taken out of their hands altogether.

C Land mismanagement and the development imperative in Fiji[1]

William Clarke, with John Morrison

1 The expansion of cultivation and degradation

1.1 A sugar-cane economy

Until after 1981, when the world sugar price went into a steep decline, sugar's share of the domestic exports of the Pacific island country of Fiji had been rising for several years, and had reached a share of 81 per cent by value

[1]We would like to thank the persons listed here for their many helpful comments on issues relevant to this paper: Mr T. Davey (NLTB, Fiji), Mr M.P. Faktaufon (Sugar Commission of Fiji), Mr F.F. Kafoa (Land Use Section, MPI, Fiji), Mr J. Martin (Native Land Development Corporation, Fiji), Mr A. Prasad (Land Use Section, MPI, Fiji), Mr N. Reddy (The University of the South Pacific), Dr S.C. Thakur (Central Planning Office, Fiji), and Professor R.G. Ward (Research School of Pacific Studies, Australian National University).

in 1980. Sugar has been the main commercial product of Fiji agriculture since late in the nineteenth century. It was initially grown by a number of plantation companies, then, after access to indentured Indian labour was withdrawn in 1920, by a system that is almost unique among the world's sugar-producing countries. Over 95 per cent of the crop is grown by smallholders, who are contracted to the four factories to provide cane. The period since national control of the industry in 1973 has seen a major expansion of sugar production, a growth which, since yields of cane have remained fairly constant, has been achieved by an expansion of the cane area from 43,804 ha in 1972 to over 70,000 ha today. Over 22 per cent of the total national labour force was employed in the industry in 1981 (Ellis 1985).

This is not the only expansion of land use that has taken place in Fiji. Ward (1985), who conducted the pioneer study of population–land relationships in Fiji twenty years ago (Ward 1965), calculates that on the main island of Viti Levu the area of land in use or committed increased by 233 per cent between 1958 and 1978, while rural population increased by only 31 per cent (1958–76). While much of this is due to the adoption of cattle raising in village agriculture, important elements arise from the extension of land settlement, especially for commercial farming.

The expansion of sugar cane is economically the most important of these changes. Until about 1950 very little sugar was grown on sloping land; it was a crop of the alluvial flats. In the 1960s and 1970s Fiji embarked on a major resource-development programme, which included planting of pines on much degraded land, but also the agricultural use of much land that had not been used, or had been little used, since pre-colonial times. A dramatic example of the process is the heavily capitalized settlement scheme at Seaqaqa (pronounced Seanggangga) in Vanua Levu. The scheme was initiated in the mid-1970s with World Bank funding, partly with the intention of bringing more indigenous Fijians into an industry dominated by the Indo-Fijian descendants of the former indentured workers. By means of expanding cane growing beyond its former perimeter it was also hoped to counteract a decline in sugar production that had taken place since the late 1960s, a decline that was in part the result of lowland drainage problems, in turn caused in part by sediments from slopes upstream. At Seaqaqa, where the land is mostly rolling or hilly, the nutrient-poor and erodible latosols have suffered serious degradation, as have many areas on Viti Levu where the cane frontier was pushed inland and upslope in the same period.

1.2 Some data on erosion

Site observations and chemical-physical analyses reveal high rates of erosion and degradation on some of this newly developed land. At Seaqaqa observations were made over five years, from initial clearing in 1978 to 1983, on a 5°–8° slope. The soil is a ferruginous latosol (Typic Haplustox). During this period 15 to 20 cm of soil were lost over the entire site, corresponding to a loss of 34 t/ha/year. Over the same five years, soil bulk density increased from 0.85 to 1.10 g/cm^3, and soil organic matter (expressed as per cent organic carbon) showed a decline from 4.43 to 3.00 per cent in the top 12 cm of soil. The most serious loss was the drop in the exchangeable bases, calcium and magnesium. While the lower layers of the soil continued to provide a friable rooting material, exchangeable bases dropped markedly from 17.9 m.e./100g in the original top 8 cm, to 1.4 m.e./100g at 16–25 cm, indicating a significant decline in fertility as the lower levels progressively became the surface soil.

Near Nadi, on Viti Levu, a lithosol (Lithic Ustorthent) on an 18°–22° slope was cleared from a grass–fern–casuarina cover. By the time the first crop of cane had been harvested 8–14 cm of soil had been lost. In some places the entire solum was removed, a loss equivalent to about 90 t/ha/year. Removal of the A, B and upper C horizons meant a drastic decrease in carbon and nitrogen, and a rise in exchangeable aluminium (present in the lower layers of the original soil) to levels toxic to most plants. Exchangeable bases did not decline because the parent material is base-rich, but the massive nature of the sub-solum meant the deterioration of the rooting medium. This site was likely to have a short production life.

Other studies show comparable results. Morrison (1981: 12) used the USLE factors to calculate erosion rates of 36.7 t/ha/year from a cane field on an 8° slope, with reasonably good cropping practices, a median value for erosion-control, and located in Fiji's drier zone. On the basis of field measurements of soil loss from six small plots in northwestern Viti Levu, Liedtke (1984) extrapolated values from 24 to 80 t/ha/year.

All these examples far exceed the 'soil-loss tolerance level' of 13.5 t/ha/year for tropical areas suggested by Hudson (1971: 192). It is evident that degradation is not only serious, but also rapid. Moving from specific sites to a wider view of cane farming in Fiji, many areas are known to suffer sheet and gully erosion as well as loss of nutrients. Similar conditions were reported twenty years ago by Twyford and Wright (1965), yet the cane area has not only increased by about 64 per cent since they wrote, but also has a higher percentage of its total area on steep slopes.

Yet the warnings were already there before this happened. Twyford and Wright (1965: 182) had already noted that, while there was only moderate deterioration after eighty years of cultivation on the alluvial lands, there was obvious deterioration on rolling land of low natural fertility and resilience. One large area of 19,425 ha in northwestern Viti Levu had already had to be retired not only from cane, but from all kinds of agricultural production as a result of bad management on the lower land coupled with overgrazing and burning on the adjacent hills. It seems arguable that, with the possibilities for further extension of the cane area now close to their limits, an overall decline in yield may soon set in as a result of the rapid deterioration of much of the land newly taken into cane since 1973.

1.3   The causes of erosion

Sugar-cane farming is an intensive form of agriculture, and in some parts of the tropics it has been practised for very long periods on the same land. Since the crop is dense, and large amounts of organic matter are conveyed to the soil in the form of trash and an extended root system (Ruthenberg 1980: 272), soil is usually exposed to erosion for only short periods. Why, then, should cane farming on sloping land in Fiji lead to such massive degradation?

The sensitivity of Fiji to erosion is high, but not remarkably high by tropical standards. Rainfall erosivity factors in the USLE scheme have been calculated as 1210 for Suva, and 930 and 885 for stations in the drier zone where cane is now concentrated (Morrison 1981; Liedtke 1984, following Roose 1977). Soil erodibility is low for over half of Fiji's soils. Slope, however, is often steep away from the alluvial lowlands: 67 per cent of Viti Levu and 72 per cent of Vanua Levu have slopes greater than 20°, while only 16 and 15 per cent, respectively, are flat (less than 2°).

The question therefore is one of management, both the manner in which sugar is cultivated, and the tenure and other conditions on which land is held and worked. Also important is the perception of erosion as a hazard in government, in the Fiji Sugar Corporation (FSC) and its multinational predecessor, and among Fiji's farmers. For over twenty years there has been an assumption of abundant land in Fiji, underlying a 'resource-frontier' policy of development (Bienefeld 1984; Ward 1985). While the average population density on land suitable for arable farming is 170/km^2, which should not encourage such a belief, the overall density is only 34/km^2 (Ward 1985). Although cries of PPR have been raised about Fiji for many years, they have always been hard to credit when standing among the large forests and grasslands away from the pockets of dense rural and urban population. Even now, much of Fiji's land is used at very low intensity, with low productivity per hectare whether measured in cash or calorific terms. The underlying causes of Fiji's degradation problem do not therefore lie in extreme sensitivity of environment, nor in PPR measured against total land. They lie elsewhere, as we shall discover below.

2   Institutional and structural factors

2.1   Land tenure

The land of Fiji is divided and allocated in a manner that is almost unique in the world, and is the product of historical decisions taken by the first generation of colonial rulers (France 1969), decisions that have become set in the concrete of Fijian multiracial politics and the country's complex political economy (Lasaqa 1984). Only some 8 per cent of the land is freehold, and much of that is now priced out of the agricultural market. A further 9 per cent is state land (Crown land), and what is known as 'native land' makes up 82.4 per cent of the total surface. Such land is allocated to landowning descent groups and to some individuals among the indigenous Fijian population, and can be used by non-Fijians only under leasehold. Some of it may not be used by non-Fijians at all.

The effect of this system is to convert a favourable person:land ratio into a distribution that excludes most of the Indo-Fijian descendants of the indentured cane workers from ownership of any land, and allows them access to land only as tenants or leaseholders. Moreover, demographic changes since the landowning descent groups and their land were defined a hundred years ago have resulted in large inequalities also among Fijians. There is therefore a paradox in the Fijian land situation. For the nation as a whole it has been possible to regard land as an abundant resource, yet for a majority of the whole rural population it is a scarce or unobtainable resource. This has important implications.

For a long time, the conditions under which land could be leased or rented were highly insecure; tenancies-at-will were common, and many leases had an extremely short term, with no security of renewal. This condition still obtained when Ward remarked that:

> lack of security results in poor farming, as the tenant is reluctant to dig or maintain drains, construct bunds or contour terraces, use fertilizer or carry out any improvement which might benefit the landlord or result in demands for higher rent. (1965: 123)

Since 1966, conditions of lease have been made more secure. The normal period is now thirty years, with some rights of renewal. However, it has taken a long time to register leases, and there remains a population of tenants-at-will. Ward's comments thus still apply to a proportion of the tenants even today.

2.2 Insecurity, poverty or neither?

What of conditions other than improvement in security of tenure, even though that improvement has been patchy and incomplete? Failures have been common among small farmers established on the land since 1960, but there have also been many successes, and individual cash incomes have generally been higher in the sugar areas than in most other parts of the country.

Ellis (1985) shows the average gross income per registered cane grower to have varied between 1300 and 6000 Fiji dollars between 1970 and 1980, and in general to have more than held its own against inflation. Using some inadequate information on costs, he went on to estimate net incomes at between $2000 and $4000 for 1981, when the Fiji dollar was close to parity with the US dollar. These are relatively high incomes for small farmers in developing countries. However, Ellis's data are averages. G. Anderson (personal communication) reports that sometimes two or three Indo-Fijian families are today living and working on a single lease, and the income is divided. At Seaqaqa, Atkins (1983) reports considerable variation among settlers in the new scheme, with some operating at a loss, and a significant group obtaining only low incomes and managing their affairs poorly. It would seem that there may be a proportion of leaseholders who have very low incomes, but there is no information on how they are distributed in terms of the quality, location and sensitivity of their land.

Among the sugar farmers, however, off-farm employment has for long been an important source of income for those with inadequate access to land, and most of this employment is in the cane-cutting gangs. Ellis (1985) shows that there are some 19,000 cane cutters who are not also growers, of whom a majority are hired workers rather than family members. Some of these hired workers are indigenous Fijians who migrate seasonally to the canefields, but most of the remainder are landless or near-landless Indo-Fijians. How far this growing population is derived from families who could not obtain land, or had to relinquish what they had, is not known. Despite the great weight of research that has gone into the conditions of the indigenous Fijians, remarkably little is yet known about the decision-making and livelihood constraints of the rural Indo-Fijians. Such research has not been fashionable.

We can therefore draw no conclusions concerning the relationship of poor farming practices either to insecurity of tenure in recent years, or to lack of resources. What evidence we have suggests that both security of tenure and income have improved rather than declined, and we have no adequate information to show that a significant minority has experienced contrary trends. For at least some of the Indo-Fijian cane farmers such contrary trends are very likely, and by no means all indigenous Fijian lease-block farmers have any longer the option of returning to villages. Not all members of either group are successful farmers, but we lack information with which to show that any growing portion of Fiji's peasantry shares the insecurity and poverty which is so common an experience elsewhere in the world. In our search for explanation we have to look elsewhere, to higher levels in the total structure of management.

3 Adaptive land management: institutional consciousness and implementation

3.1 Many words . . .

In Fiji as in much of the world it is clear that all would be well if regulations and rhetoric alone could prevent land degradation. With regard to good crop husbandry in Fiji, Ward wrote two decades ago that:

> The difference between the letter of the regulations and the reality of conditions on the farms is a major problem in Fiji. Many of the criticisms

which may be levelled against the state of agriculture would not be valid if existing regulations relating to land tenure and land use were freely observed or could be enforced. (1965:130)

One can despair at how easily Ward's comments apply to today's situation; and officials with environmental concern for Fiji often cast a rueful look at the inertia that acts against implementation of long-standing conservationist regulations. In an ex-colonial developing country such as Fiji the policy-making and investigative stages have often been artificial rather than organic; that is, they have been put in place from the outside by colonial officials or, more recently, by expatriate experts before they had come to live within the organism of the society. In such a situation, effective implementation of adaptive land management would be difficult even if other obstacles were not present. Yet the conservationist consciousness that is strongly present in certain institutional quarters may in time come alive as part of a chain reaction that permeates the whole society, as it is beginning to do in some other developing countries. We sketch here some of the expressions of that consciousness in Fiji and relate their history to changing levels of implementation.

In the 1930s the worldwide concern over soil erosion reached colonial officials in Fiji through the British Colonial Office, which had become strongly committed on paper to the prevention of soil erosion in the colonies (Anderson 1984). H.W. Jack, then Fiji's Director of Agriculture and Conservator of Forests, wrote an article in the local agricultural journal (Jack 1937) that described Fijian examples of erosion, which Jack linked to global distress about 'erosion and empire' (Anderson 1984: 327). Official expressions of concern continued through the 1940s and 1950s (Parham 1954; Whitehead 1954; 1955), culminating in two influential reports: one by Burns, Watson and Peacock (1960) on Fiji's natural resources and population and one by Spate (1959) on the economic problems and prospects of the Fijian people. Both lamented the seriousness of erosion, with the geographer Spate commenting that one of the most urgent problems facing Fiji was the:

> truly appalling state to which burning, overstocking, and bad cultivation have reduced the soil in much of western Viti Levu: I have seen some of the classic areas of erosion in India, Australia, and New Caledonia, but I do not think I have anywhere seen sheet erosion of such intensity as in parts of the hinterland of Nadi and Lautoka. (1959: 97)

Erosion and mismanagement, especially burning, continued to be decried by various authorities and scholars through the 1960s (e.g. Cochrane 1969). After Fiji's independence in 1970, institutional concern lulled in what must have been a common post-colonial response of getting on with the business of development and dismantling the 'do-gooder' restrictions of colonial governments. Aside from some consultants' reports (e.g. Swartz 1974), it is only in the past few years that an official concern with conservation has waxed again. The 1976–80 Development Plan outlined a philosophy of environmental management that was not implemented (Baines 1984). The 1981–5 plan discussed environment mainly in terms of national parks. The new 1986–90 plan (Central Planning Office, Fiji 1985) on the other hand specifically proposes action on the issue of soil erosion, for the first time in twenty years in a major national document; however, the action is not clearly specified.

3.2 . . . but little action

The history of implementation of adaptive land management from the 1930s to the present echoes the parallel history of institutional consciousness just outlined. In the 1930s agricultural field officers tried to teach 'a sounder system of agriculture, based on soil management and crop rotation' (Jack 1937: 6). The Department of Agriculture established demonstrations of contour cultivation, terraces, cover cropping, and other techniques of soil conservation.

By 1949 a full-time soil conservation officer had been appointed and a comprehensive educational programme on soil conservation had been developed. The location and extent of erosion had been surveyed; land-use surveys had been carried out 'to prepare plans for agricultural and ecological rearrangement necessary' (Whitehead 1954: 3). In 1953 the Land Conservation and Improvement Ordinance was ratified and became part of the Laws of Fiji. Under the Ordinance, a Land Conservation Board was established; some of the measures ordered by the Board were at least partially successful, particularly the contouring of cane lands, which was a practice also supported by the sugar company. Other measures, such as restrictions on burning and on the use of agricultural sledges, which initiated gully erosion, were less successful. From the 1920s, when the subdivision of sugar estates into small leaseholdings began, the company exercised considerable control over tenants' farming operations and could, if they so desired, enforce practices that maintained fertility. However, the company's principal goal was the maintenance of supplies of cane for its mills (Ward 1980: 149). Conservation as such was not their concern.

So it is that during the colonial period some fairly strong soil-conservation measures were taken; these may have slowed down but certainly did not eliminate the processes of degradation, which were especially evident on sloping grazing grounds subject to annual burning. Following independence, implementation of adaptive land management went into a decline similar to that of conservation consciousness. After national ownership in 1973, concern that tenants carry out conservationist practices diminished, partly because of the drive for production and partly because any punitive measures against damaging practices took on political overtones, associated with the sensitive issue of land. However, the possibility for improved practices on sugar lands does exist in the detailed conservationist instructions provided to

Fiji Sugar Corporation (FSC) field officers in their Field Manual (Fiji Sugar Corporation n.d.), even though the officers can only advise, not enforce.

Similarly, the Native Land Trust Board (NLTB) has potential power through a clause in its leases to ensure that tenants 'farm and manage the land in such a way as to preserve its fertility and keep it in good condition'. Rather than exercising this authority directly, however, the NLTB requires evidence in the form of a certificate of bad husbandry issued by an officer of the Department of Agriculture. To date, only one such certificate is said to have ever been issued. The three-decade-old Land Conservation Board meets rarely, and then usually only to take action on drainage.

Within the Ministry of Primary Industry is a small and competent Land Use Section; here the still lone Soil Conservation Officer is housed. This Section can only offer advice, not enforce its views. Further, its very limited resources are mainly directed at planning land use with regard to production potential rather than to longer-term degradation risk. The Land Use Section would restrict cane cultivation to slopes less than about 11°. In the interests of production, however, the FSC is said to have encouraged the use of slopes greater than 20°. By leasing land of that steepness to cane producers, the NLTB condones, if not encourages, the practice of steepland cultivation, which almost everyone agrees is one of the most immediately facilitating factors of soil erosion in Fiji.

4 The political dimension

In today's Fiji the world of action wherein land degradation takes place is easier to comprehend and describe than the world of ideas and policy wherein fragments of adaptive land management appear, then fade from view under the ever changing influences of perception of environmental threat, economic necessity and political expediency. If we are to understand this neglect, in the face of fifty years of warning and abundance of evidence on the land, we need to understand the central preoccupations of Fiji, with its ethnic plurality and emerging class conflict. The first problem is that of generating employment and income, to sustain further improvement in living standards which are among the Pacific's highest, and which are important for internal harmony. Land, in this context, has a dual role: it is a means of production, and it is the cornerstone of the political dominance of the indigenous Fijians, a minority in their own land, and economically the weakest of the three population groups in the country. It has been important to develop the land so that large areas of indigenous patrimony are not seen to be lying idle. Thus leasing has been encouraged, both to independent indigenous Fijian farmers and to Indo-Fijians, yet the land must remain the property of the thousand villages and their people, and may not be alienated in perpetuity to anyone.

The development imperative, with its twofold economic and political aims, can almost be said to have encouraged the continued misuse of land, after tenure conditions have been eased in a manner which should have encouraged better management. Fiji has even been helped in its policy of bringing marginal land into use by preferential long-term marketing arrangements, especially for sugar; much land that would otherwise be unprofitable to work has been farmed economically in consequence. Moreover, direct aid to Seaqaqa and a dozen other schemes has brought some very varied land into production.

The political aspect is critical, and it underlies the weak support given to conservation throughout a period of major expansion, and the insouciant manner in which leases over very steep land have been granted in recent years. However, the recent economic crisis in Fiji, produced by declining prices for export crops, a balance-of-payments problem and rising unemployment, has generated for the first time some serious class-based opposition to the ethnic structure of Fiji's politics. These are not circumstances in which the ills of the land are likely soon to rise high in popular awareness. Reconciliation with the environment, and the redevelopment in Fiji of an adaptive land management such as was practised in at least parts of the country during its ancient past, would seem to require that the abiding national concern with land comes to be cast in a new and more ecologically aware framework of consciousness.

10 The degradation of common property resources

A Common property resources and degradation worldwide

Piers Blaikie and Harold Brookfield

1 Common property resources – a definition

The discovery by social scientists that vast numbers of farmers, pastoralists and fishermen have been managing 'common property resources' (CPRs) in pursuing their livelihoods is akin to M. Jourdain's discovery from his elocution tutor in *Le Bourgeois Gentilhomme* that he had been speaking prose all his life. A new name for an old phenomenon, and we realize in the case of CPRs that we have been involved with it all along. However, this recognition of the widespread and important nature of CPRs has served to focus attention on a problem which is shared by seemingly very different physical and social phenomena, and therefore has brought together social and natural scientists in a qualitatively new way.

A 'common' is a resource or facility which is distinguished by three characteristics. First, it is subject to individual use but not to individual possession. Secondly, it has a number of users who have independent rights of use. This implies that a common ceases to be one if one user or some group of users, or an outside person or institution can control the use made by others without their consent. Thirdly, users constitute a collectivity and together have the right to exclude others who are not members of that collectivity.

Thus a common can be distinguished from collective consumption goods (or 'pure' public goods) which are collectively consumed but are not subject to exclusion of use outside the collectivity, since the rate of consumption is independent of the number of users (e.g. a street lamp). Also a common can be distinguished from a private good which is subject to exclusive use and possession by individuals, even though this use and possession may be subject to restriction in the collective interest of a whole society, or simply of neighbouring landholders.

These three characteristics imply that a degree of co-ordination between users is necessary to create rules of use and exclusion and to enforce them. This is achieved by institutions which perform the function of reducing the uncertainty of users by defining and stabilizing their expectations. Therefore both theoretical and empirical work on the common usually focus upon how users organize themselves in its use: how they mutually assure each other of a 'fair' (i.e. expected and agreed) system of attributing costs (if any) and benefits amongst themselves; and how they manage relationships with outsiders who are not members of the collectivity, but who have some sort of interest in the common.

CPRs are of wide domain, and even in the more limited sense of material economic resources held by groups in common, they can range from a common pathway into the fields of a number of individuals to the 'exclusive economic zones' now being claimed by whole nations over the former 'pure public good' óf the world's oceans. Some CPRs are mobile, such as fish or game, or ground water resources; others are static, such as fuelwood stands and common pastures. They may be of great areal extent such as a large forest, or exist only at a point, for example, a watering hole for livestock. There may be arable land held in common, as in Fiji (chapter 9C) where most indigenous land is registered as the collective property of the descent group whose members have the right to use it. They may alternate between use as CPRs and as private domain, as in the old European three-field system where the fallow land was thrown open to the livestock of all on a fixed date (chapter 7) or as in highland New Guinea (chapter 8B) where long-rotation land is individually worked while enclosed but becomes CPR when unfenced and fallowed. To shifting cultivators the primary forest is a CPR available for hunting and gathering by all who lay claim to it; once it is cleared it becomes private and often the one who has cleared it retains some residual rights to use the same land again. Within land that is otherwise CPR, economic trees may be individually owned. The distinction between private and common property is by no means always clear, and even in societies where 'freehold' seems to convey absolute control the mineral rights under the land often remain CPR vested in the state.

The manner of use of CPRs is of importance in understanding why they are particularly vulnerable to induced degradation. Except where they comprise water-management works of some sort, they rarely carry any landesque capital. Use is essentially extensive rather than site-specific, so that an increase in use readily affects the whole natural resource. Fish stocks in communally held lagoons may become depleted, common pastures may be overgrazed, community forests may be cut too heavily for survival, tubewells have to be deepened to pursue a falling water table that is too rapidly drained. Where CPRs are encroached upon and privatized through enclosure, the remaining areas have to carry the added displaced load of the CPR users. It remains in the rest of part A of this chapter to understand how and why these things occur.

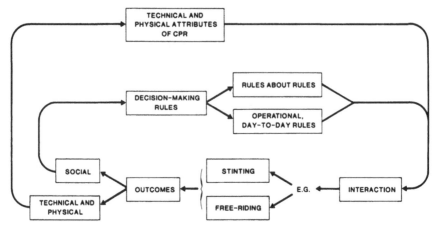

Figure 10.1 A framework for understanding CPRs (*after* Oakerson 1985)

2 Modelling the management of CPRs

In order to understand the ways in which CPRs can become degraded, it is necessary to provide a framework which links the resource itself to its management (mainly concerning 'rules' for governing its use), social interaction between users, and the outcome in terms of use, maintenance or degradation of the resource itself. The framework below was developed by Oakerson (1985), and is presented here in simplified and slightly altered form in figure 10.1.

The technical and physical attributes of the CPR will clearly determine aspects of the institutions and rules about management. By definition, a CPR is characterized by *subtractability* in use, in that each user is potentially able to subtract from the benefits accruing to other users. There are limits to the offtake which any one user can make if the resource is to be sustained and used by others. These will be technically specified by the capacity of the CPR to reproduce its capability and the technology applied in its use. For example, these may be the stocking rate of a pasture or the quantity and type of timber extracted from a forest, the rate of soil formation, the size, age and quantity of various types of fish catch, or the volume of water pumped from the water table at different times of the year.

Another set of physical attributes of the CPR will affect the *excludability* of the resource, the extent to which, or the degree of difficulty with which outsiders can be excluded. How does the collectivity keep others out? Fencing and surveillance are the two most usual ways. The size of the CPR (e.g. forest or pasture), the regularity of its boundaries, the proximity to settlements in which the collectivity of users live or to government functionaries who frequently have taken over the role of the management of the commons, are all relevant factors which have to be reflected in the institutions and rules to manage CPRs.

A basic set of issues in a study of CPR management revolves around decision-making arrangements. These can conveniently be divided into 'rules about rules' (or how decisions are made), and the practical and day-to-day management rules of how the CPR is to be used. Typically there are a number of 'rules about rules' such as the degree of personal discretion in the practice or modification of operational rules. Can, for example, a local 'big man' decide to put all his large herd of cattle on the village pasture? There are also questions about the sources of remedy in case of use which adversely affects a member of the collectivity or the maintenance of the CPR as a whole. The capacity for collective decisions, and the potential for veto are others.

The day-to-day management of the CPR revolves around operational rules which govern partitioning – how to 'share' the CPR and deal with different, and sometimes incompatible, uses. Others concern regulation of entry to and exit from use of the CPR. These rules then produce interactions between users and would-be users. To use the analogy of rules, the 'game' is now played. Users may decide to reciprocate and successfully manage the commons, or they may fail and 'free-rider' strategies prevail. The outcome (the 'result of the game') can be viewed both socially (efficiency, equity, etc.) and physically (sustainable yield, degradation, etc.). In social terms, co-operation, stinting and regulation are one set of satisfactory outcomes, which usually bring continuous benefits to users, while free-riding, increase of use and deregulation may be another outcome. If actual usage undermines the sustainability of the CPR, it implies both a change in the technical and physical attributes of the CPR and may also cause changes in the decision-making rules. Thus the framework iterates through time and history.

Some writers have tried to show that the decision-making rules will inherently tend to break down and bring about the destruction of the CPR. Garrett Hardin's formulation is the most celebrated and the most widely criticized. His thesis, named the 'tragedy of the commons', was originally published in 1972 (available also in Hardin and Baden 1977) and is essentially a restatement of Malthus' dilemma. However, Hardin also applied his general principles to the situation of a herdsman who owns livestock which graze on common lands. The herdsman derives a positive utility of one unit (or nearly one) with each additional animal he grazes on the common, but a negative utility of only a fraction of one unit is felt by the herdsman in a case of overgrazing. Adding together the component partial utilities, a rational herdsman adds one more animal, and one more ... leading to the 'tragedy' (in the sense of the 'solemnity of the remorseless working of things') of the commons.

The persuasiveness of Hardin's argument has led many to urge privatization of CPRs (e.g. Johnson 1972: Picardi 1974). Given the assumption that actors rationally pursue their short-term gain, game-theoretic approaches can be used to demonstrate the inefficiency of common property arrangements. However, Runge (1983) and others have shown that this 'free-rider' solution does not exhaust possible modes of rational behaviour; where long-term benefit is to be maximized, a voluntarist 'mutual assurance' allows for co-operation and regulation. Important in this analysis is the approach of Coase (1960) who argued that adversely affected parties can 'bribe' an instigator to reduce the effect, and that this would leave both parties better off. Coase's theorem, and associated 'voluntarist' approaches to management of use of common property, have led to a prolonged argument mostly built around the problems of pollution. It has been suggested that because the necessary bargaining between unequal parties has high transaction costs if it is to cover all sources of damage, and because a set of bargains between individuals does not necessarily lead to a socially optimal solution, the only effective response to the problems created by individual CPR users externalizing the costs of their usage is the assumption of certain property rights by a regulatory agency (Lowe and Lewis 1980). Lipton (1984) has recast this debate in terms appropriate to the Third World. The objections to 'voluntarism', and the emergent impossibility of privatization of all resources capable of being degraded by individual actions, are, however, interesting in our present context. These conclusions reinforce a growing recognition that formalized CPR arrangements are not a quaint anachronism inherited from a pre-industrial past, but can have positive and enduring benefits to all users. This recognition has led to a new wave of interest in the problems of CPR management, including an international workshop recently organized by the US National Academy of Sciences.

3   CPRs, private property resources (PR) and the state

The previous section has emphasized that CPRs are distinctive and are managed differently from private resources (PRs).[1] While this point has considerable strength, there are two issues which qualify this separateness and distinctiveness of CPRs. They must be considered first before examining the pressures on CPRs and their frequent physical degradation.

3.1   Relationships between private and common land

The first is that CPRs and PRs frequently have very close relationships, and pressures upon one set may well be transmitted to the other. Most of the formal modelling of decision-making in the use of CPRs ignores this very important dimension, particularly characteristic of rural economies but by no means unique to them. It is perhaps true to say that PRs and CPRs are more

[1] We will call private property resources 'PRs' in this chapter, and not PPRs since this abbreviation has been used in chapter 2 and elsewhere to denote pressure of population on resources.

usually found to be functionally related, and that CPRs alone rarely provide the whole livelihood of any rural people.

This is true even in the classic example of pastoralists and farmers, users of land as CPR and as PR, respectively. The migration paths of pastoralists often cross the land of farmers and are so timed as to arrive on the latter at a season when the land lies fallow; this has been amply demonstrated in many parts of the semi-arid regions of Africa and southwest Asia (e.g. Barth 1961, 1973; Bates 1973; Scott 1984; Turner 1984; Gupta 1984, 1985). The case of the Basseri of Iran (Barth 1961) is particularly apposite, and is thus summarized by Carlstein:

> By utilizing pastures *successively* over time, yet avoiding congestion and competition with other tribes using the same route ... the Basseri can optimize the use of their environment and maintain a breed of sheep which is larger and more productive than those bred by the sedentary peasants.... On the way back [to the lowlands in late summer], some groups stop temporarily and let their herds graze on the stubble in the harvested fields (simultaneously manuring these fields). (1982: 110)

Barth (1959, 1961) provides much greater detail on these mutually advantageous arrangements.

Policies to restrict nomadism, such as were later applied to the Basseri, and the privatization of 'their' land for settlement thus adversely affect the welfare not only of pastoralists but of agriculturalists as well, and the land of the latter. A persuasive example is provided by Franke and Chasin (1980, 1981). The introduction of peanuts as a cash crop in parts of Niger reduced the fallow which formerly provided forage for pastoral groups, people who also lost access to land further north by its expropriation for privately operated beef ranches. The transfer of capability through manure into both sets of land was thus lost.

More common is the situation in which holders of PR also use CPR land, so that the farming system includes both types of property and the resources found on them. Changes in the circumstances of either PRs or CPRs have impacts on the usage of the other. We have discussed this interrelationship in the middle hills of Nepal in chapter 2, section 4. Use of common forest and pasture to feed livestock which work and manure the PR fields is an essential part of the whole system, but it is threatened by degradation of the CPR resources under heavy pressure; livestock are reduced, and are unable to deliver sufficient plough-power and manure to PRs. We saw the same problem in eighteenth-century France (chapter 7), and in chapter 9, section 2.1 noted the effect of clearance and tree-crop planting in the CPR forests of Malaysia on the PRs of the *sawah*-rice fields in the valleys; the latter have been gradually driven out of production by silt and sands washed down from the surrounding former CPRs. The case study in part B of this chapter illustrates a similar interrelationship, while elsewhere Jodha (1983) makes the point that pressure on PRs is a major contributory factor to the degradation of CPRs.

3.2   The role of the state

The second issue which qualifies the notion that CPRs and their degradation are distinctive from other forms of the management of resources is the role of the state. Very frequently, especially in modern times, the state has sought to control and regulate CPRs and in effect to intervene in the decision-making rules of their management. In some extreme cases such as in many parts of India, the various institutions of state (notably the Department of Forests) have so curtailed the access of local people to certain CPRs that it is a moot point whether they can still be called CPRs at all. Blaikie, Harriss and Pain had this to say of CPR management in Tamil Nadu State, India:

> Firstly, the development of institutions for collective choice within the group involved with these 'commons' is very restricted indeed; secondly, there is extensive bureaucratic control, which, however, leaves a great deal to the discretion of field officers. (1985: 43)

In this case the decision-making rules no longer operate in the village but in the pages of official forest manuals, and through the whim of local officials.

It is usual in most countries of Latin America for forest lands to belong to the state (Lanly 1982) and it is with the bureaucracy that users of CPRs other than local communities deal in the form of private purchase, logging contracts, etc. Sometimes land reform undertaken by the state restructures feudal forms of control over CPRs (where a landlord regulated the use of CPRs by charging user fees) as in the case of India, described in part B of this chapter. This has led to a situation where the state is driven to protect CPRs against the local people. The removal of responsibility of management from the community (however unequally it used to operate in the distribution of benefits from CPRs) to bureaucratic decree has frequently been disastrous. Poaching and the 'green apple syndrome' (in the spirit of 'I must collect the apples, however green and unpalatable, to forestall others') reduces the total benefit from the harvest (Bromley and Chapagain, 1985), and is a frequent outcome. The nationalization of forests in Nepal between 1957 and 1961 had such an effect when local villagers feared that the government's new powers in taking over the forest would lead to the timber being auctioned off to logging contractors (Arnold and Campbell 1985; Mahat 1985).

4   Pressures on CPRs

There are a number of interlocking and mutually reinforcing pressures on CPRs worldwide. These are the growth of populations of humans and livestock, commercialization of CPRs, encroachment of rights on and areal extent of CPRs, and state regulation. These are illustrated in detail in the case study in part B of this chapter, and here it remains to see the way by which they interact and alter the decision-making rules and outcomes in the management of CPRs. Figure 10.2 schematically explains this. It can be seen that encroachment and an independently caused higher rate of use of CPRs by both commercial operations as well as peasants and pastoralists are mutually reinforcing. Also a higher rate of use of CPRs can lead to degradation which reduces CPR availability, which in turn encourages an even more desperate and heavy-handed use at a higher rate. A vicious circle can be established.

To understand the ways in which these interlocking pressures affect the management of CPRs requires a closer analysis of changes in decision-making and management. For example, the removal of a feudal class of landowners who used to control the management of CPRs and their replacement by a number of competing and expansionary commercial farmers has posed serious problems for communal decision-making. 'Big men' can use the law, deceit or strong-arm tactics to acquire rights on the commons themselves, and the means to remedy this and the capacity for collective decision-making are markedly reduced. Less powerful groups, hitherto content to 'stint' in the belief that others will do the same, are more tempted to free-ride. Under these circumstances to 'stint' is a mug's game. The state's involvement in controlling CPRs also puts immense strains on the community's external arrangements and in many cases has supplanted local initiative by

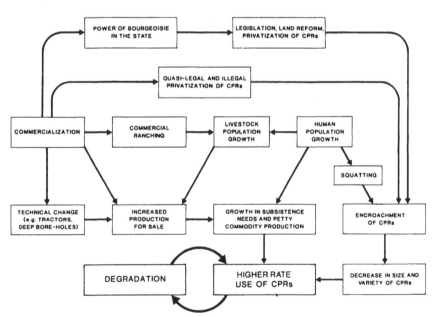

Figure 10.2 Pressures on common property resources

Model of ecological change on a sandy loam soil initially carrying bushed and wooded grassland

SOCIO-ECONOMIC SYSTEM:

SOIL AND VEGETATION SYSTEM:

IN OPEN BETWEEN WOODY PLANTS

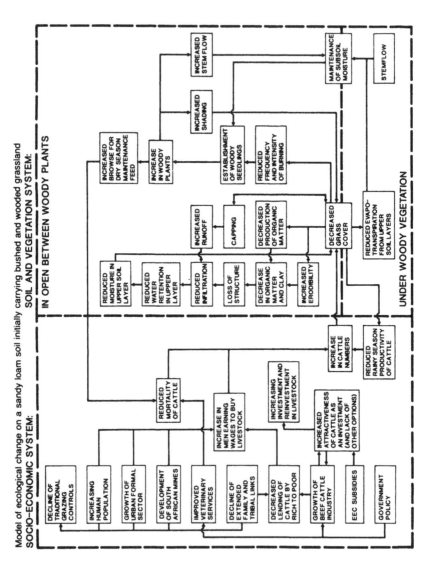

Figure 10.3 Social and physical interactions in common-property degradation in Botswana (*after* Abel *et al.* 1985)

top-down bureaucratic coercion. This process is part of what Lipton (1984: 3) calls a transition of trust. An apposite case study is provided by Beck (1981) which traces the efforts of the Shah of Iran's government to manage and control the pastures of the Qashqu'is pastoralists. Land reform, privatization of large tracts of common pasture for irrigation projects, hunting preserves for the Shah's relatives, watershed protection and other schemes all encroached upon their pastoral CPRs. The authorities disrupted the principles of reciprocity and local dispute settlement by fixing the identity and number of households using each section of the pastures. They also destroyed the flexibility of the previous arrangements which allowed a variety of migratory patterns in response to variable ecological, economic, political and social conditions.

Figure 10.3 from Abel *et al.* (1985) models the pressures and degradation on pastoral CPRs in Ngwaketse District, Botswana. Here deteriorating ecological conditions between 1963 and 1982, during which the national cattle herd nearly doubled in size, have been identified from aerial photographs. Increasing commercialization encouraged by an EEC subsidy has replaced tribal and extended family obligations and communal management with individualistic and competitive behaviour. In the study area, the increased cattle population reduces grass cover and encourages soil erosion, which in turn reduces the productivity of the grass. Fewer and cooler fires result which enables more seedlings of woody plants to survive. Reduced competition with grasses also aids the survival of seedlings. Once established, woody plants shade out grasses and intercept rainfall, so that grass cover is further reduced. The decline in grass cover reduces the productivity of cattle in the rainy season, so that more are needed for the same output. The increase in shrubs helps survival in the dry season, so that fewer cattle die. The vicious circle is completed. It is perhaps worth noting that this is a case in which it is the replacement of grassland by woody scrub which constitutes degradation; a neat example of the importance of perception and context stressed in chapter 1.

It is interesting to note that this report calls for the reintroduction of decision-making procedures and institutions which existed prior to commercialization, albeit in rather a pessimistic vein:

Traditionally grazing land *was* managed communally by the Bangwatetse tribe as a whole. But the rural economy has since become increasingly commercialized, extended family links have weakened, and social transactions are defined increasingly in contractual rather than co-operative terms. Declining tribal institutions are being replaced by an emerging class structure.... While the need and justification for communal management are strong, the effect of these trends is to make the socioeconomic environment for establishing institutions for communal management increasingly unfavourable ... if communal institutions are not viable then communal management cannot work. (Abel *et al.* 1985: 15)

In their many different forms and combinations, all these pressures affect the use of CPRs through the changes they bring about in the decision-making rules, the interactions and therefore the outcomes of the management of CPRs. The pressures are similar in general terms to those affecting private property resources, and can bring about similar degrees of degradation. However, the issues of decision-making are different. The decision-making tree discussed in chapter 4 (figure 4.1) is equally applicable to CPRs as to PRs. In the terms of figure 4.1 many CPR users seek to change the 'social data', particularly institutional structures which allow them better private access to natural resources. While the private benefits may, for a while, increase to the free-rider, the tragedy of the commons is set in motion – not, as Hardin suggests, as an inevitable law, but because of the breakdown of particular institutions of land management. The following case study in India illustrates the general issues discussed here only too well, and exposes the growing hazards which threaten the environment and the people who use it, the poor people most especially.

B   A case study of the degradation of common property resources in India
Narpat Jodha

1   Introduction
1.1   The problem

Degradation of the land, reflected through decline in its production potential, is one of the most debated but least attended problems of many developing countries. One such is India. The problem applies to most categories of land, but is particularly severe in what are described as rural common property resources (CPRs). Their nature varies from region to region, sometimes being forests and watersheds, sometimes watering points and irrigation tanks, sometimes community pastures. Their extent varies from region to region, but in the semi-arid areas they constitute a significant proportion of total land resources. This study is focused in such a semi-arid area, in western Rajasthan.

In Rajasthan, control over CPRs was exercised not through informal regulatory measures but through a landlord who could impose charges on access or produce. A land reform conducted in the early 1950s removed this system of controls with the consequence that a condition approaching the Hardin model has evolved and has not yet been regulated. This is the situation described and analysed in this chapter.

1.2 Method and material

What follows is based on detailed information collected at different points of time from selected villages in three districts of the arid region. The districts are Jaisalmer, Jodhpur and Nagaur. On the basis of rainfall and pressure on land, etc. they present three agro-ecological sub-zones within the arid region (Jodha 1982). The average annual rainfall in Jaisalmer, Jodhpur and Nagaur is 179, 264 and 310 mm, respectively. Based on 1981 census figures the density of rural population is 6, 48 and 78 persons/km^2 in the three districts. All three areas have sandy to sandy-loam soils, most of which, due to their low fertility, high erosion susceptibility and low and erratic rainfall, are suited to pasture-based livestock farming. However, mixed farming, involving crop and livestock raising, is prevalent in all the areas. The extent of the livestock component in mixed farming declines as one moves from lower to higher rainfall areas. This is partly indicated by the number of cattle and small stock (sheep and goats) which are the key categories of livestock sustained by CPRs in these areas. Expressed in terms of animal units their number is 6 per 100 ha of area and 164 per 100 persons of rural population in Jaisalmer. The corresponding figures for Jodhpur are 41 per 100 ha and 124 per 100 persons and for Nagaur 70 per 100 ha and 111 per 100 persons.

Most of the data presented in this paper were initially collected through intensive field work during 1963–5 when the author (N.S. Jodha) worked for

Table 10.1 Details indicating dependence of livestock farming on common grazing lands (CPRs) in three villages

	SITUATION DURING DIFFERENT YEARS IN THE SELECTED VILLAGES				
	NAGAUR		JODHPUR		JAISALMER
DETAILS	1963–4	1977–8	1963–5	1977–8	1963–5
Total animal days (%) involving grazing only and no stall feeding*	21	33	18	37	8
CPRs (%) in total animal grazing days†	88	84	87	85	95

Source: Field data, based on sample survey (see Jodha 1984).
*Animal days equals numbers of animals of sample households multiplied by their respective grazing days.
†CPRs for grazing include private lands available for common grazing after harvest of crop.

the Central Arid Zone Research Institute (CAZRI), Jodhpur, and lived in the villages for more than 20 days every month. These were subsequently updated and supplemented in the case of Jodhpur and Nagaur districts during field work for other projects during 1973, 1978 and 1982–4. Though the above projects had varied and multiple objectives, resources degradation and conservation were always subjects of enquiry.

The importance of CPRs, i.e. common grazing lands, for livestock rearing in the region can be seen in table 10.1. Animal rearing is not only a largely pasture-based enterprise but the bulk of grazing takes place on CPRs. Moreover, households with no land have as many or more livestock units as households with land, even as much as 15 ha or more. The presence of vast areas of CPRs for grazing ensured unrestricted mobility of livestock. This helped in adjusting to rainfall-induced fluctuations in availability of forage and water and enhanced the degree of comparative advantage to this enterprise over crop farming in the region. This comparative advantage may disappear with the decline of CPRs.

2 Decline of CPRs

2.1 The time frame

The time frame used for indicating quantitative and qualitative decline of CPRs in the study areas broadly covers the period from the early 1950s to the early 1980s. The early 1950s witnessed several changes in the agrarian history of Rajasthan. Introduction of land reforms in 1952, which abolished the feudal system of rule in the villages, transferred the bulk of the CPR-grazing lands to private ownership for cultivation and dismantled the traditional arrangements regulating the usage of CPRs, was a major event which greatly affected the status and usage of CPRs. Several post-independence development programmes, particularly the infrastructural development works which facilitated the process of commercialization in the region, were also initiated during this period as a part of the First Five Year Plan (1951–5). These events or periods of their initiation can serve both as a useful reference point for comparative study and as easy-to-remember contexts to facilitate respondents' recall of the situations before and after.

2.2 Quantitative decline

Taking grazing lands first, it is clear that, over the period under review, grazing CPRs in the arid region have been degraded because of over-exploitation. This in turn resulted from overstocking, which again was largely a consequence of shrinkage of their area. Village forest, permanent pastures, uncultivable and cultivable waste lands, and crop lands under bush fallow (long fallow) constitute the total grazing area in the villages. This is supplemented by private crop lands which acquire the character of grazing

Table 10.2 The proportion of grazing lands (CPRs) in total geographical area of selected villages in three districts of arid zone of western Rajasthan at three points of time

	PERCENTAGE AREA OF DIFFERENT LANDS IN THE STUDY VILLAGES							
	NAGAUR			JODHPUR			JAISALMER	
	1953–4	1963–4	1977–8	1953–4	1963–4	1977–8	1952–3	1963–4
Forest	2	1	0	3	2	2	0	0
Pastures	6	3	1	7	3	3	3	1
Uncultivable waste lands	17	13	11	13	10	9	38	35
Cultivable waste lands	18	10	6	15	7	3	26	12
Fallow lands (other than current fallow)	15	10	6	18	13	9	16	13

Source: From *patwari* records during different rounds of field work. The data relate to two villages in each district. Data for 1953–4 were culled out from old village records at *tehsil* headquarters. The table is adapted from Jodha (1985).

CPRs in the post-crop season when anyone can freely graze his animals there. The changing situation of crop lands including current fallows will be discussed later.

The decline in the other CPRs since 1953–4, when land reform programmes were implemented, is indicated by table 10.2. The decline in the area of most of the CPRs was greatest between 1953–4 and 1963–4. This period coincided with a land reform phase when CPRs were extensively privatized. Forest and permanent pastures, which were small in area to begin with, faced maximum decline. They had much better soils and moisture-accumulating topography than other wastelands. Local influential 'big men' managed to privatize them by various legal, quasi-legal and illegal means (Jodha 1984). The extent of fallow lands (bush fallow or long fallow) declined mainly due to the decline in the practice of crop and bush fallow rotation following the increased pressure on land.

The same tendency was also observed at a macro-level in the arid region as a whole (Jodha 1982). For all eleven districts comprising the arid zone of western Rajasthan the area of CPRs declined from 11.3 to 9.8 million ha between 1951–2 and 1961–2, then to 8.7 million ha by 1977–8. In this region, population density increased from 18/km^2 in 1901 to 26/km^2 in 1941 and 51/km^2 in 1971. The density of livestock expressed in terms of animal units per 100 ha of grazing lands increased from 30 in 1951–2 to 86 in 1961–2 and 105 in 1977–8. A part of this increase was of course due to growth in the number of livestock.

Another important group of CPRs are the ponds or tanks, filled by runoff, and important in the support of pasture-based livestock farming in the region. They are scattered all over the grazing lands and help in evenly distributed rotational grazing. They were dug by the village communities and in the past maintained (desilted) through voluntary and involuntary labour contributions of villagers as well as reinvestment of revenue generated by the auction of trees and dung collection rights around the ponds (Jodha 1967). Catchments of the ponds are used for grazing purposes as well.

Quite intensive details about the past and present of the ponds were collected from one village each in the Jodhpur and Nagaur districts. The purpose was to compare the traditional watering points with *tanka* (underground water storage tanks of more modern design and materials) tried by CAZRI in its range-land development and management experiments at 52 locations in different parts of the arid zone (Prajapati, Vangani and Ahuja 1973; Ahuja and Mann, 1975). The investigations revealed that the number of watering points in the two villages declined from 19 and 17 to 8 and 9 respectively from 1953–4 to 1972–3. The area of catchments for all the existing ponds also declined from 358 ha in 1953–4 to 181 ha in 1972–3 in Nagaur village. The corresponding figures for Jodhpur village are 411 and 275 ha. Reduced area of the catchments and disregard of their desilting requirements led to depletion of their water-retention capacities. The people's contribution to maintenance and the reinvestment of CPR revenue fell to zero by 1972–3, and their place has only partly been filled by government relief grants.

The third group of CPRs is found on private lands, but is made common only seasonally. Privately owned crop lands are designated as seasonal CPRs because they have common access for grazing only during the off-season, when there is no standing crop in the fields. This is the only group of CPRs whose area has grown. For the arid region as a whole the crop lands broadly indicated by net sown area have increased from 6.6 million ha in 1956–7 to 8.3 million ha during 1977–8. But despite an increase of over 25 per cent in the area of seasonal CPRs their contribution to grazing seems to have declined in recent decades. This has happened due to a decline in the practice of periodical fallowing of crop lands as mentioned earlier; elimination of bushes like *Ziziphus nummularia* (berbush), shelter belts and other trees from crop lands which were important sources of top feed. This happened because of large-scale use of tractors for cultivation in the arid region (Jodha 1974). The practice of post-harvest ploughing, facilitated by the introduction of the tractor, has further reduced the grazing potential of the seasonal CPRs. The crop left over, as well as undergrowth of crops available after harvest, are covered by soil before the animals have an opportunity to graze them.

2.3 Qualitative degradation

Degradation of CPRs indicated by a fall in their productivity is quite well known. However, assessment of degradation in qualitative terms is not an easy task. There are no historical records and/or other means to recreate the past to compare it with the present. One may try to capture the change through building information from scattered sources at village level. Fragmentary records from ex-*Jagirdars* (landlords) and interviews with the village elders who have seen and faced the consequences of changes in CPRs in their own lifetime are two sources used in this research. We have also employed some unconventional measures to reconstruct the past of CPR units. They are: (a) use of nomenclatures and physical locations of CPR units as sources of past information, and (b) initially anecdotal and subsequently recalled experience-based information from village elders. The importance of the former (i.e. nomenclature, etc.) as a source of past information is explained by the fact that, traditionally, areas with good soil, high vegetative productivity, etc. were differentiated from others by names which specifically connoted their high productivity status or the predominance of specific plant species. Furthermore, drinking water ponds and scattered dwellings, particularly of traditional grazers, were also located in high-productivity land parcels. High-productivity pastures were also earmarked for priority grazing for milch cattle or feudal landlords' horses. The information about the past status of selected units of CPRs, generated through this approach, is presented along with their present details in table 10.3. Whether one looks at the vegetative composition, ability to sustain a

Table 10.3 Qualitative indicators of degradation of CPRs in selected villages of western Rajasthan

CPR UNITS (LAND PARCEL/POND) (HA)	INDICATORS OF PRODUCTIVITY STATUS	
	IN THE PAST BASED ON (A) NOMENCLATURE, (B) VILLAGE ELDERS' EXPERIENCE, (C) FRAGMENTARY VILLAGE RECORDS	AT PRESENT BASED ON (A) FIELD OBSERVATION, (B) VILLAGERS' EXPERIENCE, (C) VILLAGE RECORDS
(1) Village forest (6 ha)	Thick density of *babul* and *indok* trees; nearly 6 types of trees existed; sustained about 100 camels who browsed throughout the year (based on a,b,c)	Only 2 medium-sized trees of same species plus thorny bushes are left; no camel can subsist on them; only goats are grazed; soil is dug for making bricks, mud-plaster (based on a,b)
(2) Pasture (11 ha)	Dominated by perennial *karad* grass; pasture used mainly for grazing cows and horses; poor households used to cut and collect grass (based on a,b,c)	Hardly 5 bunches of *karad*/ha are visible; mainly sheep grazing takes place; no grass worth cutting and collection; *bharoot* and other thorny, annual grasses available, lasting up to 4 months after the rains (based on a,b,c)
(3) Pasture (3 ha)	Pasture adjoining the village main tank, used for grazing young stock of cattle, sheep, etc. (based on a,b,c)	Barren land, literally no blade of grass except during the 2 rainy months (based on a,b,c)
(4) Pasture (15 ha)	Dominated by *berbush* colonies; had *matt* (shelterbelt of *ker* and other bushes/trees); used for grazing cattle in milk which needed better forage; contained 8 types of grasses including legumes (based on b,c)	Plain field with no bushes, trees and matt; only 3 types of rainy season grasses available and used largely for grazing sheep (based on a,b)
(5) Pond (catchment 40 ha)	Major pond, where water dried up only once in 4–5 years during severe droughts (based on a,b,c)	Completely silted, water dries up in every summer season (based on a,b)

Source: Field work.

specific category of animals or actual usage, the productivity status is much lower at present than about forty years ago. Areas designated as forest or permanent pastures in the revenue records (reflecting their past status) are now no more than barren, plantless patches of land. This is more so in the case of CPRs located nearer the watering points and habitations.

We managed to secure from an ex-*Jagirdar* and his accountant some records for the sale of produce from CPRs in 1945–7, when these sales were an important source of *Jagirdars'* revenue. Comparable data for 1963–5 were obtained from the village *Panchayat* (elected council). Over four plots totalling 40 ha the number of cartloads of timber declined from 43 to 4, of top-feed from 85 to 27, of fuelwood from 52 to 14, and of cut grass from 110 to 45. The sale of dung and of gum from *babok* and *indok* trees had ceased altogether.

3   Factors underlying CPR degradation

3.1   Consequences of land reform

Prior to the introduction of land reforms in the former princely states of Rajasthan, the feudal landlord (*Jagirdar*) used to be the sole custodian/owner of the village lands. All farmers except his kinsmen were the *Jagirdar's* tenants. They paid substantial rent in kind (one-quarter to one-half of farm produce) to him for the land they cultivated. Though the CPRs also belonged to the *Jagirdar*, there was a provision of common access; he also used CPRs as a source of revenue. While a fixed proportion of land revenue from cultivated land went as payment to the ruler of the princely state, the CPR revenue went to the *Jagirdar's* own exchequer. The methods of revenue generation from CPRs included a grazing tax per head of animal, auctioning of CPR products, a number of levies or cesses (*laag-baag*) on CPR users, and penalties for violation of a variety of regulations imposed on the usage of CPRs. The number of levies/taxes (*laag-baag*) including those for CPR users varied from 50 to 150 in different areas (Singh 1979). However, as a by-product of this exploitative mechanism there emerged a management system which helped in protection and maintenance of CPRs and regulated their use. The need for revenue rather than ecological concern created circumstances which insured conservation and regulated use of CPRs. With the introduction of land reform in 1952 the whole situation changed (Jodha 1985).

The land reforms abolished the *Jagirdari* system and its variants. Tenants were made owners of the lands. The land revenue payable annually to the government on such lands was drastically reduced. Vast areas of CPRs, mostly submarginal lands unsuited to cultivation, were distributed as crop lands to the landless as well as to those who already had land. Consequently, within a decade of land reforms, 3.4 million ha of CPR land in the arid region as a whole was transferred to private ownership for the purpose of arable farming. This meant an increase of nearly 50 per cent in the land put under the plough in the arid zone. This also meant in the case of most CPRs the decline of 7 to 26 per cent of land used for grazing (Jodha 1966).

Village communities represented by village *Panchayats* became the custodians of the remaining CPRs. While common access continued, the old system of management disappeared. The grazing taxes and levies from CPR users were abolished. Despite the legal provisions, the village *Panchayats*, being elected bodies and concerned about the electoral response of the CPR users, could neither impose grazing taxes nor regulate the use of CPRs. For maintenance of CPRs they depended more on government than the self-help of the villagers.

Discontinuation of a wide range of management practices after the introduction of land reforms has encouraged overexploitation and depletion. This has happened largely because there is now no private cost of using CPRs. Estimates based on detailed investigations in some villages of Nagaur district indicate that before land reforms the grazer had to pay annually in cash or kind (at 1976–7 prices) Rs. 41/household besides Rs. 1.25 *ghasmari* (grazing tax) per animal. After the land reforms this cost was reduced to zero.

3.2 Demographic factors

It is difficult to assess the impact of population growth on CPRs as no data on land utilization (indicating the area of CPRs) are available for the period prior to 1951. The analysis of information since 1951 reveals the following broad picture. The population of arid zones increased by 29.8 per cent during 1951–61 while the area of CPRs declined by 16 per cent. The corresponding figures for the decade ending 1971 are 27.9 and 7 per cent (Jodha 1985). The substantial fall in CPR area can be attributed more to privatization during land reforms than to population growth.

The micro-level evidence presented by Jodha (1985) revealed that before land reform the area of CPRs did not decline in the villages ruled by feudal landlords, despite substantial increases in rural population. The reason was that most of the CPR lands were marginal in ecological terms, and hence unsuited to crop farming, and were also made submarginal in economic terms by the very high rental charges (from a quarter to a half of farm produce) levied by the *Jagirdar* on crop lands. The crop share available to the tenant under such circumstances was insufficient to compensate for the cost and effort of raising crops on such poor lands. Hence, there was little conversion of CPRs into crop lands despite growth in village populations.

The land reforms drastically changed the economics of using submarginal land for crop production. Calculations based on comparable figures of rent payable on such lands before and after land reform in two villages of Nagaur district indicated the following. The rent (at 1976–7 prices) payable as crop share to *Jagirdar* was Rs 16/ha compared to the Rs 1.50/ha payable after the introduction of land reform. Such a low rate of rent was fixed in keeping with the low crop productivity of these submarginal lands compared to high-productivity lands in well endowed areas. The reduced cost of using land, in association with liberal land distribution policies of the government, led to a large-scale privatization of CPRs as crop lands.

3.3 Commercialization and mechanization of CPR-based activities

Hard physical conditions and poor communications resulted in isolation of arid areas from the mainstream marketing situation in the past. Post-independence infrastructural developments have helped reduce this isolation and increase the degree of monetization and commercialization of the economy in these areas (Jodha 1983). The demand for regional products, particularly animal products, is no longer guided by local requirements. The marketability and value of CPR-based products have increased substantially. This has led to greater emphasis on CPR-dependent enterprises and overexploitation of CPRs. Profitability rather than concern for upkeep of CPRs became the guiding force behind usage (Jodha 1980, 1983). This process was further accelerated by different provisions of land reform. The improved profitability of CPR-based enterprises also induced changes in the occupational pattern of the community. High-caste, rich farmers adopted sheep and goat raising, a traditional occupation of low-caste, poor people, which added to the pressure on CPRs.

The use of tractors even by hire has become extremely popular in several parts of the arid zone (Jodha 1974), because of the resulting ability to plant larger areas during the short wet period in the arid region. During the decade 1961 to 1971, the number of tractors almost tripled from 2251 to 6652 in the arid region (Mann and Kalla 1977). This has further encouraged the conversion of CPRs into crop lands. A decline in the extent of land-fallowing, partly unavoidable in the past due to the inability to plant during the wet season, can also be attributed to the use of tractors.

4 Consequences of CPR decline

4.1 Long-term consequences

The major long-term consequence is conversion to crop of marginal grazing lands, without regard to use capabilities. As expected, and in some areas already visible, the consequence is increased soil erosion. This is accentuated by the use of tractors which eliminates the roots of perennial bushes (e.g. *Z. nummularia*) which sprout along with crops, without competing with them, and are harvested annually (Jodha 1982). Yield levels of major crops have declined with successive additions of land to the cropped area (Jodha 1982).

Studies indicate that CPRs are used mainly by the rural poor (Jodha 1983). Inequality in private property resources is partly reduced due to the presence of CPRs. Decline of CPRs, both in quantitative and qualitative terms, means greater loss to the rural poor. At present we do not have data to compare losses to the poor due to decline of CPRs with the gains the poor *might* have made through acquisition of some CPR as private land. However, data reported elsewhere (Jodha 1985) indicate that the bulk of the privatized CPR lands have not reached the landless, but have gone to those who already owned land.

4.2   Short-term consequences for the livestock economy

Following the decline of CPRs, livestock farming in the study areas has undergone several changes. The average size of animal holdings, expressed in terms of animal units, has declined during the period 1963–5 to 1977–8. Moreover, this has happened both in the case of large and small farmers. The composition of animal holdings has also changed. The proportions of sheep and goats as well as buffalo have increased, while that of unproductive animals (young stock, dry milch stock) has declined. While the extent of stall feeding (except in the rainy season) has increased, dependence on CPR grazing has diminished.

There could be two competing explanations for these changes. The decline of CPRs in terms of both shrinkage of their area and decline in the quality of forage compelled the farmers to depend more on their own resources (stall feeding, etc.). The implied higher private cost of rearing animals induced a need to reduce the unproductive component of the animal holding. The shift towards sheep and goats, which are able to sustain themselves on degraded pastures where cattle cannot manage, is an important consequence of changes in the status of CPRs. Moreover, it is easier to migrate seasonally to well-watered areas in Punjab with sheep and goats than with cattle, and thus overcome the seasonal scarcity of forage and water in local CPRs.

An equally plausible explanation of the changes may relate to improved marketability and profitability of animal products in recent years. Prices of wool (net of inflation) have increased from 90 to 480 Rs/kg. Following the introduction of co-operative marketing facilities under Operation Flood Project, sale of milk has become convenient and profitable. Milk prices determined on the basis of fat content offer a comparative advantage to the sellers of buffalo milk. These factors may explain the shift towards sheep, goats and buffalo. Once animal rearing, particularly milk production, is governed by commercial considerations, it is bad economics to collect small quantities of milk from more animals subsisting on CPRs, compared to collecting substantial quantities from a small number of better yielding, better managed, stall-fed animals. Commercialization also encourages the discarding of unproductive animals.

One may add that the reduced cost of using CPRs has also contributed to the increased profitability of sheep- and goat-rearing. The comparative advantage of sheep and goats over cattle, particularly unproductive ones, sustained by degraded pastures, can explain the reduction in numbers of the latter. One way or another, the decline of CPRs forms part of any explanation. However, the comparative advantage of the arid region in livestock farming may disappear once the stall-feeding system becomes important. Well-watered areas have year-round availability of green fodder for stall feeding which the arid zone lacks.

The changes have had important redistributive consequences. A shift towards livestock farming based on private resources is of greater value to farmers well endowed with land. The poor, who had advantages under the CPR system, are likely to lose in comparison. Even on the remaining CPR land, well-to-do people can better maintain herds of sheep and goats and the poor tend to rear them as hired workers. Those who were able to gain from the process of privatization of CPR land are therefore now also those best placed to use the remaining CPR land for commercial enterprises.

The effect on the CPR lands themselves derives as much from the breakdown of former regulation of land use as it does from increased pressure, except in so far as the pressure derives from greater commercialization of the livestock economy. Meanwhile, on the land converted from pastoral to arable use under privatization, we see all the consequences of using ecologically marginal land without regard for its resilience. A focus on the social consequences of major changes both in institutions and economy, as in this chapter, also makes it possible to observe how processes which lead to land degradation also lead to a redistribution of resources to the disadvantage of the poor. The consequences of a well-intentioned land reform, and of equally well-intentioned efforts to provide the infrastructure for greater economic development, are thus deleterious both for the land and many of the people it supports. This is not an uncommon finding, but analysis of this Indian case perhaps shows better how the link between physical deterioration and socioeconomic differentiation operates in practice. The need to evolve new policies for managing the land is clearly evident, but they will inevitably involve fundamental social changes if they are to move from rhetoric to action.

11 Land degradation in socialist countries

A Socialism and the environment
Piers Blaikie and Harold Brookfield

1 Is there a distinctive socialist environmental management?

There are theoretical grounds for arguing that socialist land management may avoid the contradictions of capitalist society and the degradation that has been described in its many forms throughout this book. Levels of surplus extraction from land managers under socialism would tend to be lower, giving them more scope for investment in conservation. The mutually competitive and atomistic environment for decision-making under capitalism would tend to give way to co-operative action so that whole watersheds could be planned and worked for the collective good. Science, in the form of the rational application of agricultural technology, would be developed and applied for social ends and not for individual gain.

However, a closer analysis will recognize that socialism has to be constructed from pre-socialist patterns in society, and that all societies that call themselves socialist will therefore follow different transitional paths. A review at one point in time of the USSR, China, Mozambique, Angola, Vietnam and Cuba, for example, is bound to find a very wide variety of transitional forms. The peasantry in the USSR was eliminated in the early years of Stalin, while they remain in many other socialist countries. The levels of economic development are so disparate amongst socialist countries that the possibilities for the application of industrial inputs into agriculture (e.g. chemical fertilizers and machinery) vary greatly between countries.

There is another reason why a distinctive socialist land management may not be found in practice. There are often marked changes in policy in the agrarian sector. The following discussion on China describes the best example of abrupt changes with immediate effects upon land management. Vietnam too has recently given back much individual scope for managing land by reprivatizing a considerable part of peasant production: a form of partial 'repeasantization'. As recently as 1986, General Secretary Gorbachev was seeking wide-ranging reforms in the agricultural (and other) sectors of the economy which would give more autonomy to farm managers. To give a contrasting example, the changing levels of military security of the regimes in Angola and Mozambique have had a marked impact on the rate of transformation of the agrarian sectors there.

From these two observations it is obvious that every society will continue to have contradictions between private and collective interests, between local and national priorities over land use and the allocation of resources, between state, peasant and professional, and so on. This is not to say that the way in which these contradictions are handled will be the same as those under capitalism. We can, however, draw two general conclusions. The first is that there will be a great variety of policies and outcomes of land-use management in socialist countries; and the second is that in *all* societies there will be problem-solving attempts to cope with the opportunities and constraints which natural conditions place on those who manage land.

There is a more radical perspective on the reasons for such variety in the quality of environmental management in socialist countries. It is that none of those countries which call themselves 'socialist' deserve the description. The Soviet Union is described as state capitalist, and as never having been able to shake off the legacy of the all-pervasive and bureaucratic Tsarist State (see Lane 1985: 85, for a discussion of this view). In a completely different context, Ethiopia, according to Molyneux and Halliday (1981), did not experience a socialist revolution at all but merely a bourgeois one. China, according to some critics, has experienced a sharp reversal of socialist construction in the post-Mao era, and so on. It is because of various deviations from a true socialist path that contradictions within society and between it and the environment still exist. These are important debates, but ones which cannot be entered into here.

2 Socialist ideology and the environment

Marxian theory says little about the relationship between people and the environment, other than to develop a thorough critique of what happens under capitalism. Elsewhere in this book (chapter 2, section 2) Marx's comments on the art of robbing both the labourer and the soil have been noted (Marx 1887/1954: 505). Robinson (1977) and Ensenberger (1974) have taken the review further, and many writings both in this book and referred to by it, owe much to Marxian theory in their explanation and critique of land degradation.

However, Marx's theory of value states that value is only produced by labour, and that natural resources are considered to be 'free' inputs in production (Pryde 1972). Ecological limits to production were relegated to secondary importance behind the necessity to transcend the barriers to the full realization of the potential of natural resources under capitalism. Alier and Naredo (1982) state that ecological analysis has long been alien to Marxism, and trace the origins of this divorce to Engels' failure to incorporate energy accounting into Marxian political economy. Be that as it may, Redclift (1984) has drawn attention to two of Engels' essays which come close to a conservationist view, but points out that both Marx and Engels were primarily interested in the growth of capitalist industry, in which nature merely provided an 'enabling function'.

Socialism was seen both by Marx and Engels as well as by Lenin as the way in which nature could be transformed, 'with its conviction that Man and his modern technology are masters of the environment, and that any technology development must be beneficial' (French 1973). This led to a sort of careless cornucopianism (MacEwan 1984), and a notion of the bounty of nature at the service of economic development. A short quotation from the *Great Plan* 1929, quoted by Burke, is indicative:

> We must discover and conquer the country in which we live. It is a tremendous country (referring to the USSR) but not yet entirely ours. Our steppe will truly become ours only when we come with columns of tractors and ploughs to break the thousand-year old virgin soil. (Burke 1956: 1048)

Marxian views on the environment clearly suffer from a vacuum in another sense, in that they only analysed all too clearly what was wrong with the capitalist use of land. Marxian contemporary critique (e.g. Coates 1972; Robinson 1977) remains no more than that, with very little constructive socialist conservation. This has led to a peculiar situation where the USSR is less than able to benefit from capitalism's mistakes. For example, the ecological hazards of atomic power plants built in the West were blithely written off as 'organic defects of capitalism' (Komarov 1981: 103). This attitude echoes some of the contemporary critiques of land degradation in the Sahel, where real ecological limits are de-emphasized and reprocessed as the depredations of peripheral capitalism.

These are also more pragmatic reasons for a cavalier attitude to the use of natural resources. A drive was pursued by the Soviet Union towards rapid industrialization at all costs. This was an understandable and perhaps essential prerequisite to the survival of the socialist state on the grounds that, first, the state had to protect itself against possible external aggressors and needed an industrial base to wage war; and, second, that it created a large industrial working class upon which its internal political guarantees were based.

There is a last reason why the Soviet Union has been unable to come to grips with the environmental repercussions of its treatment of nature, which concerns the perceived need to export its own version of socialism to the rest of the world.

> As long as the objectives of spreading the USSR's political and ideological influence in the world are the prime elements in government policy, our attitude to nature will not change. Among all the announced measures to protect the environment, there is one area where these steps have good effect: it is propaganda, the showcase. (Komarov 1981: 136)

The recognition of land degradation, pollution and waste undoubtedly exists in the Soviet Union, but too trenchant criticism runs the risk of being labelled 'Sakharovism' and anti-Soviet. Where the state is so important in the management of land and other natural resources, it is essential that monitoring of environmental impacts, the admission of errors and rapid responses can find public expression.

3 Land management and degradation in the USSR

Certainly these ideological currents in Soviet thinking were, and continue to be, important in explaining how natural resources are used in practice. However, we have to look further into the political, administrative and economic nature of the management of land itself to give added dimension to the explanation.

At the outset, let us quote two conclusions from commentators:

> In sum, based on Soviet experience, there is no reason to believe that state ownership of the means of production will necessarily guarantee the elimination of environmental disruption. (Goldman 1972: 7)

> The record of the Soviet Union's first fifty years would seem to suggest that the centralized planning of an economy *per se* provides no guarantee that natural resources will be judiciously used. (Pryde 1972: 177)

A number of books and articles on environmental degradation and pollution in the USSR have become available in the West over the past fifteen years, many of them highly critical. It is difficult to be reasonably satisfied that some accounts are not as flagrantly biased or downright mendacious as those official Soviet ones mentioned at the end of the last section. Perhaps the most sensational and seemingly well-documented is *The Destruction of Nature in the Soviet Union* by Boris Komarov (1981), the name being a pseudonym for a Soviet official, still residing in the Soviet Union. However, it seems reasonably certain that there is widespread land degradation, misuse of resources, destruction of wildlife and high levels of industrial pollution.

The problem of land degradation of course did not start with the Russian Revolution. The Black Earth (loess) region of the northeastern Ukraine has long been a focus of intensive agricultural production. During the nineteenth century there was particularly serious pressure of population on resources (PPR), associated poverty and an extremely unequal distribution of landholdings. Shallow ploughing by the wooden *sokha* tended to create panformation and inhibited percolation. The peasant organization for allocating labour and land, the *obchiny*, insisted upon equality of holdings of different types of land amongst peasant households, and allocated land in thin strips down the slopes. This virtually obliged farmers to plough down the slopes too, leading to rill formation and excessive soil loss (Stebelsky 1974). This is reminiscent of France 100 years earlier (chapter 7).

After the Revolution, there *were* opportunities for socialist agriculture to transcend the constraints of an archaic, inefficient and inegalitarian agrarian structure. In many ways these were grasped in the early years. Lenin in the early years and Stalin later on both took care to have shelter belts planted, particularly in the 1930s, which reduced serious wind erosion. Mechanized ploughing was able to cut deeper into the soil creating better percolation and reducing runoff. The old tenurial system and its inefficient boundaries were swept away. New rotations which eliminated the old clean fallowing system were introduced. No doubt workers and officials felt the same expansive optimism which can be felt in the quotation from the *Great Plan* in the last section.

However, together with these liberating social and technical changes also came fresh causes of land mismanagement. Some of these appeared at the same time as the other improvements in land management. First, fields were redrawn with a rigid adherence to a laid-down minimum size of 100 ha, which frequently meant that they were criss-crossed with gullies, or boundaries were not laid out with reference to the contour. Secondly, many of the shelter-belts were planted in ways to increase ablation and to cause the drifting of snow. Thirdly, contour strips were abandoned because they were too labour intensive. Fourthly, the standard set of practices of land management were laid down inflexibly, so that, for example, the same set applied to the soils of the Tambor Lowlands as to those much more sensitive soils of the Central Russian Upland.

Some years after collectivization and especially after Stalin's death, many of the sound management practices were abandoned. State farms (*sovkhozy*) and collective farms (*kolkhoz*) were established, and both had to meet production targets set by state plans. Particularly in the case of state farms, command planning and industrial principles of management took much of the flexibility and room to manoeuvre away from local farm management (Lane 1985: 10). Both types of farm had to put up with all manner of bottlenecks in supplies of inputs, machinery and labour (Krylov 1979, chapter 16). As more pervasive problems of planning and implementation in the agricultural sector as a whole emerged, Soviet agriculture stagnated and failed to provide sufficient cereals and other basic foodstuffs. This encouraged planners to press ever greater targets upon the state and collective farms, which in turn had to endure all the shortages of inputs mentioned. The result was the extension of ploughing to ecologically marginal land, and an emphasis on wheat and other row crops instead of other rotations which provided better ground cover, and kept soil loss to tolerable levels.

The environmental impact of these assaults was not long in coming. By the late 1920s, a great increase in dust storms was observed in the Ukraine (Goldman 1972: 170), which culminated in 1 million ha of soil being 'lost' in five *oblasts* of the Ukraine between 1950 and 1960. Severe gullying of the Central Russia Upland (to the west of the River Don), which had long been a problem, was greatly accelerated. In some areas today, particularly on the west bank of the River Don near Lipetsk, gully densities of up to 3 km of gully/km^2 are reported. Also, severe sheet erosion is reported, where in many instances the entire A-horizon has been removed. In some areas the remnants of the B horizon and the parent material itself are being cultivated (Stebelsky 1974: 59). Although erosion-induced losses of soil productivity in the Soviet Union must have been estimated, it has not been possible to trace the relevant publications. However, it is reasonable to suggest that one of the causes of the disappointing performance of Soviet agriculture is land degradation itself (Brown 1982). This in turn put increased pressure upon planners to insist upon targets being met from land which was being already mismanaged. In the words of Brown 'production quotas can apparently be at least as destructive to soils as the profit motive' (1982: 23).

Perhaps one of the best known and disastrous attempts to revive Soviet agriculture's lagging production was President Khrushchev's Virgin Lands campaign when, faced with political and economic pressures, he expanded wheat production into Kazakhstan, in an area which had been largely uncultivated since the 1930s. The dust storms were so bad that lorries were obliged to keep their headlamps on at all times, and after only four years the exhausted soil often returned yields which barely covered seeding norms (Bush 1974: 17).

There are a number of underlying problems in the management of the agricultural sector which affect land management. The first is the question of price setting where the law of value operates, which in agriculture requires reimbursement of material expenditures only and the usual equal pay for equal work. This is satisfactory *only* when the prices tell the producer what products to produce, what amounts and what quality in a given agricultural environment. But prices do not provide this information, since they are planned prices, which do not necessarily represent relative scarcity. This means that farm managers cannot allocate their resources so as to reflect scarcity, nor choose their cropping or livestock production according to available resources nor according to comparative advantage of the quality and location of land. This vital aspect of flexibility of matching crop and tillage methods to land is reduced enormously (Krylov 1979: 158, 163). Other problems also hampered effective land management. Support for agriculture from the non-farm sector was inadequate, partly because of outdated and inefficient design of machinery, and partly because of standardization which could not respond to local needs. For example, the Ministry of Tractor and Agricultural Machine Building was given the responsibility for developing eleven machines for combating soil erosion, while it was only able to develop two. The same story is repeated for specialized equipment for fertilizer application, or for the production of plant protection compounds (Brown 1982: 22). The heavy and outdated design of tractors and bulldozers was particularly harmful to soil conservation, both on agricultural land and in newly felled forests (Goldman 1972: 170). Again the result was a reduction of land managers' abilities to manage the land flexibly and adapt available resources to sustain production.

In conclusion, it is possible to evaluate the social causes of land degradation in the Soviet Union. The more abstract points of Marxian theory are probably only contextual and do not provide a direct explanation since these are more or less constant and do not account for changes in policy. However, it is possible to see how the opportunities offered by the Revolution were grasped. Inefficient ploughing, archaic tenurial systems, clean fallowing and population pressure upon the land were all swept away. But other deleterious practices were substituted. More than anything else, political pressure, often amounting to coercion, forced farm managers to chase ambitious production quotas, while at the same time subjecting them to crippling bottlenecks and shortages of machinery, skills, labour and other inputs. It was not surprising that land management deteriorated, particularly in the sensitive environments of Central Russia.

In some senses, the USSR and China, the subject of the next section, have similar histories in land management. Both suffered from population pressure, grinding poverty, great inequalities and degradation. Both swept the old social order away and provided their people with tremendous opportunities to manage land better. China had an ancient tradition of transforming nature, as the opening paragraphs of the next section tell, which was continued in the garb of a new ideology. The USSR on the other hand developed the same ideology of 'transforming nature' in a Bolshevik form, every bit as ambitious as in China, although without such a long cultural tradition. Where the party or state can mobilize labour and resources with such completeness (whether through exhortation, coercion or brute force), a few decisions can have enormous implications. So it was with these two nations, and when mistakes were made, the costs were on a truly gigantic scale.

B   Land degradation in China: an ancient problem getting worse
Vaclav Smil

1   The Great Leap Forward ... to destruction
1.1   Continuity in the subjugation of nature

Chinese admiration for nature, expressed so marvellously in landscape paintings, in lively plant, insect and animal sketches and in nostalgic poetry, has always had a forthright utilitarian counterpart in bold acts of subjugation: to plant crops, to secure fuelwood, to make ink, to smelt metals, to build cities – all these ever more demanding tasks required a continuous onslaught on untamed nature. Sima Qian, the grand historian of the Han Dynasty (202

BC–AD 9) and the author of China's greatest review of ancient history, quoted approvingly the *Book of Zhou* which said that:

> without farmers, food will be scarce; without artisans, goods will be scarce, without merchants, the three precious things will disappear; without men to open up the mountains and marshes, there will be a shortage of wealth. (Sima Qian 91 BC/1979)

About eleven centuries later Lu Yu (AD 1129–1209), the leading poet of the Southern Dung Dynasty, wrote in *The Farmer's Lament* that 'If there's mountains, we'll cover it with wheat/If there's water to be found, we'll use it all to plant rice' (Liu Po Chan 1978: 106). Nearly 800 years further on, after the establishment of communal farming in new, Communist China, the provincial and nationwide monthly and annual accounts of the millions engaged in land reclamation, and the thousands of hectares of newly created fields, were among the most readily available figures; at that time, normal output and consumption figures were unavailable. The peak was reached during the winters of the early 1970s when at least 100 million peasants were engaged in off-season terracing of slopes, levelling fields, filling lakes and marshes and the opening up of grasslands (Smil 1978).

In a country where sufficient food and fuel harvests seemed to be always lagging behind the latest push into the mountains and marshes, cutting the forests and turning any convertible surface into fields has always been both an obvious necessity of everyday peasant life and a matter of state policy. Stunning consequences of this heritage can be seen in China's mountains: endless vistas of deforested slopes, often involving the elimination of every single tree leaving an emptiness the scale of which is difficult to associate with human intervention at all (Smil 1983). With marshes and swamps the changes are much less obvious; here one must consult ancient topographies and travelogues and, more recently, maps and aerial photographs to gauge the extensive disappearance of China's natural wetlands.

Not all of these changes have led to land degradation. Numerous deforested slopes have been carefully terraced, and soil behind the embankments enriched by generations of laborious application of compost and by countless baskets of alluvial mud carried up on shoulder poles. Intricate water control schemes made it possible to drain former marshes for proper cultivation and to flood the soils again for growing of rice. Such examples became the mainstays of Western admiration of China's highly productive traditional farming (Smil 1981a). But the destructive effects have caused enormous hardships.

1.2   Modern acceleration: the 'Great Leap Forward'

After the Communist Party prevailed in the long civil war two considerations kept the country on the path of expanding cultivation into the mountains and marshes. The first was an ancient one – the nation's poverty. In the early

1950s virtually no modern farming inputs were available. Production of small tractors started only in 1958, application of synthetic nitrogenous fertilizers averaged a mere 3 kg/ha by the end of the First Five Year Plan in 1957 when the power-irrigated area was only 4.4 per cent of all irrigated land (State Statistical Bureau 1984). Farmland reclamation added 3.9 million ha of new farmland (Larsen 1967: 250).

Secondly, however, these changes in land use were given strong ideological support from the ruling party which allied itself very closely with Stalinist orthodoxies, one of them being an attitude of arrogant dominion over nature translated into grandiose official schemes for 'transformation of the environment'. Growing up at that time in another Stalinist country in central Europe, I vividly recall the repeated propaganda songs about how 'we shall order rain and wind when to fall and when to blow'. Ironically, this arrogance found its most daring and most catastrophic expression in the plan to dam the Huang He at a place just before it leaves the loess region (Smil 1979). Soviet advisers proposed a grandiose construction of hundreds of thousands of silt check-dams and massive revegetation on the Loess Plateau to cut the silt influx into the Huang He, and confidently predicted that the Sanmenxia reservoir would support at least half a century of power generation in a 1,100 MW plant. Yet 45 per cent of its ultimate storage capacity was silted up in just four years after the reservoir's completion, between September 1960 and October 1964. Power generation had to be abandoned and during a costly dam reconstruction large outlets had to be opened up at the bottom of the structure to flush the silt downstream. Clearly, silting rates were much higher than anticipated and revegetation did not proceed as planned.[1]

The early 1960s were in fact the beginning of a sharply accelerated degradation of China's farm soils, grasslands, forests and wetlands, largely a result of the 'grain-first' policy which had its unmistakable roots in the greatest famine of human history which was, in turn, a predominantly man-made affair, a grievous consequence of ideological errors. To trace China's staggering land degradation of the 1960s and 1970s thus requires us to go back to the late 1950s, to Mao Zedong's 'Great Leap Forward'. China's more or less orderly and in many ways impressive economic progress during the country's first Stalinist Five Year Plan did not satisfy the 'Great

---

[1]The erosion rates on the loess regions of China are the highest in the world by common agreement. They were high 2000 years ago, when new land was already being warped from the sediment at the mouth of the Huang He (Jiang and Wu 1980: 1135). It will be recalled from chapter 7 that the loess lands of Europe have also suffered severe erosion losses in the past and present a major management problem, but in China the problem is exacerbated by the great depth of loess and the high relief of the loess regions (Wang and Zhang 1980). The extent of these Pleistocene periglacial deposits in China is undoubtedly a major environmental 'cause' of the high sensitivity of much of China to erosion, and the sort of 'development' which these regions have experienced in modern times has produced results which should occasion no surprise.

Helmsman' who dreamed of an unprecedented shortcut to a modern, industrialized society.

Obsession with iron-making led to smelting of 13.7 million tonnes of pig iron in 1958, two to three times above the 1957 level, 21.9 million tonnes in 1959 and 27.2 million tonnes in 1960. Virtually all of this huge expansion came from primitive 'backyard' furnaces fuelled with charcoal from trees indiscriminately cut in the nearest remaining forests or with poor quality coke made from coal locally extracted in small open-cut mines. Pig iron production was enormously wasteful; the recently released historical statistics for the period show that while pig iron output rose 4.6 times the output of final rolled steel products went up only 2.7 times between 1957 and 1960. Moreover, it was bought at a crippling cost of environmental destruction, deforestation and land despoilation. It demanded extraordinary human exertion as all the work was done purely by hand. Tens of millions of peasants were cutting the forests, firing charcoal, digging coal, smelting shoddy pig iron – and neglecting the crops.

1.3   The famine of 1959–61

The 1958 grain harvest was 2.5 per cent above the 1957 level but rice, wheat and corn crops were actually down and the rise was solely due to more tubers which the Chinese count as a part of grain output (at a fifth of their weight). Then, on the top of the Great Leap's wasteful industrialization and falling cereal production came poor weather including serious drought, floods, frost and typhoons. The share of the farmland seriously affected by reduced yields rose from about 7 per cent of the total arable land in 1958 to 12 per cent in 1959 and to 25 per cent in 1961 (State Statistical Bureau 1984).

Because of the Great Leap Forward, grain harvests plummeted much more deeply than they would otherwise have done. In 1980, when natural disasters affected about 22 per cent of farmland, the total grain production fell by only 3.5 per cent; in 1960, when the same proportion of cropland suffered, cereal harvests declined by nearly 16 per cent. Such a loss of staple foodstuffs, in a society which was producing no more than about 2100 kcal/day *per capita* even during the best years before the Great Leap Forward (Smil 1981b), was bound to bring widespread suffering. However, it was only in the early 1980s, with the publication of historical demographic statistics, especially the detailed age distribution from the 1964 census, and with the results from China's first truly modern and accurate 1982 census, that it became possible to judge the incredible toll of the 'three bad years' 1959–61. The best available reconstruction of this unparalleled tragedy puts the number of excess deaths at 30 million and the number of forgone and postponed births at 33 million (Ashton, Hill, Piazza and Zeitz 1984), a disaster comparable, in relation to the population, with the worst famines of medieval Europe.

1.4   The 'grain-first' policy

After the collapse of the Great Leap Forward and the sufferings from the nationwide famine China lived in the fear of repeated grain insufficiency and these concerns led to a preoccupation with cereal production. The policy of 'taking grain as the first link' was born out of the necessity to feed the nation, but its implementation has caused extensive and unprecedented land degradation.

The 'grain-first' policy called for maximum local self-sufficiency in cereal production. This stems from the country's meagre grain reserves and a transportation infrastructure inadequate to move large quantities of cereals over long distances. In only a few provinces could the self-sufficiency requirement be met, even by increasing yields and converting more grassland into grain fields. In most instances, especially in inland provinces, shortage or outright absence of subsidies led to sluggish increases in yields, and extensive conversion of grasslands, brushlands, forests and wetlands appeared to be the only way to increase the grain harvests. Its immediate cost was only the hard work of hundreds of millions of peasants making grain fields indiscriminately out of any accessible land. But worse was to follow.

1.5   Environmental consequences

The environmental consequences were not long in coming. Rapidly accelerated deforestation and grassland destruction resulted in spreading erosion. Reduction of wetlands and advancing desertification were relatively less important, although locally and regionally serious. After 1978, the Chinese news media and scientific publications began to publish numerous figures on the extent of provincial and nationwide land degradation and its consequences, so that the aggregate deterioration can now be illustrated.

Reported losses of arable land have been almost unbelievably large: for the two decades between 1957 and 1977 they were put at at least 29.3 million ha (Yi Zhi 1981), or no less than 26 per cent of the 1957 total, an alarming decline in a country which has to feed more than one-fifth of the world's population from a mere one-fifteenth of the earth's cultivated land. According to a document issued by the Central Committee of the Chinese Communist Party (1979) about 7 million ha were lost through requisitions for various construction projects, ranging from urban growth and new railways to small rural industrial enterprises and irrigation canals. While such figures may not distinguish total from partial loss, and are thus not necessarily comparable with the 'damage' figures for the United States and Australia cited in chapter 12, the important point is the attribution of so much devastation to so short a period of time.

I have come across no figures disaggregating loss due to degradation, but information published about the extent of principal processes of land reclamation and loss indicates their relative importance. According to the Policy Research Office of the Ministry of Forestry (1981) the felling of forests for grain fields destroyed at least 6 million ha of forests since 1949, most of it after 1961. This is almost one-third of all forest losses (20 million ha) sustained between 1949 and 1980. Yet most of the newly created fields had an ephemeral existence. Chinese literature abounds with hardly unexpected descriptions of rapid depletion of accumulated organic matter, plummeting yields, soaring erosion and inevitable abandonment of the fields and the beginning of a new instalment of this crippling sequence.

Desertification claimed about 6.5 million ha of farmlands and grasslands between 1949 and 1980 (Ma and Chang 1980) with nine-tenths of the loss ascribed to improper land use, mostly conversions of semi-arid grasslands to grain fields, and to overgrazing. Reclamation of lakes and wetlands created 1.3 million ha of new fields but, as in the case of the new farmlands on the slopes previously occupied by forests, yields on this new land were very often disappointing and cultivation was frequently abandoned after only a few years.

Total farmland reclamation between the late 1950s and the late 1970s was put by Yi Zhi (1981) at 17.3 million ha. With about 7 million ha of the loss coming from the conversion and subsequent destruction of forests and wetlands, the bulk had to come from conversion of grasslands and, together with deforestation, this process must have been the single largest anthropogenic contributor to China's spreading erosion which now affects 150 million ha or one-seventh of the country's whole territory (Ma and Chang 1980).[1]

The largest single cause of arable land loss would seem to have been erosion on grassland or wasteland taken into use since the late 1950s, though some may be of earlier origin especially in the loess regions. If the actual loss amounted to around 29 million ha, as suggested by Yi Zhi (1981), we can perhaps deduct 7 million ha of land lost after conversion from forest and wetland, 5 million ha lost through desertification and 7 million ha from conversion to non-farm uses to arrive at about 10 million ha due to erosion of former sloping grasslands converted into fields. More details can be found in Smil (1984). Much of this loss was predictable, and with it associated consequences including shortages of wood and wood products, diminished water-retention capacity, changes in terrain climate, aggravation of floods, silting of reservoirs and diminished crop yields; modern Chinese publications provide extensive information on these latter burdens.

[1] This huge area compares with the 159 million ha said to be seriously affected by erosion in Australia, which includes 38 per cent of land in use in the non-arid zone (Woods 1984: 53, 86), or with the 34 per cent of cropland in the United States which suffers erosion at more than 'tolerable' levels (Reichelderfer 1984). In order to evaluate such a comparison, however, we need to know more about the criteria used in each case and for China, as for Australia, this information is not readily available.

2 The new transformation of China and its consequences

2.1 The new initiatives: The baogan *system*

Policies of 'taking grain as the key link' and 'planting crops in the middle of lakes and on the tops of mountains' were officially abrogated with the arrival of a new political orientation since 1978. There are now strict edicts forbidding conversion of forests and grasslands into cropfields; all planned reclamations of wetlands should be carefully appraised and undertaken on scales compatible with preservation of essential water storage and aquatic production capacities. New forestry laws have a very strict section banning any excessive timber logging and fuelwood cutting.

Undoubtedly, with the removal of the State's blessing and active involvement, many of the most offensive forms of land degradation have been much reduced, and in many locations even completely eliminated. For example, Mongolian herdsmen readily turned their low-yielding grain fields back into pasture. Former fishermen in coastal regions and in the Yangzi lakelands responded readily to the options provided by the new household production responsibility system (*baogan*) and abandoned forced cultivation of grains for more lucrative breeding of fish and other aquatic foodstuffs. But deforestation and accelerated erosion continue.

Villagers in mountainous regions now can get special grain rations in order not to plant grains locally and to concentrate on forestry production or animal breeding, but the severe rural energy shortages are responsible for continuing overcutting. Even according to the most conservative of several available estimates, no less than one-half of all rural households is short of fuel for cooking and heating for at least half a year. The same bureaucracy which has been vigorously promoting adoption of *biogas* digesters as the best means of solving rural energy shortages had, until 1980, been forbidding private ownership of fuelwood lots on plentifully available deforested slopes and on patches of barren wasteland unsuitable for crop cultivation.

Only after the post-1978 reformist policies started to bring a degree of environmental realism to Chinese decision-making, came the recognition that private fuelwood lots are not 'the last vestiges of capitalism' but a good way to assure excellent quality and a high survival rate of plantings through assiduous care and without any state financing (Zhang Pinghua 1981). Eventually about two-thirds of China's 170 million peasant households will set up woodlots averaging 0.2 ha so that some 23 million ha, an equivalent of about 20 per cent of China's currently afforested land, can be covered by fast-growing fuelwood trees harvested, and then replanted or let to coppice in short rotations. They will provide both the much needed fuel and no less needed anti-erosion protection. But they will not become a reality until around 1990, and meanwhile deforestation motivated by desperation continues.

2.2 Commercialization and industrialization

Unfortunately, these changes have not necessarily heralded the end of widespread land degradation. The most worrying threat arises now from the rapid intensification of crop production under the *baogan* system, as hard-working incentive-driven families harvest in a few months more rice or breed more pigs than a whole production team did during a year of command communal farming. This new development must throw doubts upon agro-ecosystemic sustainability in the future.

Traditionally, Chinese farming served as a paragon of sustainable agriculture supported by complex crop rotations, planting of green manures and recycling of crop residues, human and animal wastes and many other organic materials. All of these practices have declined considerably and further large reductions appear inevitable. 'Grain-first' policies did away with many traditional rotations and also reduced the cultivation of green manures. These changes are unfortunately retained and continued in the new *baogan* initiative in which the encouragement of comparative-advantage farming leads to monocultural specialization or to simple two-crop rotations similar to continuous corn or corn-soybean practices in the American Midwest. The stress on high productivity encourages an obvious preference for applications of increasingly available synthetic fertilizers, rather than to planting of manurial leguminous crops and to organic recycling.

Recycling of organic wastes is also more difficult owing to present heavy needs for straw and stalks as household fuel; at least half of China's crop residues are now burned as fuel. Also many urban wastes are increasingly unsuitable for composting and field application. With the improvement of living standards, previously wholly organic city garbage now contains growing shares of non-biodegradable matter. Urban sewage also is usually much contaminated with heavy metals, phenols and oils, an inevitable consequence of rapid industrialization unmatched by pollution controls.

On the land, heavy applications of nitrogen have created phosphorus and potassium deficiencies and declining yield response. China now applies on average about 125 kg nitrogen per ha of arable land per year, three times as much as the United States and only 20 per cent less than Japan, Besides these macro-nutrient imbalances, other signs of widespread qualitative degradation can be noted. These include loss of organic soil matter, greater compactibility of soil and degraded quality of tilth, decreased water retention and micro-nutrient deficiencies. Also evident are consequences of inappropriate irrigation, alkalinization and salinization in dry northern provinces and formation of bog soils in overwatered fields in the south.

Correction of all these problems is a long-term process and it remains to be seen what the new long-term production contracts (for up to fifteen years) will do in these respects. Clearly, without taking care of this degradation the recent burst of high yields and high earnings will soon encounter diminishing returns. Exchanging command farming, with its inefficient production and excessive stress on grain growing, for *baogan*, with its newly found benefits of incentive performance and comparative-advantage cropping, has been, on the whole, a most beneficial move in terms of productivity, incomes and food availability. Nonetheless it has not generated an automatic cessation of China's land degradation practices. It is just a welcome abandonment of some of the most offensive practices, the moderation of others, and a shift to new concerns in company with much of the rest of the world.

2.3   Signs of hope

Yet there has been also a significant change since 1978, resulting from realization of the immediate costs and the disastrous longer term penalties to be paid for continued neglect. Most encouragingly, this realization has reached beyond the pages of scientific publications and popular periodicals. The top party and government leaders are now actively urging soil conservation, acknowledging the link between deforestation and worsening flooding, and promoting new policies banning the worst environmental offences. Perceptions have shifted and solutions are beginning to be sought in the most rewarding direction, namely, in making sure that the initial human blunders are minimized, if not eliminated. If one seeks confirmation of a thesis about social rather than natural causes of land degradation and about the importance of social as well as natural sciences in the understanding of this process, then China is a classic example.

For any situation there is a saying from the bottomless store of Chinese idioms and proverbs. One hopes that the following would apply in this case: *wu ji bi fan* – 'when things are at their worst, they begin to mend'.

12 The farmer, the state and the land in developed market economies

A New conservation crisis in the 'North'?

Piers Blaikie and Harold Brookfield

1 Reasons for concern

Except for its historical excursions, this book has been principally concerned with the land degradation problems of the developing countries. Though reference has been made to the situation in the 'developed market economies' at many points, it is only in this penultimate chapter that we focus specifically on countries where today pressure of population on resources (PPR) is never, and poverty only rarely, advanced as a reason for failure to manage the land adequately. This essay, followed by a piece on the sociology and politics of land degradation in Australia, can do no more than sketch a few outlines of a set of problems that, since the 1970s, have caused increasing concern and have begun to achieve wider public notice. Briefly, it has come to be realized that fifty years of soil-conservation awareness and action have failed to eliminate the problem of land degradation from some of the world's most productive farming regions. Indeed, there are trends which would seem to be making a worrying problem worse, notwithstanding the accumulation of an enormous fund of scientific, experimental and practical knowledge on the management of land.

Almost half of all the cultivable land in use in the world is in the so-called 'developed' countries, where the proportion of total cultivable land that is actually in use is much higher than in the 'developing' countries (Dudal 1982). The modern agricultural revolution has been far more complete in these 'developed' lands ('north') than in the 'developing' lands ('south'), with the result that enormous improvements in productivity have been achieved and substantial surpluses created, even in erstwhile deficit regions such as western Europe. The long and slow improvement in grain and other yields from medieval to modern times has taken a sharp upward turn, doubling or trebling since 1950 in most 'northern' countries. At the same time, the decline in farm labour, which began in the mid- to late-nineteenth century, has greatly accelerated, so that the productivity of labour in agriculture is now many times that obtained in most countries of the 'south'. Although total food-production growth since 1965 has been more rapid in the 'south' than in either the 'centrally planned' or 'market' economies of the 'north', this result has been achieved mainly by increasing the area under cultivation; in most market economies of the 'north', on the other hand, the area under cultivation has actually contracted (Tarrant 1980).

The contraction of area which has accompanied this remarkable success in increasing production has multiple origins. The major element is the abstraction of land from agriculture for urban and other non-agricultural use. A second element is that the cost of applying modern technology has yielded more benefits on good than on less capable land, thus reducing profit margins on the latter and leading to abandonment (Beattie, Bond and Manning 1981; Champion 1983). A third element, similar in operation to the latter, has been the poor response even to large applications of fertilizers on degraded land; while some of this land has been put down to grass and remains in production, other areas lie idle except during periods of high prices, when they are quickly, and sometimes destructively, brought back into production. Production controls, either by direct payment to withdraw land from use, or by withdrawal of price supports from crops in surplus to demand, have led to the retirement of substantial areas of low-yielding land (Tarrant 1980).

The latter problem may well grow much worse, for while modern energy-intensive agriculture can ensure a crop so long as there is a rooting medium for the plant, the cost is high and plant response deteriorates. It is estimated that if top-soil losses even from the United States corn belt continue at present rates, crop yields will be cut by 15–30 per cent by 2030 (Korsching and Nowak 1982). At the same time, there is evidence that erosion losses have reversed the downward trend of several decades, and are again on the increase (Timmons 1979). It is not necessary to join the eco-doom lobby to realize that, even were the goal only one of matching demand with supply at reasonable cost, there would be a problem. Since 34 per cent of the cropland of the United States, the world's major food exporter, loses each year more than the generous 'tolerable' 12.4 t/ha, and 23 per cent loses more than twice this level (Reichelderfer 1984; McCullough and Weiss 1985), it is clear that the problem for the United States and for the recipients of food exports or aid is a very serious one indeed.

2 Conservation in the United States of America

2.1 The genesis of effective state intervention

Although the first modern examples of state intervention in land management for conservation lie in nineteenth-century Europe, and in certain colonial territories early in this century (Hudson 1985), it was in the United States that the first effective national programme was initiated and implemented. The twin disasters of severe wind- and water-erosion in the late 1920s and early 1930s and of the Great Depression after 1929 precipitated a crisis in an American rural economy that then occupied 26 per cent of the population, with many more dependent on the spending of farm incomes that declined by 70 per cent between 1929 and 1932 (Batie 1985). To the notion of conservation, already actively advocated in the 1920s, was added massive loss of livelihood. In 1929 funds were allocated to set up the first erosion research stations in the world, and in 1933 at the beginning of Roosevelt's 'New Deal' a Soil Conservation Service was set up, using unemployed labour to undertake works. Under the charismatic leadership of H.H. Bennett, the Soil Conservation Service sold itself so well that in 1935 it was entrenched as an integral part of the US Department of Agriculture (USDA); this was followed in 1937 by the initiation of Soil Conservation Districts, voluntary associations of farmers with USDA support, that were set up state by state until by the end of the 1950s there were almost 3000 of them covering the whole country (Berg 1979).

From the beginning, conservation was not the only objective. Maintenance of farm income was also important, so that farmers were not only subsidized to undertake conservation works, but also to take threatened land out of production; given that there was overproduction of many crops in relation to demand, farmers were also assisted not to grow these crops while retaining an income, a step justified by the fact that these same crops were 'soil depleting'. The principle of farm support, which was simultaneously but in different ways developed also in several European countries, quickly became entrenched and enlarged so that the system has acquired a political inertia that governments have not found easy to shift. The consequences are described by Batie:

> Soil conservation programs have remained popular because they have lowered farmers' operating costs, improved yields, and provided for compensation for idling lands from production. The programs were not designed to achieve the most erosion control per program dollar spent, and criteria for the receipt of cost sharing funds have not (until very recently) been tied to the severity of the farmer's conservation problem. . . . Such a distribution of funds was . . . justifiable because farm income support was the chief goal. (1985: 110)

Some of the distributional results of this approach will be shown below.

The landesque capital created by the programme, together with some created without federal cost-sharing assistance, can be valued. Pavelis (1983) has standardized the data, and amortized the gross capital stocks, devalued to 1977 dollars, to show that the net value of conservation assets peaked in 1955 and has since declined, due to a declining rate of investment as successive post-Roosevelt governments have reduced contributions to the programme in real terms. A decline from a level equal to 4.8 per cent of the net land value in 1955 to 3.7 per cent in 1980 would have been greater but for the initiation in the 1950s of watershed conservation projects for flood control. The figures do not, however, take full account of the actual physical removal of terraces, grassed waterways, windbreaks and the like by a large number of land-holders as farms became consolidated and adapted to the use of large-scale machinery. Pavelis (1983: 457) argues that the decline represents consumption of capital, in consequence of 'production pressure on resources'.

2.2 The effect of changes in the farming system

During the fifty years during which the United States' soil conservation programme has proceeded very much along lines established at the outset, there have been enormous changes in the farming system as a whole. Whereas there were 7 million farms in the 1930s there are now fewer than 3 million, and while most of the remaining farm population (now only 2.4 per cent of the national population) still resides on small family farms, over 50 per cent of total commercial sales now come from only 5 per cent of farms, having average sales above $200,000/year. Most small farm owners now rely on off-farm employment for at least part of their income (Batie 1985). There have been similar trends in Canada, where the amount of land in use has declined from 70 million ha with some 600,000 farm operators in 1951 (and 732,000 in 1941) to 68.5 million ha worked by only 300,000 farm operators in the late 1970s (McCuaig and Manning 1982). As in the United States, the greatest losses have been in the east, and most of the counterbalancing gains have lain on land of lower quality but adaptable to capital-intensive operation toward the fringes of settlement in the west.

This massive farm consolidation, which is paralleled by trends in Europe though at a different order of scale (Bowler 1983), has been accompanied by the creation of a two-class system based on corporate and family ownership (Goldschmidt 1978), though some of the corporations have arisen out of successful family operations which have diversified from farming alone into the forward and backward linkages of the business (e.g. McGregor 1980). These trends have been accompanied by the use of machinery on an ever-widening scale. Thus, on a ranch in the heavily eroding Columbia plateau, by 1970:

> seven men, equipped with three self-propelled air-conditioned combines, a 'bankout' wagon (a huge diesel-powered machine similar to an earth mover, which moved with ease over steep hillsides hauling wheat from the combines to trucks waiting at the edge of the field) and three trucks handled harvest operations which fifty years earlier had required more than a hundred men and three hundred work animals. (McGregor 1980: 26)

On these semi-arid hills, summer fallow farming on slopes up to 45 per cent can result in soil losses as high as 370 t/ha (Dickinson *et al.* 1981). However, terracing and a rotated annual cropping pattern are now being adopted by at least some farmers. In California, 3 × 12 m 'land-planing' machines have been used to level land more efficiently, but with a high cost in wind-erosion loss. Efforts to 'target' the national soil conservation programme on to limited areas of high risk have, in recent years, been bedevilled both by political pressures to keep the subsidy aspect of the programme widely spread, and also by the fact that between a half and two-thirds of the crop land eroding above 'tolerance levels' is operated by farmers who are not participating in the programme (Reichelderfer 1984). At the same time, however, most of that large part of American farmland on which erosion is not regarded as a serious problem (i.e. with a soil loss below 12 t/ha/year) is also operated by farmers outside the programme.

3 Structural changes and their consequences

The diminution in the number of farms, the increasing mechanization and the emergence of a new group of very large corporate farms, have been accompanied by other changes briefly discussed in chapter 4. It is often claimed for the more 'advanced developed market economies' that agriculture has shrunk to a small proportion not only of employment, but also of gross domestic product (GDP). This conclusion arises from the habit of sectoral thinking which national accounting has engendered ever since the tripartite division into 'primary', 'secondary' and 'tertiary' industries was devised fifty years ago by Fisher (1935) and Clark (1940). In fact, if the supply of inputs to agriculture and the marketing, processing and distribution of outputs from agriculture are added to the 'agriculture sector' the proportion of GDP in the whole economy represented by the farm-associated sector often comprises 30 to 40 per cent of GDP in most 'advanced' countries; in Australia, for example, the so-called 'agri-business' sector alone contributes 30–35 per cent. It is in the organization and volume of this 'farm-related' business that the developed countries differ from those that are 'developing', where most of what 'agri-business' now covers is still integrated within the farm or village.

This change has been very fully analysed in terms of energy input, though less so in economic terms. Pimentel and Pimentel (1979: 119), for example, show that production of a 1 kg loaf of bread requires 7345 kcal of energy up to the supermarket shelf, of which only 45 per cent is accounted for by production, while only 10 per cent of the energy expended in production and marketing of a can of sweet corn is contributed by production itself. Nor is it clear that even these low proportions take full account of the composition of inputs into production. More significantly from our point of view is the fact that both the forward and backward linkages of agriculture, and sometimes both, are increasingly vertically integrated. The large companies that service agriculture have benefited from state aid to agriculture, much of which has gone to capitalize the value of farmland and hence has made investment in agriculture and agriculture-related business more attractive and more secure (Body 1982; Bowers and Cheshire 1983). But these companies participate only to a limited degree in agricultural production itself, except in some highly specialized sectors; the risk and uncertainty of agricultural production are left in the hands of the farmer, while the more securely profitable backward and forward linkages are in the hands of corporations (Wallace 1985; Smith 1986). Vogeler's (1981) analysis, discussed in chapter 4, leads him to conclude that the agri-business companies have become the main beneficiaries of the farm-support programmes, rather than the farmers themselves.

Moreover, there has been a strong tendency towards concentration of ownership of agri-business, not only on the input side where the agricultural machinery companies have been joined by petrochemical companies and their subsidiaries supplying fertilizers and other chemicals, but also on the side of marketing, processing and packaging, where major retail companies also have a share. Fewer and fewer major firms control more and more of the business. By means of contract arrangements, by the provision of technical advice, and through arrangements with the smaller locally based firms which give the major firm a large measure of control, such companies play an increasing role in actual land-use decisions (Smith 1984; Hely 1985). Moreover, they are becoming increasingly multinational, especially in certain fields of marketing and processing, with overwhelming dominance by companies based in the United States and the United Kingdom (Rastoin 1973, 1977). The effect of control on the input side is rather neatly illustrated from Australia, where the means for chemical amelioration of the acidification produced under clover, discussed in chapter 2, are not manufactured by the fertilizer-supplying companies nor held by their client dealers, and hence are not available to the farmer (CSIRO, 1985). On the output side, a farmer who has contracted to supply a certain commodity to a wholesaler who is in turn contracted to a processor, packer or retail chain, has tied his land to a particular use, and at a particular level of intensity, for the period of the contract.

4 New trends

4.1 Trends in the United States

Since the Second World War, commodity support schemes have evolved independently of conservation, in America as in Europe, and have tended to benefit all producers and also the agri-business interests, whether or not they practise conservation or assist in its practice. Indeed, the link more recently made to production control schemes has had some harmful effects from the conservation point of view, since land already in use for a crop may be regarded as 'base acreage', and thus continuous monocropping may be actually encouraged. The production from such land may be supported irrespective of the state of management; on the other hand a farmer who practises conservation but does not have 'base acreage' under a controlled crop may be denied price support (Reichelderfer 1984).

These entrenched interests have, however, lost some political support in consequence of the shift of voting strength to urban electorates, and the consequent decline in the power of the 'farm lobby', strong though it still remains. At the same time, the more general environmental-conservation movement, which has surged in strength since the early 1970s, has introduced a wholly new element into the political situation. Conservationists may well be more concerned with amenity than with the health of the land, but included in amenity considerations is water quality, and the greatest polluter of flowing water is soil erosion. In consequence there is increasingly a situation in which 'largely middle-class urban groups are in effect telling farmers what is in their and the nation's best interest to do' (Lowitt 1985: 324). While a large proportion, especially of family farmers themselves, support a new conservation effort, therefore, a powerful alliance of other interests has also emerged (Cook 1985; McCullough and Weiss 1985). The thrust is toward an integration of conservation and commodity programmes (Benbrook 1979), in such a way as to benefit the conservationist farmer, but penalize the careless or profit-hungry bad manager of the land. Such proposals, incorporated in new farm laws proposed in 1984 and implemented in 1985, have taken a long time to get off the ground.

Essentially, the new proposals take account of the fact that 70 per cent of erosion above the 'tolerable' 12 t/ha/year limit takes place on less than 9 per cent of the land (Batie 1985). Using soil survey and the Universal Soil Loss Equation, highly sensitive land can be defined in fairly precise terms, and the intention is to deny all and any programme benefits to those who cultivate such land if it now lies idle and has lain idle a number of years, unless approved soil conservation measures are adopted and, in some versions of the proposal, are also proved to be successful. An important second arm of the proposals was the creation of a 12 million ha 'conservation reserve', composed of sensitive land that has been in production for at least three of the last five years, and is to be retired from cultivation for at least ten years. The support required for this second arm is much greater than that needed for the first (Cook 1985).

The upshot of this is that almost all of this was incorporated into the new Farm Bill signed into law at the end of 1985. There was, however, a price. Farm supports were substantially enlarged, with bounties and other measures designed to increase exports. The retirement of some sensitive land, the 'sodbuster' and 'swampbuster' provisions designed to ensure the non-use of damaged land for a period of years, and the phased reduction of wheat and maize acreage are paid for by farm support measures which exceed anything previously offered. Nor do they distinguish between large and small farmers; the Deputy Secretary of Agriculture, who would be unable to claim support while a Public Servant, was reported to stand to lose about $1 million in subsidies to his 'family farm' in California and Arizona (*The Economist*, 11 January 1986: 34).

Despite the sweeteners, and their likely effect on other countries seeking to export farm produce without benefit of such support, the conservation proposals inserted in the new legislation represent an important departure in the role of the state. Recognition of the continuing seriousness and even increasing severity of land degradation are combined with the realization that, by themselves, commodity support programmes have tended to encourage rather than discourage neglect of good management. The proposals involve a new element of selectivity that was absent from the historic land conservation programme, and embody sanctions of an order that might be effective. Support for this compulsive element has, remarkably, arisen most strongly in the one country most deeply committed to 'free market' institutions.

4.2 Methods and their implementation

Behind the strivings of the powerful alliance of conservationist movements in North America lies an enormous body of research into degradation, and into more effective management of the land. Particular emphasis is now placed on techniques which involve minimum tillage of land, and maximum conservation of organic wastes, in which there is wide interest not only for conservation reasons, but also because field operations can be curtailed and costs can also be reduced. A detailed statement by Unger and McCalla (1980) provides an excellent review, and shows that the declining real cost of herbicides is now an important factor in making the adoption of conservation tillage systems more likely. While herbicides are not without their problems, it is perhaps worth noting that they are also widely attractive in the Third World as a means of reducing labour spent on weeding; for example, commercial root-crop farmers on new settlement blocks on Taveuni, Fiji, bought the equipment for herbicide spraying even before they built themselves solid houses in the mid-1970s; they were thus able to almost double the land under crop. The use of mulch-conservation methods is of great antiquity; modern research is rediscovering its advantages and providing new methods of application.

Other conservation methods are also now much more widely understood, and the need for them is also understood; however, the cost is high, not only in creating landesque capital, but also in more careful and time-consuming field operations. The sort of massive machinery now in use on the land in Europe and North America was not developed with conservation in mind, but with cost reduction of field operations as the objective. The potential, and real, conflict between economic and conservation interests changes with every new innovation, and changes also with the cost of inputs, particularly energy inputs, into farming. Without a belief by farmers in the need for conservationist management, and without some form of incentive to adopt appropriate farming systems, it would not be likely, given the structure of modern 'free market economy' agriculture, that long-term goals would receive priority over short-term interests (Dickinson *et al.* 1981).

4.3   The state, the public and the farmer

The 'free-market economies', however, embody a monstrous set of contradictions, not only, but especially, in their agriculture. While the profit motive governs production decisions and the restructuring of the industry on more capital-intensive lines, and while corporate control at the same time reduces the farmer's freedom to make his own decisions, the intervention of the state has in most countries been more concerned to prop up farm incomes than to achieve social goals other than those of retaining a viable rural economy and population. The mass of subsidies and supports within the 'Common Agricultural Policy' of the EEC is merely the most glaring example of a general trend or was until the US Farm Bill of 1985. At the same time, however, this practice of intervention opens opportunities for conservationist measures, and also opens up a means for alliances of conservationist groups, mainly urban and middle class, to exert pressure on a diminished farm sector to improve its management of the land.

The case of Australia, examined by Messer below, is different from that of America, and different again from that of Europe. There is less freehold land, and there is less private tenancy; far more land belongs to the state. The rural lobby has not yet suffered the same degree of erosion in its power, and the high concentration and peripheral geographical distribution of the urban population militates toward a lower degree of concern with rural problems among environmentalists. None the less, there are many parallels, in particular in the contrast between the ability of the state to intervene, and the low level of political will for effective intervention. In the United States, the strength of the conservation movement may be such as to force a policy change on government, but this seems less likely in other countries.

Even the policy change in the United States, however, is of a limited order. It rests on the questionable assumption that a fairly handsome level of erosion is 'tolerable', and its concentration on erosion risks the neglect of other forms of degradation. It takes no real account of the major structural change that has taken place in farming since the first programme was initiated in the 1930s, and it finds no means of incorporating some incentive to or control over the operations of the large and powerful agribusiness sector in the system of land use. Perhaps democracies are always a stage or two behind events, in their dependence on public awareness as a precursor to action. But at least the evident ability of interest groups to bring about policy changes is of importance. Conservation in the United States received its initial boost from a concern with rural poverty and a fear of major social upheaval by its association with farm-income support it became an institution which survived the disappearance of the forces which brought the programme into being and the new surge arises largely from the wider environmental concerns of the well-heeled. But the land is the beneficiary, at least in some measure, and the example is well worth more careful study in countries where the state of land management has not achieved comparable notice among those active in public affairs.

B The sociology and politics of land degradation in Australia
Judy Messer

1 An historical overview

All studies of land capability and degradation in Australia are in agreement regarding the biophysical characteristics of the land: Australia is the driest continent; the soils are ancient, of low fertility and slow reproduction, and sensitive to disturbance; rainfall is highly and inconsistently variable; and most of the water is overconcentrated in certain areas (Bureau of Agricultural Economics, BAE, 1983). Even the 'moderate' risk of desertification is as high as 69 per cent in the semi-arid zones (Mabbutt 1978: 253). Nearly 70 per cent of the land surface is composed of arid lands and semi-arid lands, and 35 per cent of these areas are too dry even for pastoralism (Wilson and Graetz 1979). Nevertheless, 65 per cent of the continent is exposed to some form of agricultural production, although only 10 per cent is arable (Woods 1984: 4).

Prior to European settlement, although grazing by 'soft-footed' animals and the use of fire by the indigenous Aborigines was widespread, and certainly greatly modified the soil–vegetation complex, the land and its users appear to have remained in dynamic equilibrium. In contrast, negative impacts of agricultural production on the natural environment were observed from the earliest days of colonization (Bolton 1981); the major forms of impact were and continue to be the loss of vegetative cover (Costin 1984) and erosion. The degradation of vegetative cover is exemplified in the transformation of native pastures in the Western Division of New South Wales from edible to inedible shrublands (Parliament of NSW, 1983). Quantitative loss of vegetation has resulted in soil erosion, adverse hydrological changes (salinity, waterlogging, and rising or declining water tables), loss of fertility and soil structure, and consequent declines in the productivity of crop and pasture land. The characteristic box-shaped form of Australian water channels reflects the high sediment loads carried by the streams. The mechanism of vegetation loss has been both intentional (land clearing, timber cutting for domestic and commercial purposes, control burning, rabbit eradication, farm management practices, and herbicides) and unintentional (overclearing, overgrazing, overcropping and intensified farm-management practices such as irrigation and chemical-based agriculture).

Since the population of Australia has always been low both absolutely and in terms of *per capita* arable land, it might appear that population can be discounted as an independent variable. However, the low population level has itself been of significance in that the consequence has always been a relative shortage of cheap labour, resulting in a reliance on a relatively high level of technological inputs. While technological innovation has given rise to 'economic efficiencies' and environmentally beneficial effects, it has also frequently and seriously impacted on the land in a negative fashion.

Government prohibitions of indiscriminate clearing occurred as early as 1803 (Bolton 1981). By the end of the nineteenth century, the combination of European farming methods, undercapitalized small farmers, ignorance, and ecological deterioration had resulted in widespread declines in productivity in both pastoral and wheat-growing areas. Various acts were implemented by state governments to protect forests and water catchments from agricultural activities. By the 1930s land degradation was so bad in the wheat belts that widespread social concern led to further intervention by the state, through facilitation of farm amalgamation and legislation to promote soil conservation practices (Condon 1978; Williams 1978; Centre for Continuing Education (CCE) 1980).

All accounts of historical land degradation in Australia draw attention to the lack of resilience of much of the land and its consequent sensitivity under intervention. An essential correlate is the widespread lack of understanding of the ecological characteristics of the land and the slow and subtle but inevitable nature of the different forms of degradation. It is generally accepted that this ignorance was due historically to a combination of inappropriate knowledge derived from European farming practices, and to cultural values and ideologies that either devalued nature or overestimated the capacity of the natural ecosystems to adapt and reproduce themselves. The failure or inability of policy-makers, professionals, and land-users to understand the nature and complexities of the land must be considered as a primary causal variable.

For example, even though salinization, as well as erosion, was understood to be a potential consequence of the clearing of native vegetation by the 1920s, little restriction has been placed on the clearing of native vegetation for agricultural purposes, either as a result of necessity or for capital gain; and the negative effects of vegetation clearing are still contested by many agricultural producers. In fact, many schemes involving vegetation clearance have been the consequence of pro-development programmes initiated by state and local government. The most striking example is the numerous 'closer settlement' schemes which indiscriminately placed urban immigrants, ex-soldiers, and the 'land hungry' on uneconomic and often marginal smallholdings (Roberts 1968; Davidson 1981). The land was said to be 'ripe' for 'closer settlement', often having been prepared for this fate by extensive pastoralism.

Even in recent times the revenue-hungry 'development' philosophies of local and state governments have generated such extensive but questionable projects. Due to the powerfully legitimizing agrarian ideology of the politically important 'Man on the Land', only one state government (that with the smallest rural electorate) has had the necessary political will to respond to public and scientific concern about the ecological implications of uncontrolled loss of vegetation on freehold land; in 1980 the South Australian government implemented the Vegetation Retention Scheme whereby vegetation clearance became a class of development requiring consent under the Planning Act (Department of Environment and Planning, South Australia, n.d.).

2   The current status of agrarian land use in Australia

2.1   On paper, a desperate situation

The critical nature of the present day situation has been described extensively in the joint federal and state government report, *A Basis for Soil Conservation Policy in Australia* (Department of Environment, Housing and Community Development, DEHCD 1978) and during the recent Senate Standing Committee Inquiry into a National Land Use Policy (Commonwealth of Australia, 1982–4). A further major statement was that of Woods, in which it was stated that 'Land degradation ... has been identified as the most significant single issue on the Australian environment scene today, affecting large areas of Australia and critically limiting the scope for sustainable development' (1984: iii). More than half of Australia's agricultural land was said to need treatment for land degradation, and the cost of repair was estimated to be $1.6 billion (Dumsday and Edwards 1984: 58). The biophysical form and degree of degradation have been extensively documented at the national level, and to some degree at the level of statistical divisions (Woods 1984). While these statements are of limited analytical value, it is recognized that the cause of degradation lies either in the type of land use or the nature of its management. This implies either that the land has always been or has become unsuitable for agricultural production or alternatively that more appropriate agronomic methods should be identified and adopted (DEHCD 1978). Unfortunately, these documents have achieved little political leverage.

2.2   The structural context of agrarian land use

There is a dearth of information on how social, economic, and institutional factors have played a causal role in these historical processes of land use and management in Australia (Blyth and Kirby 1984; Costin 1984; Woods 1984).

The lack of informed discussion is a reflection of the generally single-discipline nature of the social analysis of Australian agriculture (Messer 1984). An overemphasis on the technical aspects of causality and solutions has been compounded by the myth of the 'average' or 'family farm' (BAE 1983) which has always disguised and continues to disguise the significant internal differentiation within agriculture overall, and between and within the different commodity sectors (Davidson 1981; Messer, in progress).

In terms of causality, most Australian studies have concentrated on the identification of attitudes and beliefs. Several have demonstrated widespread community concern for rural nature conservation and environmental issues (Earle, Brownlea and Rose 1981; Newman and Cameron 1982; Stadler 1983; Hill 1984). Most studies suggest significant gaps between attitudes and behaviour. Class structure was shown to be a significant variable in a study of 490 farmers' attitudes in South Australia. Respondents who had positive attitudes to schemes promoting vegetation retention were more likely to be 'younger, better educated, less conservative, and less likely to be full-time farmers' (Craig, Smith and Sheahan 1983: 22). Similar results were found in a study of 500 Tasmanian farmers (Stadler 1983).

Studies of rates of adoption of new soil conservation practices in Queensland (Chamala and Coughenour 1982; Chamala, Keith and Quinn 1982) revealed an intricate web of structural elements underlying the behaviour of individual decision-makers. Higher levels of social participation correlated with innovation of soil conservation practices, while younger, newer full-time farmers with dependent children were less likely to have the economic flexibility necessary to implement long-term management practices. Since many well-intentioned technological innovations have been shown to have negative or ambiguous environmental impacts, hindsight also may be a significant social variable affecting decisions about relatively novel land-management practices.

2.3 Land tenure

Landlord/tenant relationships are generally thought to have negative implications for land management. In the context of the extensive system of public leasehold tenancy in the pastoral zones in Australia, it has been argued that freehold tenure would increase the landholder's motivation to care for the land (Young 1981; Parliament of NSW 1983). However, the evidence suggests that it is security of tenure rather than the type of tenure system that is important. In reality, the worst forms and degree of land degradation in New South Wales are not occurring on the Crown lands, but in the freehold grain-growing areas which are virtually untrammelled by statutory controls (Soil Conservation Service of NSW, n.d.). In contrast, the extensive, semi-arid and sensitive pastoral region that comprises the leasehold lands of the Western Division of NSW is subject to strict controls over property size, stocking rates, pest eradication, and other management strategies (Condon 1978).

2.4 Land use – who decides?

The historical evidence suggests that much of the land should never have been subjected to conventional agriculture. Given that it has, the role of the state has been highly influential (Davidson 1981). Australia has a land tenure system that is in contrast to those of most advanced capitalist societies in that most of the land is owned by the state, less than 15 per cent being alienated from the Crown by freehold title (BAE 1983: 23). Constitutionally, control over land use is vested in the state governments, but the federal government can exercise a controlling influence by means of its foreign affairs powers, where Australia is a signatory of international conservationist agreements, and through its power to grant conditional loans (Prineas 1984). Despite the differences in political philosophy, the land settlement policies of all state governments have reflected an ongoing concern to promote 'progress and development'. The ecological capabilities of the land and the economic and technical capacities of the land-users were seldom taken into account. The history of land settlement in Australia is not only one of frequently inappropriate land use but also a repetitious cycle of the failure of small-scale farming (Davidson 1981). Whether the suffering of the land or that of the small settlers was greater is a moot point.

Condon (1978) noted that there have been two periods in the history of European land use in Australia when severe drought and economic recession coincided, following a period of new settlement, with severe negative implications for the land. The resulting crises gave rise to widespread public concern, attendant inquiries, and the gradual evolution of land-use theory and policy (Anon 1901; Slatyer and Perry 1969; DEHCD 1978; Young 1981; Oates et al. 1981; Commonwealth of Australia 1982–4). The excesses of nineteenth-century pastoralism led to disasters which precipitated the 1901 Royal Commission into the condition of the Crown tenants, and subsequently to the Western Lands Act which regulated land use and management in the Western Division of NSW. The first Soil Conservation Act in Australia was passed in New South Wales in 1938, as a consequence of the severe land degradation problems of the thirties.

Despite the fact that the NSW Service did much to repair historical damage, as well as implementing controls over vegetation clearing on slopes and along water courses, by the 1970s NSW still had the highest level of land degradation in Australia with more than 80 per cent of the land in use requiring treatment (Woods 1984: 53). The Service has powers only in specified stituations, with the result that its role has been more remedial than preventative. More stringent statutory controls on rural land use are gradually evolving under the aegis of recent legislation, but the issue remains politically sensitive and administratively complex due to conflicts between private, institutional, and public interests (Dick 1985).

Although the degree to which constraints on Crown tenants are exercised is always subject to political will, the public interest is advantaged in that the state does have juridical power to ensure the long-term viability of the land and to invoke periodic reviews. On the other hand, resolution or amelioration of the decline in productivity of freehold agricultural land has been dependent on substantial public funds and scientific and technical expertise (Wagner 1978; Soil Conservation Service of NSW 1985). The enormity of the problem has led to increasing debate on the conflict between the public and private interest, and the rights and obligations which are attached to private property rights (Wills 1984; Gilmour, Hamer and Bourchier 1984; Fowler 1985).

2.5 Economic aspects of land degradation

The highly variable nature of Australian climatic conditions gives rise to unpredictable instabilities in production, and due to the relatively infertile soils yields are low. Agriculture in Australia has only been economically successful in the long term when it has utilized large areas of cheap land and very little labour to produce commodities that can be sold on the world market (Davidson 1981). Thus agricultural producers are also subject to the recurrent instabilities of both world market demand and commodity prices. Returns to labour and capital are volatile. The 1978 Joint Commonwealth –State Soil Study was criticized for failing to take any real account of the degree to which socioeconomic forces determine the way investment and disinvestment in land occurs, or 'how arbitrarily determined conservation goals might be achieved' (Campbell 1982: 239). There has been almost no analysis of the influence of the powerful agribusiness and finance institutions (McIntyre 1975) on the agricultural land-user's capacity to make short- or long-term decisions, let alone how this influence is exerted unequally by sector, farm size and equity criteria.

A number of recent economic analyses have focused on the failure of market and of political or bureaucratic resource allocation mechanisms to allocate land in the best interests of the community (Dumsday 1983; Wills 1984). The land-users' emphasis on short-term productivity, and their failure to budget for externalities have been identified as important underlying economic factors (e.g. Kerin 1982; Dumsday and Edwards 1984; Edwards and Lumley 1985). Recent analysis has also drawn attention to the economic influences of government policy on land management in consequence of subsidies (fertilizer side-effects), water pricing (adverse hydrological outcomes), taxation concessions (overclearing of vegetation, wetlands destruction, etc), inappropriate drought relief measures (delayed destocking), and interventionary commodity pricing policies (encouraging expansion into marginal lands) (Blyth and Kirby 1984; Parliament of NSW 1984). Yet economic theorists have not attempted to differentiate between land degradation that has occurred as a result of the opportunity for capital gain and that due to income maintenance, or simply survival. The socioeconomic factors underlying the processes of land degradation in the Mallee in the

1930s are not strictly comparable to those underlying the rapid expansion of broad-acre cropping in northern NSW in the 1970s.

3 Conclusion

This brief account of 'capitalist' agriculture in Australia indicates that sensitive ecosystems and fragile soils have been degraded as an interactive consequence of the witting and unwitting short-term social priorities of individuals, corporate and state institutions, and the general society. Despite the *ad hoc* nature of the social analyses of agrarian land degradation in Australia, they support the contention that similar systematic structural relationships underlie the apparently diverse forms of land degradation that have occurred historically and continue to occur.

Agriculture in capitalist societies is subject to an inherent tendency to exploit the land as a result of the necessity for both household and capitalist producers to reproduce their economic conditions of production or to maximize their profits. The degree to which specific cases of land exploitation are influenced by ecological variability, the world market, finance institutions, state policies and juridical controls, or the internal social relations of agriculture, can however only be determined by empirical analysis. It is essential that both macro- and micro-variables be taken into account.

The tendencies towards agrarian land degradation in Australia have generated counter tendencies that are gradually leading to increased social controls over the private (individual and corporate) land-user's exploitation of the land as a natural resource. Land use and management is therefore increasingly the subject of collective decision-making processes, although the mechanisms for ensuring public participation in such decision-making have yet to be fully articulated. The emergent emphasis appears to be on such strategies as achieving economic viability of farms by amalgamation, better farm-management practices by means of technological innovation, and a juridical framework that will integrate national, state, regional, and on-farm decision-making by means of a judicious mixture of 'carrot and stick' laws, regulations, economic incentives, and ideological 'shifts' (Oates *et al.* 1981; Department of Primary Industry 1984; Fowler 1985; Gilmour, Hamer and Bourchier 1984). Increasingly, this is occurring in the context of a multidisciplinary approach which recognizes that the necessary policy guidelines can only be identified by acknowledging that the structural forces underlying decision-making processes are as significant as the technical problems and solutions. To what extent these efforts will overcome the intractable obstacles generated by anarchic production, economic necessity, constitutional constraints, bureaucratic resistance, insufficient social and scientific understanding, and the enduring sensitivity of the land itself remains to be seen.

13 Retrospect and prospect

Piers Blaikie and Harold Brookfield

1 Concluding comments on the book's approach

We have organized the material in this book by putting the land managers 'central stage' and studying their interactive relationship with a changing environment. Under the summary label of regional political ecology we have then put this interactive relationship in a historical, political and economic context.

Of course this approach was developed partly *a priori* to the case study material of this book and other empirical work, and partly in response to it. Nonetheless, it is perhaps useful to review how the approach has fared in the light of the case studies. The variety of social and environmental causes of land degradation is clearly enormous, and explanation of any one instance is usually complicated and beset by uncertainty. It is commonplace to say that each instance of land degradation is highly conjunctural, which implies that it cannot be satisfactorily theorized. Our approach attempts to build a theory which allows for complexity, and identifies the sources of that complexity. The contrasting outcomes of the three Indonesian communities' efforts to manage their land in chapter 9B defy single hypothesis explanations. The bewildering complexity produced by inter-relating environmental variability in resilience and sensitivity with social and economic variations in agricultural production, as evidenced in New Guinea (chapter 8B) or Nepal (chapter 2, section 4), must point to conditional and multiple hypotheses, not universal nor single ones.

The other component of our approach is to pursue the quest for better measurement, both of technical relationships as well as of costs and benefits of degradation and conservation. Such measurements can dispel uncertainty, even if they cannot reduce complexity. The case study of measuring costs and benefits in Indonesia in chapter 5 indicates both the importance and difficulties of this effort.

Stocking's review of measurement in chapter 3 demonstrated that we are not dealing with a situation in which 'soft' social science is merely trying to explain phenomena which 'hard' natural science has firmly established. Nor, as Seckler showed in chapter 5, is there 'hard' economics of land degradation. In all areas of diagnosis, explanation and prescription we still have to rely heavily on logical deduction from facts and principles which are established by locality-bound empirical research, but must recognize that both the 'facts' and the logical system reflect uncertainty in measurement and different theories of social processes. Even the most basic of our questions, that concerning the relative roles of natural forces and of human activity in bringing about degradation, can seldom be regarded as unequivocal (the case studies of eighteenth-century France and contemporary Nepal are examples).

There seem to be no short-cuts to an explanation of land degradation. Simply we must accept that explanation will be difficult and costly for policy-makers to achieve as a first step to effective action. Whether our approach has made the path to explanation any easier is for the reader to judge. It has been our aim to clarify complexity, but not to reduce it. Maybe that is too modest an aim, but it is conditioned by the nature of land degradation itself.

2 Some conclusions from case material

Ignorance of the consequences of actions on the land, the reckless quest for profit, poverty and deprivation leading to 'desperate ecocide', pressure of population on resources (PPR) – on which we remain somewhat ambivalent – and population decline all emerge as underlying causal agents of degradation in these pages. Except where a production system is in collapse, however, there does seem to be one common element, which is a 'pressure of production on resources', to use the term coined by Pavelis (1983) and introduced in chapter 12. We identify it in situations as far apart as the modern United States and USSR, pre-colonial Papua New Guinea, eighteenth-century France and colonial Africa. While this pressure can lead to innovation in landesque capital, it can also lead to the consumption of such capital, and to the unwise or undercapitalized use of sensitive and unresilient land. The origins of this pressure, however, vary enormously, as also does the nature of its impact. We move from the satisfying discovery of an apparently general conclusion to a set of conditional hypotheses, which we then try to disentangle.

'Pressure of production on resources' can arise in a number of ways: from the nature of the crops or livestock themselves in relation to the capability and sensitivity of the land; from high or rising population or from heavy extraction of surpluses from the land-manager. In Kalimantan (chapter 9B) shortage of land leads both to degradation and to innovation. In Nepal (chapter 2, section 4) the lack of capital and labour resources available to farmers emerges as an important factor in combination with PPR; farmers must undertake tasks requiring stonework and specialized tools with their own labour alone, and without the means to take more efficient conservation measures of which they are well aware. In India, Jodha (chapter 10B) interprets production pressure as due to encroachment of common property resources (CPRs), mechanization, commercialization and population growth. In Fiji (chapter 9C) the pressure arises from a resource-based development policy combined with a rigid land tenure system. In early-modern France (chapter 7) it arose from rising population and commercialization, but if rural PPR itself was also a cause it must be recalled that there was greater PPR in the early nineteenth century, when there was less degradation.

Strong market or social demand for production can arise in a variety of ways, and at all times. Extensive cattle ranching on a scale involving overgrazing of grasslands created from the tropical rain forest (chapter 9A) can clearly do major damage; and so could the demands of 'big men' in the pre-contact New Guinea highlands for production of large herds of pigs and other prestation items (chapter 8B).

Pressure to produce is, however, also heightened by social relations of production involving onerous rates of surplus extraction. Heavy exactions of tax and tribute, very low wages which do not provide enough for the worker and family, and denial of access to CPRs all create such pressure, at the same time as they deprive farmers of the resources they need to manage the land. Perhaps this is best described as 'pressure of deprivation', clearly a factor where CPRs are reduced, whether by engrossment as in early-modern France or by a well-intentioned but ill-conceived land reform in India. Pressure leads to a breakdown of management that affects not only the CPRs themselves but also private lands with which their use is linked in the farming system as a whole; it is almost always associated with degradation.

Surpluses can also be extracted in the sphere of exchange. Low prices for commodities produced by farmers can be imposed by state pricing policies or other forces distorting the market in favour of particular nations or groups (see chapters 6, section 2 and 6, section 3). Small producers may have to compete with larger capitalist producers in areas of the market where the latter have an advantage. Monopsonist local traders may be able to reap large profits while leaving the producers without the capital, labour or tools to manage the land effectively. Indebtedness through usury, or offering goods on credit, brings about the same effect (see chapter 9B, section 2).

Degradation is thus often encapsulated in a web of surplus-extracting relationships, in feudal systems, in colonial/capitalist economic relationships, and almost wherever there is marked inequality. Neither centrally planned nor 'market' economies are free from such relationships. The surplus product in a socialist system is theoretically 'at the disposal of society as a whole' (Maksimovic, reported in Kaser 1968: 279), but often goes elsewhere in the USSR (chapter 11A) as it does in the revolutionary states of Africa (Williams 1976).

3 Ways forward

3.1 The world conservation strategy (WCS)

This major initiative was taken by the International Union for Conservation of Nature and Natural Resources (IUCN) or other international institutions. Its aims cover a wider field than those of this book, but there are also shared concerns and it is therefore worth considering in what ways each can inform the other.

The first point concerns a difference in the criteria used by WCS and ourselves. Whereas our approach is utilitarian in that it focuses on the economic and social aspects of land degradation and conservation, the WCS necessarily deals with important ethical considerations relating to areas of special scientific interest and the preservation of wildlife and endangered species.

The second concerns one of the central concepts in WCS: that of 'sustainable development' and 'sustainable utilization'. Here we feel that our thematic and case study material may be of use in producing a clearer definition of the above. 'Sustainable utilization' of land is both a technical and political economic concept and it can be achieved in a number of different ways, for example, with high inputs of chemical fertilizer and considerable earthworks, or substituting labour inputs and locally derived nutrients. Therefore, we argue for the necessity of including the future prospects of land managers for *access* to land, capital, extension, state assistance and so on, in any assessment of sustainable production. Clearly this covers potentially dangerous political ground and is subject to conjecture, but it pursues the issue in a realistic and practical direction. Perhaps some of these views are implicit in this statement:

> The efficiency of traditional cropping systems can often be increased not by introducing completely different ones but by identifying those elements which could be improved and making the appropriate improvement. (IUCN 1980: 14, para 11)

This is not all that the IUCN advocates, but it is a message that nowadays finds a growing number of supporters. Following on from this, our view is that sustainable land utilization often requires fundamental social and political changes in the process of development. 'Short cuts' such as coercive state action on a massive scale are unsuccessful, whatever the arguments for an elite guardianship of the biosphere.

The last point concerns the audience for WCS and the institutional and political means of achieving its objectives, by aiming at an audience of government policy-makers, conservationists and development practioners (as specified on their page iv). However, the writers disregard the importance of land managers at grass roots level. Emphasis on the education, political will and international conventions for policy-makers and administrators may be useful in improving national and international policies, but we feel that the person who is ultimately responsible for making the decisions as to how the land should be used, and the context in which that person makes those decisions, could receive more consideration. Some decisions can be directly influenced by policy changes, as in the US case outlined in chapter 12A, and in the Common Agricultural Policy of the EEC (not discussed in this book), but for the majority of land managers, the technical and social choices open to them need to be clearly emphasized and undestood and the policies evolved more wide-ranging. If this emphasis is not given the policies remain 'top-down' and will neither respond to the complexity of the problem nor be acceptable to the land managers themselves.

3.2 Starting with the land manager

For developing countries, it is tempting to call for a small-farmer-orientated set of policies. Already it is claimed that there is a new consensus as far as research and development for agriculture in Africa is concerned (Richards 1986). This involves research and development (R&D) for resource-poor farmers (especially food producers and women) which would be highly specific to local ecological conditions, would be founded in a thorough knowledge of local agricultural practice, and would be participatory in style. Chapter 6 has shown how different this approach is from historical practice in colonial regions, and there is a good deal of evidence from the case studies and examples in this book to lend weight to these views. Land management needs on-the-spot and on-the-dot attention involving local knowledge, continuous monitoring and sometimes urgent action. Surely then, bureaucrats, policy-makers and research workers should listen to and learn from farmers? Too often government policies have greatly aggravated the problems of land degradation, and there is also evidence that pressure (and policies) on resource-poor farmers have encouraged them to rob the soil, pastures and forest merely to survive. All this provides a sound basis for supporting a 'bottom-up', locally specific, participatory approach to research into and diffusion of land management and repair. Certainly the authors would endorse this direction of state-sponsored research, however difficult it may be to change attitudes and working practice of soil conservation experts, research stations, agricultural extension officers and the like. As Richards has written in the West African context:

> Rather than concentrate on selling a 'package' of soil 'medicine' to farmers, the challenge for 'populist' agricultural development is to establish a thorough-going and self-sustaining programme of improved soil management drawing strength and initial impetus from the skills, experience and experimental ability already present within the farming community. (1985: 61)

3.3 Inadequacies of the populist approach

When we place this initiative into the wider considerations of land management, leaving aside R&D, it quickly becomes apparent that knowledge of appropriate methods is not by itself enough. Land management also requires resources of labour, capital, sometimes special tools, food reserves to fill the gap created by conservation works, and alternative sources of materials, crop residues, grazing areas and fuelwood. These resources are often not forthcoming because of pressures of various types which have been identified at many points in this book, but to which must be added the pressure from land degradation itself, which completes the vicious circle. Such pressure can, if experienced for a generation or more tend to 'de-skill' farmers and pastoralists. The provision of resources and incentives to make land management and repair possible for poor farmers requires considerable state involvement, but one that goes against most political economic currents of the contemporary world.

Moreover, most ethno-scientific knowledge has been evolved in pre-capitalist and/or non-market situations. Sustainable production for subsistence needs, perhaps also to produce materials for petty-commodity production and trade, remains a major objective. However, the market penetrates virtually all land use in the world, either directly because crops, livestock or timber are produced for intended sale, or indirectly because the resources allocated to subsistence production are affected in type, quantity and timing of allocation by the market sector of the farm and the labour of the family. There have been *tremendous* changes in the societies which nurtured and developed these systems of land management. Beinart notes:

> although there is considerable evidence of self-regulatory practices in land use in pre-colonial African societies, it cannot necessarily be assumed that peasant communities in the colonial period, their old systems of authority eroded, and faced with a shortage of land, increasing population densities, new opportunities and new constraints in their battle for survival, had the capacity to regulate themselves. (1984:84)

Market opportunities add to population pressures and surplus extraction as a strong additional force which puts pressure upon resources. In many cases, locally enforced authority has been broken down. The state is forced to step in to protect resources, not only from pressured small farmers and pastoralists, but from contractors and opportunist farmers. The state is pulled into a vacuum created by pressures of the market, and the dissolution of informal, traditional, local authority which the market and capitalist relations have helped to dissolve (see the case of the commons in India, chapter 10B). Simply, the problems which small farmers and pastoralists face today are an overlapping but a different and shifting set from those which 'ethno-scientific' land management set out to solve.

The question of adapting indigenous land-mangement technology to the needs of market agriculture is complex. The process is inevitably painful, since it involves contradictions which are irreconcilable in the short run: mechanization (and its consequences for employment) and yield maximization versus work spreading and risk minimization; short-term versus long-term goals; tenurial systems which increase access to land versus consolidation and greater inequalities. Already the market is an essential element in the reproduction of peasantries throughout the developing world, the strains (Bernstein 1979). Also, despite these strains, most of the world's peasantry would not wish to withdraw from the market.

3.4 Means of resolution

The importance of achieving means of land management which are within local competence is not, however, diminished by these difficulties. Ultimately, the task of effective land management must lie with those who work the land, and who continuously have to make decisions concerning agronomic practice and repair. Yet local knowledge is clearly not enough, otherwise there would be no problems. In practical terms there would seem to be two main avenues by which the contradictions might be resolved.

In the first place, land management must be conceived of as it is in practice: an integral part of farm management as a whole. This means that the full context of farm management must be taken into account, including all its social, cultural, economic and political aspects. There must be particular emphasis on *access* to the resources needed for effective management and repair. The farm, then, must remain the logical unit for management on private land.

For CPRs, including those such as water and drainage which run across both private and public land, the question is one of strengthening local institutions where they still operate, of handing power back to them where it has been removed, or even of creating such institutions anew. However, to hand precious resources to a local power structure which cannot give mutual assurance to all users is a charter for rascals and a recipe for accelerated degradation.

The second part of the approach brings us towards the role of the state. The appropriate technology of management and repair will tend to be that already known, whether still used or not at the site, or used elsewhere under comparable conditions of sensitivity and resilience. Most of the conservation must be done by farmers themselves, but this will involve mainly the 'cheaper' methods – agronomic techniques rather than major earth works. Where larger works are necessary, demanding resources beyond those available to a local council, *Panchayat* or chief, the role of the state becomes more direct, but there is constant likelihood that it will be seen either as coercion or as something which government should do and is not the business of, or even in the interests of, the farmers themselves.

Coercion of the kind which limits or denies access to resources is the most delicate aspect of intervention, though it has often been handled so crudely as to be wholly counter-productive, or has failed because of lack of will to enforce regulations, as we saw in chapter 6. Yet coercion may be necessary in cases where failure to protect remaining resources would lead to rapid overuse and degradation. There are no instant cornucopias awaiting even the most egalitarian reforms in such matters. Undue inequalities in access to resources lead to poaching and expensive protection, and simply diverts pressure on to resources elsewhere. Draconian measures by the state *may* be necessary but seldom work to the advantage of the poorer local people. The ideology of the Nepalese and Indian approaches to forest conservation may usefully be compared (the full discussion being found in chapters 2, section 4 and 6, section 5.2).

In the case of the Nepalese community forest projects there was no pressure on villagers. *Panchayat* leaders volunteered their villages to join the programme, had to find a small area to set aside as a nursery, and people to run it and the means to keep all livestock off areas which they chose to have replanted. No wire was provided, only 'psychological' barriers consisting of a perimeter of white painted stones. A relatively egalitarian social structure and surviving traditions of communal organization help, but even so, notwith-standing successful cases (Nepal–Australia Forestry Project 1985), results are patchy. In the Indian case state forestry departments have protected the remaining resources with something like a 'laager' mentality. While this approach and its recent extension do protect forests from indiscriminate felling for the market there is also much corruption and poaching. Local people come to view the situation as a zero-sum game where their loss of income from the prohibition is someone else's gain (Thomson 1977). Sometimes, what Porter (1986) terms a 'rip-off upward' takes place, in which the size of the individual reward from corruption increases through successive 'percentages' accumulating in the few hands of the higher levels of officialdom. Attempts have been made to minimize this 'rip-off upward' of benefits. Some have succeeded in the sense that government-financed trees are poached by local people at a sufficient rate to ensure that worthwhile resources for large-scale commercial poaching do not accumulate, and environmental damage is limited by its small scale.

The promotion of a 'bottom-up' and participatory approach to the repair and prevention of degradation is thus perhaps the only way in which success can ultimately be achieved, but it has many problems and the coercive as well as supportive roles of the state are both indispensable. We now turn to the role of the state in more specific terms, suggesting ways in which it might be exercised in a less clumsy and counterproductive manner than has so often been the case in the past. The state clearly has it in its power to accelerate land degradation greatly by its policies, but also to make a much more positive impact on the problem.

3.5 The need for selectivity in state intervention

The cases in which there has been positive intervention by the state in the interests of better management of the land are few, although the amount of regulation and advice has been substantial. The outstanding case of such intervention, in the United States in the 1930s, is shown by commentators to have had more to do with the support of farm incomes than with the health of the land itself, important though the latter was (chapter 12A). Other interventions have been more spasmodic and selective. Against successful examples must be set many failures, such as the series initiated by the leadership of the Communist Party in China since the early 1950s (chapter 11B).

It falls to governments to decide *where* to concentrate their participatory and farm-based initiatives. There are a number of criteria which can be used, and it is perhaps feasible to prioritize these in terms of an ordinal scale of measurement. First, areas have to be identified where improved land management would reduce land degradation, and create successful models for demonstration (IUCN 1980). Many areas with massive damage owe this to natural processes which cannot readily be controlled. A second criterion is that offered in chapters 5A and B. Will people need land under threat of degradation in the future? Are there other lands to which they could move; could they find gainful employment in cities, or do they have a reasonable chance of having access to capital and savings to reduce or repair damage themselves? The plight of marginal people in marginal areas discussed in chapter 6 is particularly important here, since the common answer is that people *will* need the land in the future, *cannot* go elsewhere to be gainfully employed, and *are* in the grips of a vicious circle of resource deprivation which reduces their prospects of coping with land degradation to vanishing point.

A third criterion addresses the economic question of returns to investments in soil conservation. As chapters 3 and 5 argue, it seems that only high potential land, not yet badly degraded but under imminent threat, provides an unequivocally good return on investment. Already degraded land, particularly in the tropics, is very expensive indeed to repair. This may steer funds away from poor farmers, marginal and depressed areas subject to drought, and outmigration. In these circumstances it is all the more necessary to give due weight to option values of land in the future, particularly when it looks likely that there will be no alternative means of livelihood for land-users in the foreseeable future. These three criteria offer conflicting choices, so that often social or political priorities dominate in any national programme of land management.

3.6 The state, society and the land

There are, however, other pervasive sets of social relations in production and exchange which the state can and does affect, and the nature of which even moulds the character of the state itself. One of the most important is the distribution of land. Great inequality of landholding, absent from very few developing regions outside the socialist countries, encourages the creation and reproduction of 'functional dualism'. One side of this coin is pressure on the land, often steep and ecologically marginal, by small peasant farmers. The other is commercial farming, often rapacious and short-sighted. Land reform is no sufficient condition for reducing land degradation, but it may be a necessary condition, hard to grasp though this nettle is for a great many governments. Even limited tenancy reform can, if actually implemented, discourage mining of the soil held under short leases and encourage tenants to invest in the land, though if such reform is combined with pressures to produce more crops it can be ineffective from the point of view of conservation, as Clarke (chapter 9C) showed in Fiji.

Pricing policies can also put additional weight on the 'reproduction squeeze' (Bernstein 1979) facing peasants and pastoralists. Parastatal marketing boards have by various means of coercion forced small farmers to grow food for the towns at low prices, or export crops for prices which such boards are notoriously poor at successfully negotiating. All these factors increase pressure upon people and their use of resources. All can be alleviated or accentuated by state action. It is not only a matter of policy, but also of the local 'climate' of implementation. It may, for example, be that friends or relatives of the president can acquire the means of exacting surplus from producers whose finances and marketing were given into their hands, a situation familiar to many citizens of the Philippines under the recently departed Marcos. It may be that bureaucrats look the other way, or are induced to do so, rather than vigorously implement a land-reform programme. It may be that squatters, or even villagers, are forcibly evicted while those with power to do so annex thousands of fertile hectares. All these things happen and seventy-five years of revolution and reform have not halted them.

The state may be more benign, or more successful in protecting the weak from the strong. The state may act with awareness for the consequences of its action, and agility in changing direction when harmful consequences occur. However, it would be idle to deny that there are few such states in the developing world, whether 'free market' or 'socialist' in complexion. Even in states which enjoy government responsible to an electorate free to exercise its majority will, politicans and bureaucracies can often be subverted to act in favour of powerful sectional interests. In widely differing circumstances the case studies have illustrated the impact of state action upon land managers. On the one hand whenever that action has been coercive, or has taken away resources (whether in the name of conservation or not), or *detracted* from the range of choice in land management, it has led to more degradation, not less (in Africa and India in chapters 6 and 10, in Indonesia, chapter 9, and China, chapter 11). On the other hand, state action can increase the freedom of land managers, but in such a way that it encourages 'ruining' of the soil or vegetation, as in the early history of the United States, in contemporary Brazil, sometimes in Indonesia, and in the case of CPRs for a few more commercially orientated and politically powerful farmers in India.

4 The earth does not abide

To detail policy recommendations would be a futile exercise in the light of this discussion; it has been consistently argued that there are no universal explanations nor universal solutions. There are no blueprints for success. None the less, we suggest that our emphasis on the land managers, on the economic, social and political conditions under which they operate, and on the dynamism of the environment in which they work, does indicate some ways forward for planners, and for scholars who seek to influence planners.

The decision-making context of land management is of primary importance at all times and places; almost equally important are the conditions of access to resources, and the ability as well as the willingness of land managers to apply what they know or can learn from others. Sanctions, where they are necessary, are most effectively applied through local institutions where such remain viable or can become viable; coercion by a remote authority lacking local knowledge is a poor second best, though often it is unavoidable. The better role of the state is to be found in creating or reinforcing the conditions which make good management worthwhile and economically feasible, and in making available the results of research. At the same time, such research needs to imbued with a willingness to learn from the experience of those who work the land.

It is, however, idealistic to expect even the most reformist of states to apply its resources equally for the benefit of all. Not only do the larger social conditions of production and exchange determine that some groups of land managers, often neither the very rich nor the very poor, are more amenable than others to successful intervention, but there are difficult political trade-offs between selecting for conservation effort either capable land at risk, or already degraded land that is needed for future use. Moreover the decision of how to allocate resources is rendered harder by the imperfect state of knowledge concerning the relationship between land quality, its future option value, and the nature and rate of the natural processes which either act to degrade it, or could be harnessed to aid in its rehabilitation.

Clearly, where natural rates of degradation are high, as in the gorges of Nepal, remedial works are likely to be unrewarding, but where degradation is only a spasmodic phenomenon, as in the loess-lands of Europe, or where it is clearly induced by human activity, as in the case of rangeland overgrazing, there is much that can usefully be done provided that the present and likely future value of the land make the investment worthwhile. In the present state of knowledge these relationships are not always easy to determine and the relative contributions of natural forces and human interference cannot be evaluated with certainty. However, our case studies suggest that wherever there is the greatest doubt concerning a human or natural causation of degradation problems, skilful and sustained management are most critical for amelioration and control, and management failure is most likely to expose the land to damage. In such cases, changes in the social conditions of land management are likely to exert a high degree of leverage. The option value of such land, which may be as different as that of Himalayan interfluves European loess, or tropical *Imperata* grassland, is however a matter for evaluation in the political, social and economic sphere.

There is, moreover, always a considerable 'slack' between policy levers and any improvement in land management. Land management involves so many variables, and a reduction in the constraints such as improved credit facilities elimination of a tax, improved conditions of tenure, or more flexible terms of contract with an agribusiness company may be no solution. More positive steps such as bench-terracing aided by grants or soft loans, which could be quite the wrong technical solution, or even the provision of 'food-for-work' aid to encourage farmers to divert inputs from production into land management may also not provide the answer. The whole complex of employment and income-gaining opportunity is involved, and while good management of the land can be achieved ultimately only at the interface where a spade or plough is dug into the soil, the whole of society and economy are involved in determining what will and what will not work.

Our final appeal, then, is to our fellow scientists, both natural and social, to realize not only the seriousness of the land degradation problem but also its complexity. For social scientists especially it is important to realize that no only do society and polity evolve and change, but so also does the land on which the system of production basically depends; the natural conditions of production are not a 'free gift' and the earth does not abide. Natural scientists know this well, but social scientists have been quite remarkably slow to include explanation and amelioration of the waste of the most basic of all resources in their agenda. It is time for a new beginning.

References

Abel, N.O.J., Flint, M., Hunter, N., Chandler, D. and Merafe, Y. (1985) 'The problems and po _sibilities of communal land management in Ngwaketse District – a case study', Summary Report, International Livestock Centre for Africa, 32 pp, mimeo.

Abel, W. (1980) *Agricultural Fluctuations in Europe from the Thirteenth to the Twentieth Centuries*, translated by O. Ordish from *Agrarkrisen und Agrarkonjunktur* (1978, Hamburg, Paul Parey), London, Methuen.

Acocks, J.P.H. (1955) *Veld Types of South Africa*, Union of South Africa, Department of Agriculture, Botanical Survey Memoir No. 28, Pretoria, Government Printer.

Adler, E.D. (1957) 'Soil conservation', in *Natal Regional Survey, Agriculture in Natal: Recent Developments, A Symposium*, Cape Town, Oxford University Press, 123–34.

Agulhon, M. and Désert, G. (1976) 'L'essor de la paysannerie 1789–1852', in Agulhon, M., Désert, G. and Specklin, R. (eds) *Histoire de la France Rurale, Tome III, Apogée et Crise de la Civilisation Paysanne* (Juillard, E., general editor), Paris, Editions du Seuil, 19–175.

Ahuja, L.D. and Mann, H.S. (1975) 'Rangeland development and management in western Rajasthan', *Annals of Arid Zone*, 14, 29–44.

Aina, P.O., Lal, R. and Taylor, G.S. (1976) 'Soil and crop management in relation to soil erosion in the rainforest of western Nigeria', *Soil Erosion, Prediction and Control*, Soil Conservation Society of America Special Publication 21, 75–84.

Alexander, D. (1982) 'Difference between "calanchi" and "biancane" badlands in Italy', in Bryan, R. and Yair, A. (eds) *Badland Geomorphology and Piping*, Norwich, Geo Books, 71–87.

Alier, J.M. (1984) *A History of Ecological Economics*, Barcelona, Facultat de Ciencies Economiques. (English version of part of a book in Spanish, to be published by la Fundacion Juan March, Madrid, mimeo.)

Alier, J.M. and Naredo, J.M. (1982) 'A Marxist precursor of energy economics: Podolinsky', *Journal of Peasant Studies*, 9, 207–23.

Allan, W. (1965) *The African Husbandman*, Edinburgh, Oliver and Boyd.

Allen, B.J. (1982) 'Subsistence agriculture: three case studies', in Carrad, B., Lea, D.A.M. and Talyaga K.T. (eds) *Enga: Foundations for Development*, Armidale, NSW, Department of Geography, University of New England, 93–127.

Allen, B.J. (ed.) (1984) *Agricultural and Nutritional Studies on the Nembi Plateau, Southern Highlands*, Department of Geography Occasional Paper (New Series), 4, Waigani, University of Papua New Guinea and Southern Highlands Regional Development Plan.

Allen, B.J. (1985) 'Dynamics of fallow successions and introduction of Robusta coffee in shifting cultivation areas of the lowlands of Papua New Guinea', *Agroforestry Systems*, 3, 227–38.

Allen, B.J., Bourke, R.M., Clarke, L.J., Cogill, B., Pain, C.F. and Wood, A.W. (1980) 'Child malnutrition and agriculture on the Nembi Plateau, Southern Highlands, Papua New Guinea', *Social Science and Medicine*, 14D, 127–32.

Allen, R. (1980) *How to Save the World. A Strategy for World Conservation*, London, Kogan Page.

Amos, M.J. (1982) 'Economics of soil conservation', unpublished MSc dissertation, Silsoe, Bedfordshire, National College of Agricultural Engineering.

Anderson, D. (1984) 'Depression, dust bowl, demography, and drought: the colonial state and soil conservation in East Africa during the 1930s', *African Affairs*, 83 (332), 321–43.

Anderson, D. (1985) 'Managing the forest: the conservation history of Lembus, Kenya 1904–1963', paper presented at workshop on The Scramble for Resources: Conservation Policies in Africa 1884–1984, April, Cambridge.

Anderson, D. and Millington, A.C. (1985) 'Historical aspects of soil erosion studies', paper read at African Studies Association of the UK Symposium, 18 September 1985, London, University College.

Anderson, D. and Millington, A.C. (1986) 'Political ecology of soil conservation in Anglophone Africa', paper presented at the Annual General Meeting, Institute of British Geographers, January, Reading, mimeo.

Anon. (1901) *Royal Commission to Inquire into the Condition of Crown Tenants, Western Division of New South Wales*, Sydney, Government Printer.

Arndt, H.W. (1983) 'Transmigration: achievements, problems, prospects', *Bulletin of Indonesian Economic Studies*, 19, 50–73.

Arnold, J.E.M. and Campbell, J.G. (1985) 'Collective management of hill forests in Nepal: the Community Forestry Development Project', paper for Common Property Resource Management Conference, Annapolis, Maryland, 21–27 April 1985, National Academy of Sciences, USA.

Ashton, B., Hill, K., Piazza, A. and Zeitz, A. (1984) 'Famine in China, 1958–61', *Population and Development Review*, 10, 613–45.

Asian Development Bank (1982) *Nepal Agriculture Sector Strategy* (2 vols), Manila, Asian Development Bank.

Atkins 'Land and water management' (1983) *Western Vanua Levu Regional Plan* (2 vols), Cambridge, Atkins, for the Ministry of Economic Planning and Development, Fiji.

Axinn, N.W. and Axinn, G.H. (1983) *Small Farms in Nepal: A Farming Systems Approach to Description*, Kathmandu, Rural Life Associates.

BAE (Bureau of Agricultural Economics) (1983) *Rural Industry in Australia*, Canberra, Australian Government Printing Service.

Bahrin, Tunku S. and Perera, P.D.A. (1977) *FELDA, 21 Years of Land Development*, Kuala Lumpur, Federal Land Development Authority.

Bailey, F.G. (1966) 'The peasant view of the bad life', *Advancement of Science*, 399–409. Also in Shanin T. (ed.) (1971) *Peasants and Peasant Societies*, Harmondsworth, Penguin.

Baines, G. (1984) 'Environment and resources – managing the South Pacific's future', *Ambio*, 13, 355–8.

Baines, J. (1983) 'Dietary patterns of pregnant women and birthweights on the Nembi Plateau, Papua New Guinea', unpublished MSc thesis in Medicine, University of London.

Bajracharya, D. (1983) 'Deforestation in the food/fuel context: historical and political perspectives from Nepal', *Mountain Research and Development*, 3, 227–40.

Bandarage, A. (1983) *Colonialism in Sri Lanka: The Political Economy of the Kandyan Highlands 1833–1886*, Berlin, Mouton.

Banister, J. and Thapa, S. (1980) *The Population Dynamics of Nepal*, Honolulu, East–West Center.

Barkley, P.W. and Seckler, D.W. (1972) *Economic Growth and Environmental Decay – The Solution Becomes the Problem*, New York, Harcourt, Brace Jovanovich.

Barrau, J. (1956) *L'Agriculture Vivrière Autochtone en Nouvelle-Calédonie*, Noumea, South Pacific Commission.

Barth, F. (1959) 'The land use pattern of migratory tribes in South Persia', *Norsk Geografisk Tidskrift*, 17, 1–11.

Barth, F. (1961) *Nomads of South Persia: the Basseri Tribe of the Kamseh Confederacy*, Boston, Little Brown.

Barth, F. (1973).'A general perspective on nomad–sedentary relations in the Middle East', in Nelson, C. (ed.) *The Desert and the Sown*, Berkeley, Institute of International Studies, University of California, 11–21.

Bates, D. (1973) *Nomads and Farmers: A Study of the Yoruk of Southeastern Turkey*, Ann Arbor, University of Michigan.

Batie, S.S. (1985) 'Soil conservation in the 1980s: a historical perspective', *Agricultural History*, 59, 107–23.

Bayliss-Smith, T.P. (1979) 'Prehistoric soil erosion in Britain: evidence from fen, bog and mire deposits', unpublished paper.

Bayliss-Smith, T.P. (1985) 'Pre-Ipomoean agriculture in the Papua New Guinea highlands above 2000 metres: some experimental data on taro cultivation', in Farrington, I.S. (ed.) *Prehistoric Intensive Agriculture in the Tropics*, Oxford, BAR International Series 232, 285–320.

Beattie, K.G., Bond, W.K. and Manning, E.W. (1981) *The Agricultural Use of Marginal Lands: Review and Bibliography*, working paper no. 13, Ottawa, Lands Directorate, Environment Canada.

Beck, L. (1981) 'Government policy and pastoral land use in southwest Iran', *Journal of Arid Environments*, 4, 253–67.

Beckerman, W. (1974) *In Defence of Economic Growth*, London, Jonathan Cape.

Beets, W.C. (1982) Multiple Cropping and Tropical Farming Systems, London, Gower, Westview Press.

Beinart, W. (1984) 'Soil erosion, conservation and ideas about development: a Southern African exploration, 1900–1960', *Journal of Southern African Studies*, 11, 1, 52–84.

Bell, M. (1981) 'Valley sediments and environmental change', in Jones, M. and Dimbleby, G. (eds) *The Environment of Man: The Iron Age to the Anglo-Saxon Period*, London, British Archaeological Research British Series 87, 75–91.

Bellwood, P. (1978) *Man's Conquest of the Pacific: The Prehistory of Southeast Asia and Oceania*, Auckland, Collins.

Belshaw, D.G.R. (1979) 'Taking indigenous technology seriously: the case of inter-cropping techniques in East Africa', *IDS Bulletin*, 10, 2, 24–7.

Benbrook, C. (1979) 'Integrating soil conservation and commodity programs: a policy proposal', *Journal of Soil and Water Conservation*, 34, 160–7.

Berg, N.A. (1979) 'Soil conservation: the physical resource setting', in Soil Conservation Society of America, *Soil Conservation Policies, an Assessment*, Ankeny, Iowa, Soil Conservation Society of America, 8–17.

254 LAND DEGRADATION AND SOCIETY

Bernstein, H. (1979) 'African peasantries: a theoretical framework', *Journal of Peasant Studies*, 6, 4, 420–44.

Berry, L. and Townshend, J. (1972) 'Soil conservation policies in the semi-arid regions of Tanzania, an historical perspective', *Geografiska Annaler* 54A, 241–53.

Berry, L. and Townshend, J. (1973) 'Soil conservation practices in Tanzania: an historical perspective', in Rapp, A., Berry, L. and Temple, P. (eds) *Studies of Soil Erosion and Sedimentation in Tanzania*, Research Monograph No. 1, BRALUP, Tanzania, University of Dar es Salaam, 241–53.

Best, S. (1985) 'Lakeba: the Prehistory of a Fijian Island', unpublished PhD thesis in Prehistory, New Zealand, Auckland, University of Auckland.

Bidwell, O.W. and Hole, F.D. (1965), 'Man as a factor of soil formation', *Soil Science* 99, 65–72.

Bienefeld, M.A. (1984) *Fiji Employment and Development Mission: Final Report to the Government of Fiji*, Parliament of Fiji, Parliamentary Paper no. 66 of 1984, Suva, Government Printer.

Birkenhauer, J. (1980) 'Rezente Bodenerosion und periglaziale Vorgänge', *Geographische Rundschau*, 32, 488–96.

Birmingham, D. and Martin, P.M. (eds) (1983) *History of Central Africa*, vol. 2, London, Longman.

Blaikie, P.M. (1979) 'Poor peasants', in Seddon J.D., Blaikie P.M. and Cameron J. (eds), *Peasants and Workers in Nepal*, Warminster, Aris and Phillips, 48–75.

Blaikie, P.M. (1985a) *The Political Economy of Soil Erosion in Developing Countries*, London, Longman.

Blaikie, P.M. (1985b) 'Soil slides south', *Inside Asia*, 2, Feb./March, 45–7.

Blaikie, P.M. (1985c) 'Natural resources and social change', Unit 7, II (Analysis: aspects of the geography of society), in Open University, *Changing Britain, Changing World: Geographical Perspectives* (course D 205), Milton Keynes, The Open University.

Blaikie, P.M. (1986) 'Natural resource use in developing countries', in Johnston, R.J. and Taylor, P.J. (eds) *A World in Crisis?, Geographical Perspectives*, Oxford, Blackwell, 105–26.

Blaikie, P.M., Cameron, J., Fleming, R. and Seddon, J.D. (1977) *Centre, Periphery and Access: Social and Spatial Relations of Inequality in West-Central Nepal*, Norwich, Development Studies Monograph No. 5.

Blaikie, P.M., Cameron, J. and Seddon, J.D. (1979) *The Struggle for Basic Needs in Nepal*, Paris, Development Centre of the Organization for Economic Cooperation and Development.

Blaikie, P.M., Cameron, J. and Seddon, D. (1980) *Nepal in Crisis: Growth and Stagnation of the Periphery*, New Delhi and London, Oxford University Press.

Blaikie, P.M., Harriss, J.C. and Pain, A. (1985) *Public Policy and the Utilisation of Common Property Resources in Tamil Nadu, India*. Report to Overseas Development Administration, Research Scheme R3988. Also reproduced in shorter form as 'The management and use of common property resources in Tamil Nadu, India', paper for Common Property Resource Management Conference, Annapolis, Maryland, 21–27 April 1985, National Academy of Sciences, USA.

Blaikie, P.M. and Seddon, J.D. (1978) 'A map of the Nepalese political economy', *Area*, 10, 1, 30–1.

Blanco, H. (1972) *Land or Death: The Peasant Struggle in Peru* (translated by Allen, N. from an original text in Spanish) New York, Pathfinder Press.

Bleeker, P. (1983) *Soils of Papua New Guinea*, Canberra, CSIRO and ANU Press.

Blyth, M.J. and Kirby, M.G. (1984) *The Impact of Government Policy on Land Degradation in the Rural Sector*, working paper, Canberra, Bureau of Agricultural Economics.

Body, R. (1982) *Agriculture: The Triumph and the Shame*, London, Temple Smith.

Bohannan, P. (1964) *Africa and the Africans*, London, Natural History Press.

Bolline, A. (1979) 'L'érosion en région limoneuse; ses causes, ses conséquences', in Vogt, H. and Vogt, T. (eds) *Colloque sur L'Erosion Agricole des Sols en Milieu Tempéré non-Mediterranéen, Strasbourg-Colmar, 20–23 September 1979*, Strasbourg, Laboratoire de Géographie, 95–100.

Bolton, G. (1981) *Spoils and Spoilers*, Sydney, Allen and Unwin.

Borah, W. and Cook, S.F. (1963) *The Aboriginal Population of Central Mexico on the Eve of the Spanish Conquest*, Ibero-Americana: 45, Berkeley and Los Angeles, University of California Press.

Bork, H.-R. (1983) 'Die holozäne Relief- und Bodenentwicklung in Lössgebieten: Beispiele aus dem südöstlichen Niedersachsen', in Bork, H.-R. and Ricken, W. *Bodenerosion, Holozaene und Pleistozaene Bodenentwicklung*, Catena Supplement 3, Braunschweig, Catena Verlag, 1–93.

Bork, H.-R. (1986) 'Pedogenesis and soil erosion during younger Holocene in Southern Lower Saxony (W. Germany)', paper presented at the XIII Congress of the International Society of Soil Science in Hamburg, W. Germany, 13–20 August 1986.

Boserup, E. (1965) *The Conditions of Agricultural Growth: The Economics of Agrarian Change Under Population Pressure*, Chicago, Aldine.

Boserup, E. (1970) 'Present and potential food production in developing countries', in Zelinsky, W., Kosinski, L.A. and Prothero, R.M. (eds) *Geography and a Crowding World: A Symposium on Population Pressures upon Physical and Social Resources in the Developing Lands*, New York, Oxford University Press, 100–110.

Boserup, E. (1981) *Population and Technology*, Oxford, Blackwell.

Boulet, J. (1975) 'Ma gourmaz, pays Mafa (Nord-Cameroun)', *Atlas de Structures au Sud du Sahara*, 11, Paris, ORSTOM.

Boutrais, J. (1973) 'La colonisation des plaines par les montagnards du nord du Cameroun', Travaux et documents de l'ORSTOM, 24, Paris, ORSTOM.

Bowers, J.K. and Cheshire, P. (1983) *Agriculture, the Countryside and Land Use*, London, Methuen.

Bowers, N. (1968) 'The ascending grasslands: an anthropological study of ecological succession in a high mountain valley of New Guinea', unpublished PhD thesis in Anthropology, New York, Columbia University.

Bowler, I.R. (1983) 'Structural change in agriculture', in Pacione, M. (ed.) *Progress in Rural Geography*, London, Croom Helm, 46–73.

Braudel, F. (1972–73) *The Mediterranean and the Mediterranean World in the Age of Philip II*, translated from the French by Reynolds, S., London, Collins.

Brett, E.A. (1973) *Colonialism and Underdevelopment in East Africa: the Politics of Economic Change 1919–1939*, London, Heinemann.

Bridgman, H.A. (1983) 'Could climatic change have had an influence on the Polynesian migrations?', *Palaeogeography, Palaeoclimatology, Palaeoecology*, 41, 193–206.

Bromfield, S.M., Cumming, R.W., David, D.J. and Williams, C.H. (1983) 'Change in soil pH, manganese and aluminium under subterranean clover pasture', *Australian Journal of Experimental Agriculture and Animal Husbandry*, 23, 181–91.

Bromley, D.W. and Chapagain, D.P. (1985) 'The village against the center: resource depletion in South Asia', 15 pp, mimeo, Department of Agricultural Economics, University of Wisconsin-Madison and Ministry of Agriculture, HMG, Kathmandu.

Brookfield, H.C. (1972) 'Intensification and disintensification in Pacific agriculture: a theoretical approach', *Pacific Viewpoint*, 13, 30–48.

Brookfield, H.C. (1973) 'Full circle in Chimbu: a study of trends and cycles', in Brookfield, H.C. (ed.) *The Pacific in Transition: Geographical Perspectives on Adaptation and Change*, London, Edward Arnold, 127–60.

Brookfield, H.C. (1984a) 'Intensification revisited', *Pacific Viewpoint*, 25, 15–44.

Brookfield, H.C. (1984b) 'Report on a Mission to advise the Nepal National Committee for Man and the Biosphere, 28 December 1983–21 January 1984', Kathmandu, National Commission for Unesco, mimeo.

Brookfield, H.C. and Brown, P. (1963) *Struggle for Land: Agriculture and Group Territories among the Chimbu of the New Guinea Highlands*, Melbourne, Oxford University Press.

Brookfield, H.C. and White, J.P. (1968) 'Revolution or evolution in the prehistory of the New Guinea Highlands: a seminar report', *Ethnology*, 7, 43–52.

Brookfield, M. (1979) 'Resource-use, economy and society: island at the cross-roads', in Brookfield, H.C. (ed.) *Lakeba: Environmental Change, Population Dynamics and Resource Use*, UNESCO/UNFPA Fiji Island Reports 5, Canberra, ANU for UNESCO, 127–97.

Brown, L.R. (1978) *The Worldwide Loss of Cropland*, Worldwatch Paper 24, Washington DC, Worldwatch Institute.

Brown, L.R. (1981) 'Eroding the base of civilisation', *Journal of Soil and Water Conservation*, October, 36, 255–60.

Brown, L.R. (1982) *US and Soviet Agriculture: The Shifting Balance of Power*, Worldwatch Paper 51, Washington DC, Worldwatch Institute.

Burbridge, P., Dixon, J. and Soewardi, B. (1980) 'Forestry and agricultural options for resource allocation in choosing lands for transmigration development', *Applied Geography*, 7, 237–58.

Burke, A.E. (1956) 'Influence of man upon Nature – the Russian view', in Thomas, W.L. (ed.), *Man's Role in Changing the Face of the Earth*, Chicago, Ill., University of Chicago Press, 1035–51.

Burkill, I.H. (1910) 'Notes from a journey to Nepal', *Records of the Botanical Survey of India*, 4, Calcutta, Superintendent of Government Printing, 59–140 (separately issued).

Burns, A., Watson, T.Y. and Peacock, A.T. (1960) *Report of the Commission of Enquiry into the Natural Resources and Population Trends of the Colony of Fiji 1959*, Council Paper No. 1 of 1960, Suva, Legislative Council of Fiji.

Burrin, P.J. and Scaife, R.G. (1984) 'Aspects of Holocene valley sedimentation and floodplain development in southern England', *Proceedings of the Geological Association*, 95, 81–96.

Bush, K. (1974) 'The Soviet response to environmental disruption', in Volgyes, I. (ed.) *Environmental Deterioration in the Soviet Union and Eastern Europe*, New York, Praeger, 8–36.

Caine, N. and Mool, P.K. (1982) 'Landslides in the Kolpu Khola drainage, middle mountains, Nepal', *Mountain Research and Development*, 2, 157–73.

Campbell, K. (1982) 'Land policy', in Williams, D.B. (ed.) *Agriculture in the*

Australian Economy, Sydney, Sydney University Press.

Caplan, L. (1970) *Land and Social Change in East Nepal*, London, Routledge and Kegan Paul.

Carlstein, T. (1982) *Time, Resources and Ecology: On the Capacity for Human Interaction in Space and Time in Preindustrial Societies*, Lund Studies in Geography, Series B, Human Geography, No. 49, London, Allen and Unwin for the Royal University of Lund.

Carson, B. (1985) *Erosion and Sedimentation Processes in the Nepalese Himalaya*, ICIMOD occasional paper no. 1, Kathmandu, International Centre for Integrated Mountain Development (ICIMOD).

Carson, R.L. (1962) *Silent Spring*, Boston, Houghton Mifflin.

Casanova, P.G. (1970) *Democracy in Mexico* (translated by Salti, D. from *La Democracia en Mexico* (Mexico City, Ediciones Era (1965)) New York, Oxford University Press.

Cassen, R.H. (1976) 'Population and development: a survey', *World Development*, 4, 785–830.

Central Committee of the Chinese Communist Party (1979) *Zhongfa 4. Issues and Studies*, 15, 105–6.

Central Planning Office, Fiji (1985) *Fiji's Ninth Development Plan 1986–1990*, Suva, Government Printer.

Centre for Continuing Education (CCE) (1980) *Australia's Marginal Lands*, Proceedings of the Wodonga Conference, November 1980, Canberra, Australian National University.

Centre for Science and Environment (1982, 1985) *The State of India's Environment: A Citizen's Report*, Delhi.

Chamala, S. and Coughenour, M. (1982) 'Using farming systems research to strengthen the soil conservation education program in Darling Downs, Australia', *Tillage Systems and Social Science*, 2, 2, 1–4.

Chamala, S., Keith, K.J. and Quinn, P. (1982) *Adoption of Commercial and Soil Conservation Innovations in Queensland*, Brisbane, Department of Agriculture and the University of Queensland.

Chambers, R. (1984) *Rural Development: Putting the Last First*, London, Longman.

Champion, A.G. (1983) 'Land use and competition', in Pacione, M. (ed.) *Progress in Rural Geography*, London, Croom Helm, 21–45.

Charley, J.L. (1983) 'Tropical highland agricultural development in a monsoonal climate: the utilization of *Imperata* grassland in Northern Thailand', *Mountain Research and Development*, 3, 389–96.

Chevalier, F. (1963) *Land and Society in Colonial Mexico: the Great Hacienda* (translated by Eustis, A. from *La Formation des Grands Domaines au Mexique*, Paris, Institut d'Ethnologie (1952)) Berkeley and Los Angeles, University of California Press.

Chorley, R.J. (1973) 'Geography as human ecology', in Chorley, R.J. (ed.) *Directions in Geography*, London, Methuen, 155–69.

Clark, C. (1940) *The Conditions of Economic Progress*, London, Macmillan.

Clarke, W.C. (1977a) 'A change of subsistence staple in prehistoric New Guinea', in Leakey, C.L.A. (ed.) *Proceedings of the Third Symposium of the International Society for Tropical Root Crops*, Ibadan, 159–63.

Clarke, W.C. (1977b) 'The structure of permanence: the relevance of self-subsistence communities for world ecosystem management', in Bayliss-Smith, T. P. and

Feachem, R. (eds) *Subsistence and Survival: Rural Ecology in the Pacific*, London, Academic Press, 363–84.

Clarkson, J.D. (1968) *The Cultural Ecology of a Chinese Village: Cameron Highlands, Malaysia*, Research Paper 114, Chicago, University of Chicago, Department of Geography.

Clay, C.G.A. (1984) *Economic Expansion and Social Change: England 1500–1700, vol 1, People, Land and Towns*, Cambridge, Cambridge University Press.

Clay, E.J. and Schaffer, B.P. (eds) (1984) *Room for Manoeuvre: An Exploration of Public Policy Planning in Agricultural and Rural Development*, London, Heinemann.

Clayton, E.S. (1964) *Agrarian Development in Peasant Economies*, Oxford, Pergamon Press.

Clout, H.D. (1977) 'Agricultural changes in the eighteenth and nineteenth centuries', in Clout, H.D. (ed.) *Themes in the Historical Geography of France*, London, Academic Press, 407–46.

Coase, R. (1960) 'The problem of social cost', *Journal of Law and Economics*, 3, 1–44.

Coates, K. (ed) (1972) *Socialism and the Environment*, Nottingham, Spokesman.

Cochrane, G.R. (1969) 'Problems of vegetation change in western Viti Levu, Fiji', in Gale, F. and Lawton, G.H. (eds) *Settlement and Encounter: Geographical Studies Presented to Sir Grenfell Price*, Melbourne, Oxford University Press.

Cohen, B.J. (1973) *The Question of Imperialism: The Political Economy of Dominance and Dependence*, New York, Basic Books Inc.

Collier, J.V. (1928/1976) 'Forestry in Nepal', in Landon, P. (ed.) *Nepal*, reprinted in Kuløy, H.K. (ed.) *Bibliotheca Himalayica*, series 1, vol. 16, Kathmandu, Ratna Pustak Bhandar, 251–55 (first printed, 1928).

Commoner, B. (1972) *The Closing Circle: Confronting the Environmental Crisis*, London, Jonathan Cape.

Commonwealth of Australia (1982–4) 'Senate Standing Committee on Science, Technology and the Environment inquiry into land use policy', *Official Hansard Report*, 1–2430.

Condon, R.W. (1978) 'Land tenure and desertification in Australia's arid lands', *Search*, 9, 261–4.

Conklin, H.C. (1954) 'An ethnoecological approach to shifting agriculture', *Transactions of the New York Academy of Sciences*, 77, 133–42.

Cook, K. (1982) 'Soil loss: a question of values', *Journal of Soil and Water Conservation*, 37, 89–92.

Cook, K. (1985) 'The 1985 farm bill: a turning point for soil conservation', *Journal of Soil and Water Conservation*, 40, 218–20.

Cook, S.F. (1949) *Soil Erosion and Population in Central Mexico*, Ibero-Americana: 34, Berkeley and Los Angeles, University of California Press.

Cool, J.C. (1983) *Factors Affecting Pressure on Mountain Resource Systems*, First International Symposium and Inauguration; Mountain Development 2000, Challenges and Opportunities, Kathmandu, International Centre for Integrated Mountain Development, mimeo.

Corbel, J. (1964) 'L'érosion terrestre, étude quantitative (méthodes–techniques–résultats)', *Annales de Géographie*, 73, 385–412.

Costin, A.B. (1984) 'Submission to the Senate Standing Committee on Science, Technology, and the Environment inquiry into land use policy', *Official Hansard Report*, 2397–430.

Coulson, A. (1981) 'Agricultural policies in mainland Tanzania 1946–76', in Heyer, J., Roberts, P. and Williams, G. (eds) *Rural Development in Tropical Africa*, London, Macmillan, 52–89.

Council on Environmental Quality (1982) *The Global 2000 Report to the President: entering the Twenty-first Century*, Harmondsworth, Penguin.

Craig, R.A., Smith N.M. and Sheahan, B.T. (1983) *Landholders and Native Vegetation: Attitudes to Retention and Clearing*, South Australia, Roseworthy Agricultural College.

Crittenden, R. (1982) 'Sustenance, seasonality and social cycles on the Nembi Plateau, Papua New Guinea', unpublished PhD thesis in Geography, Canberra, Australian National University.

Crittenden, R. (1984) 'Pigs, women, colonialism and climate', in Allen, B.J. (ed.) *Agricultural and Nutritional Studies on the Nembi Plateau, Southern Highlands*, Department of Geography Occasional Paper (New Series) 4, Waigani, University of Papua New Guinea and Southern Highlands, Regional Development Plan, 121–72.

Crush, J.W. (1980) 'The genesis of colonial land policy in Swaziland', *South African Geographical Journal*, 62, 73–88.

CSIRO (Commonwealth Scientific and Industrial Research Organisation) (1985) Press Release, 30 September, Canberra, Division of Plant Industry, CSIRO.

Davidson, B. (1981) *European Farming in Australia: an Economic History of Australian Farming*, Amsterdam, Elsevier.

Davidson, D.A. (1980) 'Erosion in Greece during the first and second milennia BC', in Cullinford, R.A., Davidson, D.A. and Lewin, J. (eds) *Timescales in Geomorphology*, Chichester, John Wiley, 143–58.

de Boodt, M. and Gabriels, D. (1979) 'Mechanics and control of losses from sensitive soils in a temperate region (Belgium)', in Vogt, H. and Vogt T. (eds) *Colloque sur l'Erosion Agricole en Milieu Tempéré non-Mediteranéen, Strasbourg-Colmar, 20–23 September 1979*, Strasbourg, Laboratoire de Géographie, 165–68.

DEHCD (Department of Environment, Housing and Community Development) (1978) *A Basis for Soil Conservation Policy in Australia*, Canberra, Australian Government Printing Service.

Department of Environment and Planning (South Australia) (n.d.) *Native Vegetation Clearance Controls*, Adelaide, Government Printer.

Department of Primary Industry (1984) 'National soil conservation program: background paper for conservation and the economy conference', Sydney, NSW, Government Printer.

de Wilde, J.C. (1967) *Experiences with Agricultural Development in Tropical Afrcia*, 2 vols, Baltimore, Johns Hopkins University Press.

Dick, A. (1985) 'Carr defends government subdivision policy', *The Land*, 6 June, 15.

Dickinson, R., Ellis, A.K., Eyvind, F., Fisher, R., Irwin, J., Scarborough, M. and Schmidt, R. (1981) 'Soil conservation problems and practices: a farmer view', *Journal of Soil and Water Conservation*, 36, 186–93.

Dinham, B. and Hines, C. (1983) *Agribusiness in Africa*, London, Earth Resource Research Ltd.

Doran, M.H., Low, A.R.C. and Kemp, R.L. (1979) 'Cattle as a store of wealth in Swaziland: implications for livestock development and overgrazing in eastern and southern Africa', *American Journal of Agricultural Economics*, 61, 41–7.

Dove, M.R. (1980) 'Development of tribal land rights in Borneo', *Borneo Research*

Bulletin, 12, 3–19.

Dove, M.R. (1984) 'Government versus peasant beliefs concerning *Imperata* and *Eupatorium*: a structural analysis of knowledge, myth, and agricultural ecology in Indonesia', Honolulu, Environment and Policy Institute, East–West Center, 39 pp, mimeo.

Dove, M.R. (1985) 'The agroecological mythology of the Javanese and the political economy of Indonesia', *Indonesia*, 39, 1–36.

Dudal, R. (1982) 'Land degradation in a world perspective', *Journal of Soil and Water Conservation*, 37, 245–9.

Dumsday, R.G. (1983) 'Agricultural resource management', *Australian Journal of Agricultural Economics*, 27, 157–63.

Dumsday, R.G. and Edwards, G.W. (1984) 'Economics of conservation in agriculture', in Gilmour, D. *et al.* (eds) *Agriculture and Conservation – Achieving a Balance*, Proceedings of a Conference held at Wodonga, 10–11 September 1984, Benalla, Victoria, Institute of Agricultural Science, 58–64.

Dunne, T., Dietrich, W.E. and Brunengo, M.J. (1978) 'Recent and past erosion rates in semi-arid Kenya', *Zeitschrift für Geomorphologie Suppl.*, 29, 130–40.

Earle, T.R., Brownlea, A.A. and Rose, C.W. (1981) 'Beliefs of a community with respect to environmental management: a case study of soil conservation beliefs on the Darling Downs', *Journal of Environmental Management*, 12, 197–219.

Eckholm, E.P. (1976) *Losing Ground: Environmental Stress and World Food Prospects*, New York, Norton.

Edwards, G. and Lumley, S. (1985) 'Dryland salting: conceptual characterization and policy preamble', paper presented to the 29th Annual Conference of the Australian Agricultural Economics Society, 12–14 February, 1985, Armidale, University of New England (unpublished).

Ehrlich, P.R. and Ehrlich, A.H. (1970) *Population. Resources, Environment: Issues in Human Ecology*, San Francisco, W.H. Freeman.

Ellenberger, D. (1969) *History of the Basuto: Ancient and Modern*, Reprint of 1912 ed., New York, Negro University Press, Greenwood, Westport, Conn.

Ellis, F. (1985) 'Employment and incomes in the Fiji sugar economy', in Brookfield, H.C., Ellis, F. and Ward, R.G., *Land, Cane and Coconuts: Papers on the Rural Economy of Fiji*, Department of Human Geography Publication 17, Canberra, Australian National University, 65–110.

Elwell, H.A. and Stocking, M.A. (1982) 'Developing a simple yet practical method of soil loss estimation', *Tropical Agriculture*, 59, 43–8.

Ensenberger, H.M. (1974) 'A critique of political economy', *New Left Review*, 8, 4, 3–32.

Escobar, G. and Beal, C.M. (1982) 'Contemporary patterns of migration in the Central Andes', *Mountain Research and Development*, 2, 63–80.

FAO (Food and Agriculture Organization of the United Nations) (1974) 'Shifting cultivation and soil conservation in Africa', *Soils Bulletin 24*, Rome, FAO.

FAO (Food and Agriculture Organization of the United Nations) (1977) 'Soil conservation and management in developing countries, *Soils Bulletin 33*, Rome, FAO

FAO (Food and Agriculture Organization of the United Nations) (1979) *A Provisional Methodology for Soil Degradation Assessment* (accompanied by mapping of North Africa), Rome, FAO.

FAO (Food and Agriculture Organization of the United Nations) (1980) *Natural*

Resources and the Human Environment for Food and Agriculture, environment paper no. 1, Rome, FAO.

FAO (Food and Agriculture Organization of the United Nations) (1982) *Potential Population Supporting Capacities of Lands in the Developing World*, technical report on FAO/UNFPA project INT 75/P13, Rome, FAO.

Feder, E. (1977) 'Agribusiness and the elimination of the Latin American rural proletariat', *World Development*, 5, 5–7.

Fel, A. (1977) 'Petite culture 1750–1850', in Clout, H.D. (ed.) *Themes in the Historical Geography of France*, London, Academic Press, 215–46.

Festy, O. (1947) *L'Agriculture pendant la Révolution Française: Les Conditions de Production et de Récolte des Céréales, Etude d'Histoire Economique, 1789–1795*, Paris, Gallimard.

Fiji Sugar Corporation (n.d.) *Field Manual*, Suva, Fiji Sugar Corporation Limited.

Fisher, A.G.B. (1935) *The Clash of Progress and Security*, London, Macmillan.

Flenley, J.R. and King, S.M. (1984) 'Late quaternary pollen records from Easter Island', *Nature*, 307 (5946), 47–50.

Flohn, H. (1949/50) 'Klimaschwankungen im Mittelalter und ihre historisch-geographische Bedeutung', *Berichte zur Deutsche Landeskunde*, 7, 347–57.

Forster, R. (1970) 'Obstacles to agricultural growth in eighteenth-century France', *American Historical Review*, 75, 1600–15.

Fosbrooke, H. and Young, R. (1976) *Land and Politics among the Lugurn of Tanganyika*, London, Routledge and Kegan Paul.

Foster, G.R., Meyer, L.D. and Onstad, C.A. (1977) 'An erosion equation derived from basic erosion principles', *Transactions of the American Society of Agricultural Engineers*, 20, 678–82.

Fournier, F. (1960) *Climat et Erosion*, Paris, Presses Universitaires de France.

Fournier, F. (1967), 'Research on soil erosion in Africa', *African Soils*, 12, 53–96.

Fowler, R.J. (1985) 'Property rights of farmers', paper delivered to Australian Farm Management Society, 11th National Conference, Roseworthy Agricultural College, 18 February (unpublished).

France, P. (1969) *The Charter of the Land: Custom and Colonization in Fiji*, Melbourne, Oxford University Press.

Frank, A.G. (1980) *Crisis in the Third World*, London, Heinemann.

Franke, R. and Chasin, B.H. (1980) *Seeds of Famine: Ecological Destruction and the Development Dilemma in the Western Sahel*, New Jersey, Allenheld, Osmun and Co.

Franke, R. and Chasin, B.H. (1981) 'Peasants, peanuts, profits and pastoralists', *The Ecologist*, 11, 4, 156–68.

French, R.A. (1973) 'Conservation and pollution in the USSR', *The Geographical Journal*, 139, 521–4.

Furnivall, J.S. (1939) *Netherlands India: A Study of Plural Economy*, Cambridge, Cambridge University Press.

Furon, R. (1947) *L'Erosion du Sol*, Paris, Payot.

Gabriels, D., Pauwels, J.M. and de Boodt, M. (1977) 'A quantitative rill erosion study on a loamy sand in the hilly region of Flanders', *Earth Surface Processes and Landforms*, 2, 257–59.

Garcia, R.V. and Escudero, J.C. (1982) *The Constant Catastrophe: Malnutrition, Famines and Drought*, Oxford, Pergamon Press.

Gardner, J. (1969) 'Observation of surficial talus movement', *Zeitschrift für*

Geomorphologie, 13, 317–23.

Geertz, C. (1963) *Agricultural Involution: the Process of Ecological Change in Indonesia*, Berkeley and Los Angeles, University of California Press.

Gerlach, J.C. (1938) 'Bevolkingsmethoden van ontginning van alang-alang-terreinen in de Zuider en Oosterafdeeling van Borneo', *Landbouw*, XIV, 7, 446–50.

Gillman, G.P. (1984) 'Nutrient availability in acid soils of the tropics following clearing and cultivation', in ACIAR (Australian Centre for International Agricultural Research), *Proceedings of the International Workshop on Soils: Research to Resolve Selected Problems of Soils in the Tropics*, Townsville, Queensland, Australia, 12–16 September 1983, Canberra, ACIAR, 39–46.

Gilmour, D., Hamer, I. and Bourchier, J. (eds) (1984) 'Agriculture and conservation achieving a balance', *Proceedings of a Conference held at Wodonga*, 10–11 September, 1984, Benalla, Victoria, Australian Institute of Agricultural Science.

Ginting, M. and Daroesman, R. (1982) 'An economic survey of North Sumatra', *Bulletin of Indonesian Economic Studies*, 8, 52–83.

Glaser, G. (1983) 'Unstable and vulnerable ecosystems: a comment based on MAB research in island ecosystems', *Mountain Research and Development*, 3, 121–23.

Glass, D.V. (ed.) (1953) *Introduction to Malthus*, London, Watts.

Glover, Sir H. (1946) *Erosion in the Punjab: its Causes and Cure*, Lahore, Feroz Printing Works.

Gluckman, M. (1941) *The Economy of the Central Barotse Plain*, Rhodes–Livingstone papers, no. 7, Manchester, Manchester University Press for Rhodes–Livingstone Institute.

Goh, K.C. (1982) 'Land use and soil erosion problems in Malaysia', in Consumers' Association of Penang, *Development and the Environmental Crisis: Proceedings of the Symposium 'The Malaysian Environment in Crisis*, 16–20 September 1978, Penang, Malaysia, Consumers' Association of Penang, 109–19.

Goldman, M.I. (1972) *The Spoils of Progress: Environmental Pollution in the Soviet Union*, Cambridge, Mass., MIT Press.

Goldschmidt, W. (1978) 'Large-scale farming and the rural social structure', *Rural Society*, 43, 362–66.

Goldstein, M.C., Ross, J.L. and Schuler, S. (1983) 'From a mountain–rural to a plains–urban society: implications of the 1981 Nepalese census', *Mountain Research and Development*, 3, 61–4.

Golson, J. (1977) 'No room at the top: agricultural intensification in the New Guinea highlands', in Allen, J., Golson, J. and Jones, R. (eds) *Sunda and Sahul: Prehistoric Studies in Southeast Asia, Melanesia and Australia*, London, Academic Press, 601–38.

Golson, J. (1981) 'New Guinea agricultural history: a case study', in Denoon, D. and Snowden, C. (eds) *A Time to Plant and A Time to Uproot: A History of Agriculture in Papua New Guinea*, Port Moresby, Institute of Papua New Guinea Studies for the Department of Primary Industry, 55–64.

Golson, J. (1982) 'The Ipomoean revolution revisited: society and the sweet potato in the upper Wahgi valley', in Strathern, A. (ed.), *Inequality in New Guinea Highlands Societies*, Cambridge, Cambridge University Press, 109–36.

Goodland, R.J. and Irwin, H.S. (1975) *Amazon Jungle: Green Hell to Red Desert?*, Amsterdam, Elsevier.

Goveia, E.V. (1969) *Slave Society in the British Leeward Islands at the End of the Eighteenth Century*, New Haven, Yale University Press.

Grandstaff, T. (1978) 'The development of swidden agriculture (shifting cultivation)', *Development and Change*, 9, 547–79.

Grantham, G.W. (1980) 'The persistence of open-field farming in nineteenth-century France', *Journal of Economic History*, 40, 515–31.

Gras, F. (1979) 'L'érosion des sols "lessivés" de Lorraine et son incidence sur les projets de remembrement rural', in Vogt, H. and Vogt T. (eds) *Colloque sur L'Erosion Agricole des Sols en Milieu Tempéré non-Mediteranéen*, *Strasbourg-Colmar*, 20–23 September 1979, Strasbourg, Laboratoire de Géographie, 89–94.

Greenland, D. and Lal, R. (eds) (1977) *Soil Conservation and Management in the Humid Tropics*, Chichester, John Wiley.

Gregory, K.J. and Walling, D.E. (1973) *Drainage Basin Form and Process: A Geomorphological Approach*, London, Edward Arnold.

Guinness, P. (ed.) (1977) *Transmigrants in South Kalimantan and South Sulawesi: Inter-island Government Sponsored Migration in Indonesia*, Yogyakarta, Indonesia, Population Institute, Gadjah Mada University.

Gupta, A. (1984) 'Managing common properties: some issues of institutional design', Vastrapur, Ahmedabad, India, Centre for Management in Agriculture, Indian Institute of Management, 12 pp, mimeo.

Gupta, A. (1985) 'The socio-ecology of stress: why do CPR management projects fail?', paper for Common Property Resource Management Conference, Annapolis, Maryland, 21–27 April 1985, National Academy of Sciences, USA.

Gurung, H. (1982) *The Himalaya: Perspective on Change*, Kathmandu, New Era.

Gurung, H. (forthcoming) *Regional Patterns of Migration in Nepal, 1952/54–1981*, Honolulu, East–West Center.

Gutman, P. (1985) 'Teona economica y problematica ambrento: un diologo difical', *Desarrollo Economico: Revista de Ciencias Sociales*, April–June, 25 (97), 47–70, with English summary.

Hall, P. (1966) *Von Thunen's Isolated State*, an English edition of *Der Isolierte Staat* by Johann Heinrich von Thunen, translated by Warenberg, C.M., edited with an introduction by Peter Hall, London, Pergamon.

Hamer, W.I. (1980, 1982) 'Soil conservation consultant reports', Center for Soil Research, Bogor (Ministry of Agriculture, Government of Indonesia and FAO).

Hanson, N.R. (1973) 'Observation', in Grandy, R.E. (ed.) *Theories and Observation in Science*, Englewood Cliffs, NJ, Prentice-Hall, 129–46.

Hard, G. (1970) 'Excessive Bodenerosion um und nach 1800', *Erdkunde*, 24, 290–308.

Hardin, G.J. (1977) 'The tragedy of the commons', in Hardin, G.J. and Baden, J. (eds) *Managing the Commons*, San Francisco, W.H. Freeman, 16–30.

Hardin, G.J. and Baden, J. (eds) (1977) *Managing the Commons*, San Francisco, W.H. Freeman.

Hardjono, J.M. (1977) *Transmigration in Indonesia*, Kuala Lumpur, Oxford University Press.

Hare, F.K. (1977) 'Connections between climate and desertification', *Environmental Conservation*, 4, 81–90.

Harris, D.R. and Vita-Finzi, C. (1968) 'Kokkinopilos – a Greek badland', *Geographical Journal*, 134, 537–45.

Hart, D.M. (1976) *The Aït Waryaghan of the Moroccan Rif: An Ethnography and History*, Viking Fund Publications no. 55, Tempe, University of Arizona Press.

Hartmann, H. (1976) 'Capitalism, patriarchy and job segregation by sex', in Blaxall,

M. and Reagan, B. (eds), *Women and the Workplace*, Chicago, University of Chicago Press, 137–69.

Headrick, D.R. (1981) *The Tools of Empire: Technology and European Imperialism in the Nineteenth Century*, New York, Oxford University Press.

Healey, C.J. (1978) 'The adaptive significance of systems of ceremonial exchange and trade in the New Guinea Highlands', *Mankind*, 11, 198–207.

Heathcote, R.L. (1965) *Back of Bourke: A Study of Land Appraisal and Settlement in Semi-arid Australia*, Melbourne, Melbourne University Press.

Heathcote, R.L. (1969) 'Drought in Australia: a problem of perception', *Geographical Review*, 59, 175–94.

Heimann, E. (1945) *History of Economic Doctrines: An Introduction to Economic Theory*, New York, Oxford University Press.

Hellen, J.A. (1968) *Rural Economic Development in Zambia, 1890–1964*, Munich, African Studies Centre of the Ifo Institute for Economic Research.

Hely, R. (1985) 'Change and challenge: a lean new elite is emerging to run a much tougher agribusiness', *Inside Australia* (Magazine of the Rural Development Centre, Armidale, NSW, Australia), 1 (2), 21–3.

Hempel, L. (1968) 'Bodenerosion in Süddeutschland; Erläuterungen zu Karten von Baden-Württemberg, Bayern, Hessen, Rheinland-Pfalz und Saarland', *Forschungen zur Deutschen Landeskunde*, 179, Bad Godesberg, Bundes-forschungsanstalt für Landeskunde und Raumordnung.

Henin, S. (1979) 'L'érosion liée à l'activité agricole en France', in Vogt, H. and Vogt, T. (eds) *Colloque: Erosion Agricole des Sols en Milieu Tempéré non Méditerranéen, Strasbourg-Colmar*, 20–23 September 1979, Strasbourg, Laboratoire de Géographie Physique en Milieu Tempéré, 9–12.

Hewitt, K. (1983) 'The idea of calamity in a technocratic age', in Hewitt, K. (ed.) *Interpretations of Calamity, from the Viewpoint of Human Ecology*, Boston, Allen and Unwin, 3–32.

Heyer, J., Roberts, P. and Williams, G. (1981) *Rural Development in Tropical Africa*, London, Macmillan.

Higgins, G.M., Kassam, A.H. and Shah, M. (1984) 'Land, food and population in the developing world', *Nature and Resources*, 20, 2–10.

Higgs, E.S. and Vita-Finzi, C. (1966) 'The climate, environment and industries of stone age Greece, Part II', *Proceedings of the Prehistoric Society*, 32, 1–29.

Higgs, E.S., Vita-Finzi, C., Harris, D.R. and Fagg, A.E. (1967) 'The climate, environment and industries of stone age Greece, Part III', *Proceedings of the Prehistoric Society*, 33, 1–29.

Hill, D.G. (1984) 'Member survey on priorities', *Australian Conservation Foundation Newsletter*, 16, 10, 2.

Hiraoka, M. and Yamamoto, S. (1980) 'Agricultural development in the upper Amazon of Ecuador', *Geographical Review*, 70, 423–45.

Hoffman, P.T. (1982) 'Sharecropping and investment in agriculture in early modern France', *Journal of Economic History*, 42, 155–59.

Holling, C.S. (1978), *Adaptive Environmental Assessment and Management*, Chichester, John Wiley.

Homewood, K. (1985) 'Swamp grazing in Baringo', paper read at the African Studies Association of the UK Symposium, 18 September 1985, London, University College.

Hooze, J.A. (1893) 'Topografische, geologische, mineralogische en mijnbouwkundige beschrijving van een gedeelte der afdeling Martapoera in de residentie Zuider- en Oosterafdeling van Borneo', *Jaarboek van het Mijnwezen in Nederlandsch-Indie*, Batavia, part 2, 1–431.

Hopkins, A. (1973) *An Economic History of West Africa*, London, Longman.

Hudson, N.W. (1971) *Soil Conservation*, London, Batsford.

Hudson, N.W. (1981) 'Non-technical constraints on soil conservation', in Tingsanchali, T. and Eggers, H. (eds) *South East Asian Regional Symposium on Problems of Soil Erosion and Sedimentation*, Bangkok, Asian Institute of Technology, 15–26.

Hudson, N.W. (1983) 'Soil conservation strategies in the Third World', *Journal of Soil and Water Conservation*, 38, 446–50.

Hudson, N.W. (1985) 'A world view of the development of soil conservation', *Agricultural History*, 59, 327–39.

Hufschmidt, M.M., James, D.E., Meister, A.D., Bower, B.T. and Dixon, J.A. (1983) *Environment, Natural Systems and Development: An Economic Valuation Guide*, Baltimore, Johns Hopkins University Press.

Hughes, P.J., Hope, C., Latham, M. and Brookfield, M. (1979) 'Man-induced degradation of the Lakeba landscape: evidence from two inland swamps', in Brookfield, H.C. (ed.) *Lakeba: Environmental Change, Population Dynamics and Resource Use*, UNESCO/UNFPA Fiji Island reports, no. 5, Canberra, Australian National University for UNESCO, 93–110.

Hurni, H. (1983) 'Soil erosion and soil formation in agricultural systems, Ethiopia and northern Thailand', *Mountain Research and Development*, 3, 131–42.

Hurni, H. and Messerli, B. (1981) 'Mountain research for conservation and development in Simen, Ethiopia', Mountain Research and Development, 1, 49–54.

Hurwitz, N. (1957) *Agriculture in Natal, 1860–1950*, Cape Town, Oxford University Press.

Hutchinson, Sir J. (1969) 'Erosion and land use: the influence of agriculture on the Epirus region of Greece', *Agriculture History Review*, 17, 85–90.

Huxley, E. (1937) 'The menace of soil erosion', *Journal of the Royal Africa Society*, 36, 365–70.

Hyams, E. (1952) *Soil and Civilisation*, London, John Murray (2nd edn 1976).

Hyden, G. (1980) *Beyond Ujamaa in Tanzania; Underdevelopment and an Uncaptured Peasantry*, London, Heinemann.

IBSRAM (International Board for Soil Research and Management) (1985) 'Report of the inaugural workshop and proposal for implementation of the tropical land clearing for sustainable agriculture network', 27 August–2 September, Jakarta and Bukittinggi, Indonesia, Bangkok, IBSRAM.

IFAD (International Fund for Agricultural Development) (1985) 'Soil and water conservation in sub-Saharan Africa: issues and options', prepared by the Centre for Development Co-operation Services, Free University, Amsterdam.

Imeson, A.C., Kwaad, F.J.M.P. and Mücher, H.J. (1980) 'Hillslope processes and deposits in forested areas of Luxembourg', in Cullingford, R.A., Davidson, D.A. and Lewin J. (eds) *Timescales in Geomorphology*, Chichester, John Wiley, 31–42.

IUCN (International Union for Conservation of Nature and Natural Resources) (1980) *World Conservation Strategy: Living Resource Conservation for Sustainable Development*, Gland, Switzerland, International Union for Conservation of

Nature and Natural Resources.

Ives, J.D. (1981) 'The Heavenly Mountains: an excursion to the Tien Shan, Peoples' Republic of China, 1–24 June 1981', *Mountain Research and Development*, 1, 293–98.

Ives, J.D. (1984) 'Does deforestation cause soil erosion?', *International Union for the Conservation of Nature Bulletin*, Suppl. 2, 4–5.

Ives, J.D. (1985) 'Mountain environments', *Progress in Physical Geography*, 9, 425–33.

Ives, J.D. and Messerli, B. (1981) 'Mountain hazards mapping in Nepal: introduction to an applied mountain research project', *Mountain Research and Development*, 1, 223–30.

Iwata, S.T., Sharma, K. and Yamanaka, H. (1984) 'A preliminary report of central Nepal and Himalayan uplift', *Journal of the Nepal Geological Society*, 4, 141–49.

Jack, H.W. (1937) 'Soil erosion', *Agricultural Journal* (Department of Agriculture, Fiji), 8 (4), 4–7.

Jacks, G.V. and Whyte, R.O. (1939) *The Rape of the Earth: A World Survey of Soil Erosion*, London, Faber and Faber.

Jackson, J.C. (1968) *Planters and Speculators: Chinese and European Agricultural Enterprise in Malaya, 1786–1921*, Kuala Lumpur, University of Malaya Press.

Jackson, P. (1983) 'The tragedy of our tropical rainforests', *Ambio*, 12, 252–4.

Jamieson, N.L. (1984) 'Multiple perceptions of environmental issues in rural Southeast Asia: the implications of cultural categories, beliefs and values for environmental communication', Honolulu, Environmental and Policy Institute, East–West Center, 26 pp, mimeo.

Japan International Co-operation Agency (1982) *Draft Feasibility Report on the Riam Kiwa River Hydro-Electric Power Development Project*, Republic of Indonesia, Perusahaan Umum Listrik Negara.

Jiang, D. and Wu, Y. (1980) 'Sediment yield and utilization', in Chinese Society of Hydraulic Engineering (ed.) *Proceedings of the International Symposium on River Sedimentation*, 24–29 March 1980, vol. 1, Beijing, Guanghua Press, 1127–36.

Jodha, N.S. (1966) 'Scarcity oriented pattern of arid agriculture', *Indian Journal of Agricultural Economics*, 21, 238–46.

Jodha, N.S. (1967) 'Capital formation in arid agriculture: a study of resource conservation and reclamation measures applied to arid agriculture', unpublished PhD thesis, University of Jodhpur, Rajasthan.

Jodha, N.S. (1974) 'A case of process of tractorization', *Economic and Political Weekly (Quarterly Review of Agriculture)*, 9, A.111–A.118.

Jodha, N.S. (1980) 'The operating mechanism of desertification and choice of interventions', in Mann, H.S. (ed.), *Arid Zone Research and Development*, Jodhpur, Scientific Publishers, 429–38.

Jodha, N.S. (1982) 'The role of administration in desertification: land tenure as a factor in the historical ecology of western Rajasthan', in Spooner, B. and Mann, H.S. (eds) *Desertification and Development: Dryland Ecology in Social Perspective*, London, Academic Press, 333–50.

Jodha, N.S. (1983) 'Market forces and erosion of common property resources', paper presented at the International Workshop on Agricultural Markets in Semi-Arid Tropics, ICRISAT Center, Patancheru (AP), India, 24–28 October.

Jodha, N.S. (1984) *Causes and Consequences of Decline of Common Property Resources in the Arid Region of Rajasthan*, Progress Report, Economics Program, ICRISAT, Patancheru (AP), India.

Jodha, N.S. (1985) 'Population growth and decline of common property resources in Rajasthan, India', *Population and Development Review*, 11, 247–64.

Johnson, K., Olson, E.A. and Manandhar, S. (1982) 'Environmental knowledge and response to natural hazards in mountainous Nepal', *Mountain Research and Development*, 2:175–88.

Johnson, O.E.G. (1972) 'Economic analysis, the legal framework and land tenure systems', *Journal of Law and Economics*, 15, 259–76.

Juo, A.S.R. and Lal, R. (1977) 'Effect of fallow and continuous cultivation on the chemical and physical properties of an alfisol in western Nigeria', *Plant and Soil*, 47, 567–84.

Kahn, H., Brown, W. and Martel, L. (eds) (1976) *The Next 2000 Years*, New York, William Morrow.

Kartawinata, K. (1979) 'An overview of the environmental consequences of tree removal from the forest in Indonesia', in Boyce, S.G. (ed.) *Biological and Sociological Basis for a Rational Use of Forest Resources for Energy and Organics*, Washington, US Department of Agriculture.

Kartawinata, K. and Vayda, A.P. (1984) 'Forest conversion in East Kalimantan, Indonesia: the activities and impact of timber companies, shifting cultivators, migrant pepper-farmers and others', in di Castri, F., Baker, F.W.G. and Hadley, M. (eds) *Ecology in Practice, I. Ecosystem Management*, Dublin and Paris, Tycooly International and Unesco, 98–126.

Kaser, M. (ed.) (1968) *Economic Development for Eastern Europe*, Proceedings of a Conference held by the International Economic Association at Plovdiv, Bulgaria, London, Macmillan.

Keeton, C.L. (1974) *King Thebaw and the Ecological Rape of Burma*, Delhi, Manohar Book Service.

Kellman, M. (1974) 'Some implications of biotic interactions for sustained tropical agriculture', *Proceedings of the Association of American Geographers*, 6, 142–45.

Kerin, J. (1982) 'Senate Standing Committee on Science and the Environment (land use policy)', Canberra, *Official Hansard Report*, 9 March, 263–324.

Khanal, N.R. (1981) 'The causes and consequences of hill migration in Nepal: a case study of Aruchaur Panchayat, Syangja', unpublished MA thesis in Geography, Lalitpur, Nepal, Tribhuvan University.

Khor, K.P. (1983) *The Malaysian Economy: Structures and Dependence*, Kuala Lumpur, Maricans and Institut Masyarakat.

Kienholz, H., Hafner, H., Schneider, G. and Tamrakar, R. (1983) 'Mountain hazards mapping in Nepal's middle mountains: maps of land use and geomorphic damages (Kathmandu – Kakani area)', *Mountain Research and Development*, 3, 195–220.

Kikuchi, M., Hafid, A., Saleh, C., Hartoyo, S. and Hayami, Y. (1980) *Changes in Community Institutions and Income Distribution in a West Java Village*, IRRI Research Papers Series 50, Manila, International Rice Research Institute.

Kington, J.A. (1980) 'Daily weather mapping from 1781: a detailed synoptic examination of weather and climate during the decade leading up to the French Revolution', *Climatic Change*, 3, 7–36.

Kirch, P.V. and Yen, D.E. (1982) 'Tikopia: the prehistory and ecology of a Polynesian outlier', *Bernice P. Bishop Museum Bulletin*, 238, Honolulu, Bishop Museum Press.

Kirkby, A.V.T. [Whyte, A.V.T.] (1973) *The Use of Land and Water Resources in the Past and Present Valley of Oaxaca, Mexico*, Memoirs of the Museum of

Anthropology, no. 5, Ann Arbor, University of Michigan.

Kirkby, M.J. (1972) *The Physical Environment of the Nochixtlan Valley, Oaxaca*, Publications in Anthropology 2, Nashville, Tenn., Vanderbilt University.

Kirkby, M.J. (1980) 'the problem' in Kirkby, M.J. and Morgan, R.P.C. (eds) *Soil Erosion*, Chichester, Wiley, 1–16.

Kitching, G. (1980) *Class and Economic Change in Kenya: The Making of an African Petite Bourgeoisie*, London and New Haven, Yale University Press.

Kitching, G. (1982) *Development and Underdevelopment in Historical Perspective: Populism, Nationalism and Industrialization*, London, Methuen.

Kjekshus, H. (1977) *Ecology Control and Economic Development in East African History: The Case of Tanganyika 1850–1950*, London, Heinemann.

Klepper, R. (1980) 'The state and peasant differentiation in Zambia', Sussex, Institute of Development Studies, 40 pp, mimeo.

Kolonisatie Bulletin (1938) nos 2, 3; (1939) nos 4, 6, 7; (1940) no. 10; (1941) no. 11, Batavia, Netherlands East Indies Government.

Komarov, B. (1981) *The Destruction of Nature in the Soviet Union*, London, Pluto Press.

Korsching, P.F. and Nowak, P.J. (1982) 'Farmer acceptance of alternative conservation policies', *Agriculture and Environment*, 7, 1–14.

Krauss, H.A. and Allmaras, R.R. (1982) 'Technology masks the effects of soil erosion on wheat yields – a case study in Whiteman County, Washington', in American Society of Agronomy, *Determinants in Soil Loss Tolerance*, special publication no. 45, Madison, Wisconsin, American Society of Agronomy, 75–86.

Kriedte, P. (1983) *Peasants, Landlords and Merchant Capitalists: Europe and the World Economy, 1500–1800* (translated by Berghahn, V.R. from *Spätfeudalismus und Handelskapital: Grundlinien der Europäischen Wirtschaftsgeschichte vom 16. bis zum Ausgang des 18. Jahrhunderts*, Göttingen, Vandenhoeck und Ruprecht (1980)) Leamington Spa, Berg Publishers.

Krylov, G.A. (1979) *The Soviet Economy*, Lexington, Lexington Books.

Labrousse, E., Leon, P., Goubert, P., Bouvier, J., Carriere, C. and Harsin, P. (1970) *Histoire Economique et Sociale de la France, Tome II: Des derniers Temps de l'Age seigneurial aux preludes de l'Age industriel (1660–1789)*, Paris, Presses Universitaires de France.

Lamarche, V.C. (1968) 'Rates of slope degradation as determined from botanical evidence, White Mountains, California', US Geological Survey professional paper 352–L.

Lamartine-Yates, P. (1940) *Food Production in Western Europe: An Economic Survey of Agriculture in Six Countries*, London, Longman Green.

Lamb, H.H. (1977) *Climate: Present, Past and Future*, vol. 2, London, Methuen.

Lamb, H.H. (1982) *Climate History and the Modern World*, London, Methuen.

Lamb, H.H. (1984) 'Climate in the last thousand years: natural climatic fluctuations and change', in Flohn, H. and Fantechi, R. (eds) *The Climate of Europe: Past, Present and Future*, Dordrecht, D. Reidel, 25–64.

Lane, D. (1985) *Soviet Economy and Society*, London, Basil Blackwell.

Lanly, J.P. (1982) *Tropical Forest Resources*, FAO Forestry paper no. 30, Rome, FAO.

Larsen, M.R. (1967) 'China's agriculture under communism', *An Economic Profile of Mainland China*, vol. 1, studies prepared for the Joint Economic Committee, Congress of the United States, Washington, US Government Printing Office,

197–267.

Larson, W.E., Pierce, F.J. and Dowdy, R.H. (1983) 'The threat of soil erosion to long-term crop production', *Science*, 219, 458–65.

Lasaqa, I. (1984) *The Fijian People Before and After Independence*, Canberra, Australian National University Press.

Latham, M. (1983) 'Le milieu natural et son utilisation sur les îles, A, Lakeba', in Latham, M. and Brookfield, H.C. (eds) *Iles Fidji Orientales: Etude du Milieu Naturel et de son Utilisation sous l'Influence humaine*, Paris, ORSTOM, Travaux et Documents de l'Orstom, no. 162, 13–62.

Lefebvre, G. (1929/1977) 'The place of the revolution in the agrarian history of France', translated by Forster, E., in Forster, R. and Ranum, O. (eds) *Rural Society in France: Selections from the Annales, Economies, Societés Civilisations*, 31–49 (first published in *Annales d'Histoire Economique et Sociale*, 1 (1929)), Baltimore, Johns Hopkins University Press, 506–23.

Leprun, J.C. (1981) 'A erosão, a conservação e o manejo do solo no nordeste Brasileiro', *Recursos de Solos*, 15, Recife, SUDENE.

Lericollais, A. (1970) 'La déterioration d'un terroir: Sob, en pays Sérèr (Sénégal)', *Etudes Rurales*, 37–9, 113–28.

Le Roy Ladurie, E. (1975) 'De la crise ultime à la vraie croissance', in Neveux, H., Jacquart, J. and Le Roy Ladurie, E. (eds) *Histoire de la France Rurale, Tome II, L'Age Classique des Paysans* (Le Roy Ladurie, E. general editor), Paris, Editions du Seuil, 359–599.

Levy, T.E. (1983) 'The emergence of specialized pastoralism in the southern Levant', *World Archaeology*, 15, 15–36.

Liedtke, H. (1984) 'Soil erosion and general denudation in northwest Fiji', Bochum, Federal Republic of Germany, unpublished, 27 pp.

Lim, T.G. (1977) *Peasants and their Agricultural Economy in Colonial Malaya 1874–1941*, East Asian Historical Monographs Series, Kuala Lumpur, Oxford University Press.

Lipton, M. (1968) 'The theory of the optimizing peasant', *Journal of Development Studies*, 4, 327–51.

Lipton, M. (1984) 'Private interest and the good of all: types of conflict in the development process', paper presented to the British Association for the Advancement of Science, Norwich, University of East Anglia, 10–14 September, 72 pp, mimeo

Liu Po Chen (1978) *Ancient China's Poets* (translated by Robinson, L.S.) Hong Kong, Commercial Press, 85–107.

Löffler, E. (1977) *Geomorphology of Papua New Guinea*, Canberra, CSIRO and ANU Press.

Lowe, J. and Lewis, D. (1980) *The Economics of Environmental Management*, Deddington (Oxfordshire), Philip Allen.

Lowitt, R. (1985) 'Agricultural policy and soil conservation: comment', *Agricultural History*, 59, 320–25.

Luxemburg, R. (1951) *The Accumulation of Capital*, London, Routledge and Kegan Paul.

Ma Shijun and Chang Shuzhang (1980) 'It is of immediate urgency to protect our environment and natural resources', *Jingji Guanli (Economic Management)*, 10, 28–39.

Mabbutt, J.A. (1978) 'Desertification of Australia in its global context', *Search*, 9,

252–6.

McAlpine, J.R., Keig, G. and Falls, R. (1983) *Climate of Papua New Guinea*, Canberra, CSIRO and ANU Press.

Macar, P. (1974) 'Etude en Belgique de phénomènes d'érosion et de sedimentation récentes en terres limoneuses', in Poser, H. (ed.) *Geomorphologische Prozessen und Prozesskombinationen in der Gegenwart unter verschiedenen Klimabedingungen (Report of the Commission on Present-day Geomorphological Processes, International Geographical Union)*, Göttingen, Vandenhoeck und Ruprecht, 334–71.

McCracken, J. (1985) 'Colonialism, capitalism and the ecological crisis in Malawi', paper presented at a workshop on The Scramble for Resources: Conservation Policies in Africa 1884–1984, April, Cambridge.

McCuaig, J.D. and Manning, E.W. (1982) *Agricultural Land-use Change in Canada: Process and Consequences*, Land Use in Canada series no. 21, Ottawa, Lands Directorate, Environment Canada.

McCullough, R. and Weiss, D. (1985) 'An environmental look at the 1985 farm bill', *Journal of Soil and Water Conservation*, 40, 267–70.

MacEwan, M. (1984) 'The greening of Britain', *Marxism Today*, July, 23–7.

Macfarlane, A. (1976) *Resources and Population: a Study of the Gurungs of Nepal*, Cambridge, Cambridge University Press.

McGregor, A.C. (1980) 'From sheep range to agribusiness: a case history of agricultural transformation on the Columbia Plateau', *Agricultural History*, 54, 11–27.

McIntyre, J. (1975) 'The farmer's dilemma', *Search*, 6, 207–11.

McVean, D.N. and Lockie, J.D. (1969) *Ecology and Land Use in Upland Scotland*, Edinburgh, Edinburgh University Press.

Maddison, A. (1971) *Class Structure and Economic Growth: India and Pakistan since the Moghuls*, London, Allen and Unwin.

Mahat, T.B.S. (1985) 'Human impact on forests in the middle hills of Nepal', PhD thesis in Forestry, Canberra, Australian National University.

Malcolm, D.W. (1938) *Sukumuland – An African People and their Country*, Oxford, International African Institute and Oxford University Press.

Mann, H.S. and Kalla, J.C. (1977) 'Asset-liability imbalances in agricultural sector of the Indian arid zone', in Anon. (ed.) *Desertification and its Control*, paper presented to UN Conference on Desertification at Nairobi, New Delhi, Indian Council of Agricultural Research, 324–34.

Mariategui, J.C. (1971) *Seven Interpretive Essays on Peruvian Reality* (translated by Urquidi, M. from *Siete Ensayos sobre la Realidad Peruana*, Lima, N.S. (1928)) with an introduction by Basadre, J., Austin, Texas, University of Texas Press.

Martin, P.M. (1983) 'The violence of empire', in Birmingham, D. and Martin, P.M. (eds) *The History of Central Africa*, London, Longman, 1–26.

Marx, K. (1887/1954) *Capital: a Critique of Political Economy*, vol. 1, Moscow, Progress Publishers, from the English edition of 1887, Engels, F. (ed.).

Mather, A.S. (1983) 'Land deterioration in upland Britain', *Progress in Physical Geography*, 7, 210–28.

Matheson, V. (1981) 'Borneo history: past lessons, current challenges', Asian Studies Association of Australia, Malaysia Society, Third Colloquium, University of Adelaide, 22–24 August.

Matheson, V. (1982) 'Conflict without resolution. The Banjarmasin War 1859–1905', paper presented to Asian Studies Association of Australia, Fourth National Conference, Monash University, 10–14 May.

Meggitt, M.J. (1965) *The Lineage System of the Mae Enga of New Guinea*, New York, Barnes and Noble.

Meggitt, M.J. (1977) *Blood is their Argument: Warfare among the Mae-Enga Tribesmen of the New Guinea Highlands*, Palo Alto, Calif., Mayfield Publishing Co.

Meinig, D. (1962) *On the Margins of the Good Earth: the South Australian Wheat Frontier 1869–1884*, Association of American Geographers, monograph series no. 2, Chicago, Rand-McNally.

Messer, J. (1984) 'The political ecology of land degradation in Australia', paper presented at the Conference on Soil Degradation: The Future of Our Land?, Canberra, Australian National University, 25–27 November (unpublished).

Messer, J. (in progress) 'The political ecology of Australian agriculture: the case of Australia', PhD thesis in Sociology, Sydney, University of New South Wales.

Meuvret, J. (1946) 'Les crises de subsistances et la démographie de la France d'ancien régime', *Population*, 1, 643–50.

Mikesell, M.W. (1969) 'The deforestation of Mount Lebanon', *Geographical Review*, 59, 1–28.

Millington, A.C. (1981) 'Relationship between three scales of erosion measurement on two small basins in Sierra Leone', in International Association of Hydrological Sciences, *Erosion and Sediment Transport Measurement*, publication no. 133, Wallingford, UK, IAHS, 485–92.

Millington, A.C. (1985) 'Environmental degradation, soil conservation and agricultural policies in Sierra Leone 1895–1984', paper presented at a workshop on The Scramble for Resources: Conservation Policies in Africa 1884–1984', April, Cambridge.

Modjeska, N. (1982) 'Production and inequality: perspectives from central New Guinea', in Strathern, A. (ed.) *Inequality in New Guinea Highlands Societies*, Cambridge, Cambridge University Press, 50–108.

Molyneux, M.E. and Halliday, F. (1981) *The Ethiopian Revolution*, London, New Left Books.

Moran, E.F. (1982) 'Ecological, anthropological and agronomic research in the Amazon Basin', *Latin American Research Review*, 17, 3–41.

Morgan, R.P.C. (1977) *Soil Erosion in the United Kingdom: Field Studies in the Silsoe Area 1973–75*, National College of Agricultural Engineering occasional paper 4.

Morgan, R.P.C. (1980a) 'Soil erosion in Britain: is it an enigma?' *Progress in Physical Geography*, 4, 24–7.

Morgan, R.P.C. (1980b) 'Implications', in Kirkby, M.J. and Morgan, R.P.C. (eds) *Soil Erosion*, Chichester, Wiley, 253–302.

Morgan, R.P.C. (1985) 'Soil erosion measurement and soil conservation research in cultivated areas of the UK', *Geographical Journal*, 151, 11–20.

Morineau, M. (1970) 'Was there an agricultural revolution in 18th century France?', in Cameron, R. (ed.) *Essays in French Economic History*, Homewood, Ill., Irwin, 170–82.

Morren, G.E.B. (1977) 'From hunting to herding: pigs and the control of energy in montane New Guinea', in Bayliss-Smith, T.P. and Feachem, R.G. (eds) *Subsistence and Survival: Rural Ecology in the Pacific*, London, Academic Press, 273–316.

Morrison, R.J. (1981) 'Factors determining the extent of soil erosion in Fiji', Institute of Natural Resources Environmental Studies report no. 7, Suva, The University of the South Pacific, 17 pp.

Moss, R.P. (1978) 'Concept and theory in land evaluation for rural land use planning',

Occasional Publication no. 6, Department of Geography, University of Birmingham.

Mueller, M.B. (1977) 'Women and men in rural Lesotho: the periphery of the periphery', PhD thesis, Department of Politics, Brandeis University, microfilm.

Murphey, R. (1951) 'The decline of North Africa since the Roman occupation: climatic or human?', *Association of American Geographers Annals*, 41, 116–32.

Murray, C. (1981) *Families Divided: the Impact of Migrant Labour in Lesotho*, Cambridge, Cambridge University Press.

MvO (various dates) *Memories van Overgave*. (Handing-over memoranda, prepared by sub-district officers in the Dutch colonial service. Originals held in the Algemeen Rijksarchief, The Hague; some other microfiche collections exist. Most references here are to *Memories van Overgave van de Onderafdeling Martapoera, Zuider- en Ooster-afdeling Borneo* for dates between 1906 and 1938.)

Myers, N. (1985) *The Primary Source: Tropical Forests and Our Future*, New York, W.W. Norton.

National Planning Commission (Nepal) (1978) *A Survey of Employment, Income Distribution and Consumption Patterns of Nepal*, Kathmandu, National Planning Commission.

Nations, J.D. and Komer, D.I. (1983) 'Central America's tropical rainforests: positive steps for survival', *Ambio*, 12, 232–8.

Naveh, Z. and Dan, J. (1973) 'The human degradation of Mediterranean landscapes in Israel', in di Castri, F. and Mooney, H.A. (eds) *Mediterranean Type Ecosystems: Origin and Structure*, London, Chapman and Hall, 373–90.

Nepal–Australia Forestry Project (1985) *Project Document: Phase Three*, Canberra, Nepal-Australia Forestry Project.

Nepali, S.B. and Regmi, I.R. (1981) 'Technological innovations for hill agricultural development' in Ong, S.E. (ed.) *Nepal's Experience in Hill Agricultural Development*, Kathmandu, Ministry of Food and Agriculture.

Neumann, J. (1977) 'Great historical events that were significantly affected by the weather: 2, The year leading to the revolution of 1789 in France', *Bulletin of the American Meteorological Society*, 58, 163–8.

Newman, P. and Cameron, I. (1982) *Attitudes to Conservation and the Environment in Western Australia*, a report for the Department of Conservation and Environment, State Conservation Strategy, Perth, Murdoch University.

Nisbett, J. (1901) *Burma under British Rule and Before*, vols I and II, London, Constable.

Nye, P.H. and Greenland, D.J. (1960) *The Soil under Shifting Cultivation*, Technical Communication 51, Farnham Royal, Bucks., Commonwealth Agricultural Bureaux.

Oakerson, R.J. (1985) 'A model for the analysis of common property resources', prepared for workshops in preparation for the Common Property Resource Management Conference, Annapolis, Maryland, 21–27 April, National Academy of Sciences, USA.

Oates, N.M., Greig, P.J., Hill, D.G., Langley, P.A. and Reid, A.J. (eds) (1981) *Focus on Farm Trees: The Decline of Trees in the Rural Landscape*, Proceedings of a National Conference, University of Melbourne, 23–26 November, 1980, Melbourne, University of Melbourne.

O'Keefe, P. (1975) *African Drought: A Review*, Disaster Research Unit, University of Bradford, occasional paper no. 8.

Oldfield, F., Appleby, P.G., Brown, A. and Thompson, R. (1980) 'Palaeoecological studies of lakes in the Highlands of Papua New Guinea. 1: the chronology of sedimentation', *Journal of Ecology*, 68, 457–77.

O'Riordan, T. (1981) *Environmentalism*, 2nd edn, London, Pion.

Ormsby-Gore, W.G.A. (1928) *Report by W.G.A. Ormsby-Gore on his Visit to Malaya, Ceylon and Java, 1928*, command paper 3235, London, HMSO.

Osborn, F. (1948) *Our Plundered Planet*, London, Faber and Faber.

Palmer, R. (1977) 'The agricultural history of Rhodesia', in Palmer, R. and Parsons, N. (eds) *The Roots of Rural Poverty in Central and Southern Africa*, London, Heinemann, 221–55.

Palmer, R. and Parsons, N. (eds) (1977) *The Roots of Rural Poverty in Central and Southern Africa*, London, Heinemann.

Parham, B.E.V. (1954) 'Two examples of land use in Fiji', *Agricultural Journal* (Department of Agriculture, Fiji), 25, 109–12.

Parliament of NSW (1983) *First and Second Reports of the Joint Select Committee of the Legislative Council and Legislative Assembly to Enquire into the Western Division of New South Wales*, March 1983 and December 1983, Sydney, NSW Government Printer.

Parliament of NSW (1984) *Fourth Report of the Joint Select Committee of the Legislative Council and Legislative Assembly to Enquire into the Western Division of New South Wales*, Sydney, NSW Government Printer.

Parra, V.R. (1972) 'Marginalidad y Sudesarrollo', in Caradone, R. (ed.) *Las Migraciones Internas*, Bogota, Editorial Andes, chapter 5.

Passmore, J. (1974) *Man's Responsibility for Nature: Ecological Problems and Western Traditions*, London, Duckworth.

Pavelis, G.A. (1983) 'Conservation capital in the United States, 1935–1980', *Journal of Soil and Water Conservation*, 38, 455–8.

Pelzer, K.J. (1945) *Pioneer Settlement in the Asiatic Tropics: Studies in Land Utilization and Agricultural Colonization in Southeastern Asia*, special publication 29, New York, American Geographical Society.

Pelzer, K.J. (1978) *Planter and Peasant: Colonial Policy and the Agrarian Struggle in East Sumatra, 1863–1947*, Verhandelingen van het Koninklijk Instituut voor Taal-, Land- en Volkenkunde 84, 's-Gravenhage, Martinus Nijhoff.

Pelzer, K.J. (1982) *Planters against Peasants: the Agrarian Struggle in East Sumatra, 1947–1958*, Verhandelingen van het Koninklijk Instituut voor Taal-, Land- en Volkenkunde 97, 's-Gravenhage, Martinus Nijhoff.

Perrens, S.J. and Trustrum, N.A. (1983) 'Assessment and evaluation for soil conservation policy', workshop on Policies for Soil and Water Conservation, 25–27 January, Honolulu, East–West Environment and Policy Institute.

Pfister, C. (1981) 'An analysis of the Little Ice Age climate in Switzerland and its consequences for agricultural production', in Wigley, T.M.L., Ingram, M.J. and Farmer, G. (eds) *Climate and History: Studies in Past Climates and their Impact on Man*, Cambridge, Cambridge University Press, 214–48.

Picardi, A.C. (1974) 'A systems analysis of pastoralism in the West African Sahel', in MIT Center for Policy Alternatives, *Framework for Evaluating Long-Term Strategies for the Development of the Sahel-Sudan Region*, Annex 5, Cambridge, Mass., MIT Press.

Pihan, J. (1979) 'Risques climatiques d'érosion hydrique des sols en France', in Vogt, H. and Vogt, T. (eds) *Colloque sur L'Erosion Agricole des Sols en Milieu Tempéré non-Méditerranéen, Strasbourg-Colmar*, 20–23 September, Strasbourg, Laboratoire de Géographie, 13–18.

Pimentel, D. and Pimentel, M. (1979) *Food, Energy and Society*, London, Edward Arnold.

Plumwood, V. and Routley, R. (1982) 'World rainforest destruction – the social factors', *The Ecologist*, 12 (1), 4–22.

Polanyi, K. (1944) *The Great Transformation*, Boston, Beacon Press.

Polanyi, K., Arensberg, C.M. and Pearson, H.W. (eds) (1957) *Trade and Market in the Early Empires: Economies in History and Theory*, Glencoe, Ill., Free Press.

Policy Research Office of the Ministry of Forestry (1981) 'Run forestry work according to law', *Hongqi (Red Flag)*, 5, 27–31.

Porter, D.J. (1986) 'Rationalisation, resistance and development practice: a question of authority', PhD thesis in Human Geography, Canberra, Australian National University.

Porter, P.W. (1970) 'The concept of environmental potential as exemplified by tropical African research', in Zelinsky, W., Kosinski, L.A. and Prothero, R.M. (eds) *Geography and a Crowding World: a Symposium on Population Pressures upon Physical and Social Resources in the Developing Lands*, New York, Oxford University Press, 187–217.

Posner, J.L. and MacPherson, M.F. (1981) 'The steep-sloped areas of tropical America: current situation and prospects for the year 2000', Agricultural Sciences Division, New York, Rockefeller Foundation, 21 pp, mimeo.

Posner, J.L. and MacPherson, M.F. (1982) 'Agriculture on the steep slopes of tropical America: the current situation and prospects', *World Development*, May, 341–54.

Post, J.D. (1977) *The Last Great Subsistence Crisis in the Western World*, Baltimore, Johns Hopkins University Press.

Postan, M.M. (1959) 'Note (in response to W.C. Robinson on "Money, population and economic change in late medieval Europe")', *Economic History Review*, 12, 77–82.

Powell, J.M. (1982) 'The history of plant use and man's impact on the vegetation', *Monographiae Biologicae*, 42, 207–25.

Prajapati, M.C., Vangani, N.S. and Ahuja, L.D. (1973) 'In the dry range lands on western Rajasthan "Tanka" can be the answer', *Indian Farming* 22 (11), 27–31.

Prineas, P. (1984) 'National Parks Association submission to the Senate Standing Committee on Science, Technology, and the Environment inquiry into land use policy', *Official Hansard Report*, 1574–95.

Pryde, P.R. (1972) *Conservation in the Soviet Union*, Cambridge, Cambridge University Press.

Rahman, A (ed. and trans. M. Munshi) (1900) *The Life of Abdur Rahman, Amin of Afghanistan* (2 vols), London, John Murray.

Rajbhandari, H.B. and Shah, S.B. (1981) 'Trends and projections of livestock production in the hills', in Ong, S.E. (ed.) *Nepal's Experience in Hill Agricultural Development*, Kathmandu, Ministry of Food and Agriculture, 43–58.

Ramsay, W.J.H. (1985) 'Erosion in the Middle Himalaya, Nepal, with a case study of the Phewa Valley', MSc dissertation at the Department of Forest Resources Management, University of British Columbia.

Ranger, T.O. (1971) *The Agricultural History of Zambia*, Lusaka, National Educational Company of Zambia.

Rapp, A. (1976) 'The Sudan' in Rapp, A., Le, Houerou, H.N. and Lundholm, B.

(eds) *Can Desert Encroachment be Stopped?*, Swedish Natural Science Research Council Ecological Bulletin 24, Stockholm, National Science Research Council, 155–64.

Rapp, A., Berry, L. and Temple, P.H. (1972) 'Soil erosion and sedimentation in Tanzania – the project', *Geografiska Annaler*, 54A, 105–9.

Rapp, A., Murray-Rust, D.H., Christiansson, C. and Berry, L. (1972) 'Soil erosion and sedimentation in four catchments near Dodoma, Tanzania', *Geografiska Annaler*, 54A, 255–318.

Rastoin, J.-L. (1973) 'Croissance des firmes agro-alimentaires multinationales', *Economies et Sociétés*, 7, 2275–305.

Rastoin, J.-L. (1977) 'The world strategy employed by agri-food multinationals in the Euro-Mediterranean area', *World Agriculture*, 26 (2/3), 18–21.

Redclift, M.R. (1984) *Development and the Environmental Crisis: Green and Red Alternatives*, London, Methuen.

Regmi, M.C. (1971) *A Study of Nepali Economic History 1768–1845*, New Delhi, Manjusri.

Regmi, M.C. (1976) *Land Ownership in Nepal*, Berkeley, University of California Press.

Regmi, M.C. (1978) *Land Tenure and Taxation in Nepal*, vols I–IV, Kathmandu, Ratna Pustak Bhandar.

Reichelderfer, K. (1984) 'Will agricultural program consistency save more soil?', *Journal of Soil and Water Conservation*, 39, 229–31.

Ricardo, D. (1951) *On the Principles of Political Economy and Taxation*, 3rd edn, London, John Murray (1821), reprinted with an Introduction by Straffa, P. with the collaboration of Dobb, M.H. as *The Works and Correspondence of David Ricardo*, vol. I, Cambridge, Cambridge University Press.

Richards, P. (1985) *Indigenous Agricultural Revolution*, London, Hutchinson.

Richards, P. (1986) 'Indigenous agricultural revolutions: new approaches to the agricultural crisis in Africa', paper presented to the Annual General Meeting of the Institute of British Geographers, Reading (unpublished).

Richter, G. and Negendank, J.F.W. (1977) 'Soil erosion processes and their measurement in the German area of the Moselle river', *Earth Surface Processes and Landforms*, 2, 261–78.

Richter, G. and Sperling, W. (eds) (1976) *Bodenerosion in der Bundesrepublik Deutschland*, Darmstadt, Wissenschaftliche Buchges.

Roberts, Sir S. (1968) *History of Australian Land Settlement*, Melbourne, Macmillan.

Robinson, D.A. (1978) *Soil Erosion and Soil Conservation in Zambia: A Geographical Appraisal*, Zambia Geographical Association, occasional study no. 9.

Robinson, J. (1963) *Economic Philosophy*, Chicago, Aldine.

Robinson, P. (1977) *The Environment Crisis: a Communist View*, London, Communist Party Headquarters.

Roose, E.J. (1976) 'Use of the universal soil loss equation in West Africa', in Soil Conservation Society of America, *Soil Erosion: Prediction and Control*, Ankeny, Iowa, SCSA, 60–74.

Roose, E.J. (1977) 'Application of the universal soil loss equation of Wischmeier and Smith in West Africa', in Greenland, D.J. and Lal, R. (eds) *Soil Conservation and Management in the Humid Tropics*, Chichester, John Wiley.

Rounce, N.V. (1949) *The Agriculture of the Cultivation Steppe of the Lake, Western and Central Provinces* (Uganda), Cape Town, Longman.

Runge, C.F. (1983) 'Common property and collective action in economic development', paper prepared for Board on Science and Technology for International Development (BOSTID), Washington, National Academy of Sciences.

Ruthenberg, H. (1980) *Farming Systems in the Tropics*, 3rd edn, Oxford, Clarendon Press.

Sahlins, M. (1974) *Stone Age Economics*, London, Tavistock.

Saint, S.J. (1934) 'The coral limestone soils of Barbados', *Journal of the Department of Science and Agriculture, Barbados*, 3, 1–37.

Sanchez, P.A., Bandy, D.E., Villachica, J.H. and Nicholaides, J.J. (1982) 'Amazon basin soils: management for continuous crop production', *Science*, 216, 821–7.

Sauer, C.O. (1941) 'Foreword to historical geography', *Annals of the Association of American Geographers*, 31, 1–24.

Sayer, A. (1980) *Epistemology and Regional Science*, Brighton, University of Sussex, School of Social Science.

Schertz, D.L. (1983) 'The basis for soil loss tolerances', *Journal of Soil and Water Conservation*, 38, 10–14.

Schlich, P. (1889/90) 'Forestry in the colonies and in India', *Proceedings of the Royal Colonial Institute*, 21, 187–238.

Schreuder, W.G. (1923) *Verslag van een globale verkenning van eenige voor grootrijstbedrijven vestemde terreinen in de Zuider-en-Oosterafdeeling van Borneo*, Jakarta, Arsip Nasional Indonesia.

Schultze, J.H. (1952) 'Die Bodenerosion in Thüringen, Wesen, Stärke und Abwehrmöglichkeiten', *Petermann's Geographischen Mitteilungen Erganzungsheft*, 247, Gotha, Justus Perthes.

Schumm, S.A. (1956) 'The evolution of drainage systems and slopes in badlands at Perth Amboy, New Jersey', *Bulletin of the Geological Society of America*, 67, 597–646.

Scott, E.P. (1984) 'Life before the drought: a human ecological perspective', in Scott, E.P. (ed.) *Life Before the Drought*, Boston and London, Allen and Unwin, 49–76.

Scott, G. and Pain, C.F. (1982) 'Land potential', in Carrad, B., Lea, D.A.M. and Talyaga, K.T. (eds) *Enga: Foundations for Development*, Armidale, University of New England, 128–45.

Scott, J.C. (1976) *The Moral Economy of the Peasant: Rebellion and Subsistence in Southeast Asia*, New Haven, Yale University Press.

Seckler, D.W. (1975) *Thorstein Veblen and the Institutionalists – A Study in the Philosophy of Economics*, London, Macmillan.

Seddon, J.D. (1981) *Moroccan Peasants: A Century of Change in the Eastern Rif*, London, Dawson and Sons.

Seddon, J.D. (ed.) with Blaikie, P.M. and Cameron, J. (1979) *Peasants and Workers in Nepal*, Warminster, Aris and Phillips.

Selby, M.J. (1982) *Hillslope Materials and Processes*, Oxford, Oxford University Press.

Semple, E.C. (1907) 'The Anglo-Saxons of the Kentucky Mountains', *Geographical Journal*, 17, 588–623.

Semple, E.C. (1911) *Influences of Geographic Environment*, New York, Henry Holt.

Shaha, R. (1975) *Nepali Politics: Retrospect and Prospect*, New Delhi, Oxford University Press.

Shaw, B.D. (1981) 'Climate, environment and history: the case of Roman North Africa', in Wigley, T.M.L., Ingram, M.J. and Farmer, G. (eds) *Climate and*

History: Studies in Past Climates and their Impact on Man, Cambridge, Cambridge University Press, 379–403.

Sheng, T.C. (1981) 'Protection of cultivated slopes – terracing steep slopes in humid regions', Rome, FAO.

Sherman, D.G. (1980a) 'What "green desert"? The ecology of Batak grassland farming in Indonesia', *Indonesia*, 29, 113–49.

Sherman, D.G. (1980b) 'The culture-bound notion of "soil fertility" in interpreting non-western criteria of selecting land for cultivation', in *Blowing in the Wind: Deforestation and Long Range Implications, Studies in Third World Societies*, 14, Williamsburg, Virginia, Department of Anthropology, College of William and Mary, 487–511.

Sherratt, A. (1980) 'Water, soil and seasonality in early cereal cultivation', *World Archaeology*, 11, 313–30.

Sherratt, A. (1981) 'Plough and pastoralism: aspects of the secondary products revolution', in Hodder, I., Isaac, G. and Hammond, N. (eds) *Patterns of the Past: Studies in Honour of David Clarke*, Cambridge, Cambridge University Press, 261–305.

Silberfein, M. (1984) 'Differential development in Machakos District, Kenya', in Scott, E.P. (ed.) *Life Before the Drought*, Boston, London, Allen and Unwin, 101–23.

Sima Qian (1979) *Records of the Historian* (translated by Yang Hsien-yi and Gladys Yang), Beijing, Foreign Languages Press, 411 (first written, 91 BC).

Simon, J.L. (1981) *The Ultimate Resource*, Princeton, NJ, Princeton University Press.

Singh, G., Babu, R. and Chandra, S. (1985) 'Research on the universal soil loss equation in India', in El-Swaify, S.A. *et al.* (eds) *Soil Erosion and Conservation*, Ankeny, Iowa, Soil Conservation Society of America, 496–508.

Singh, H. (1979) 'Caste and Kisan movement in Marwar: some questions to the conventional sociology of kin and caste', *Journal of Peasant Studies*, 7, 103–18.

Slatyer, R.O. and Perry, R.A. (eds) (1969) *Arid Lands of Australia*, Canberra, Australian National University Press.

Slicher van Bath, B.H. (1963) *The Agrarian History of Western Europe AD 500–1850*, London, Edward Arnold.

Smil, V. (1978) 'China's energetics: a system analysis', in *Chinese Economy Post-Mao*, Washington DC, US Government Printing Office, 324–69.

Smil, V. (1979) 'Controlling the Yellow River', *Geographical Review*, 69, 253–72.

Smil, V. (1981a) 'China's agroecosystem', *Agro-ecosystems*, 7, 27–46.

Smil, V. (1981b) 'China's food', *Food Policy*, 6, 67–77.

Smil, V. (1983) 'Deforestation in China', *Ambio*, 12, 226–31.

Smil, V. (1984) *The Bad Earth: Environmental Degradation in China*, Armonk, New York, M.S. Sharpe.

Smith, N.J.H. (1978) 'Agricultural productivity along Brazil's transamazon highway', *Agro-Ecosystems*, 4, 415–32.

Smith, N.J.H. (1980) 'Anthrosols and human carrying capacity in Amazonia', *Annals of the Association of American Geographers*, 70, 553–66.

Smith, R.T. (1975) 'Early agriculture and soil degradation', in Evans, J.G., Limbrey, S. and Cleere, H. (eds) *The Effect of Man on the Landscape: The Highland Zone*, research report no. 11, London, Council for British Archaeology, 27–36.

Smith, W. (1984) 'The "vortex model" and the changing agricultural landscape of Quebec', *Canadian Geographer*, 28, 358–72.

Smith, W. (1986) 'Agricultural marketing and distribution', in Pacione, M. (ed.) *Progress in Agricultural Geography* (II), London, Croom Helm, 219–38.

Snowy Mountains Engineering Corporation (1982) 'Report on the Serang River Project', Cooma, SMEC.

Soemitro, A., Anwar, A. and Pawobo, D. (1983) 'Natural resources and environmental management' [Draft final report on Agricultural Policy Study funded by USAID and IDRC (Canada)], Bogor, Center for Soil Research.

Soerjani, M., Eussen, H.J.J. and Titrosudirdjo, S. (1983) '*Imperata* research and management in Indonesia', *Mountain Research and Development*, 3, 397–404.

Soewardi, B., Burbridge, P. and Djokosudardjo, M.J. (1980) 'Improving the choice of resource systems for transmigration', *Prisma* (English), 18, 56–70.

Soil Conservation Service of NSW (n.d.) *The Northern Crisis – Soil Erosion*, Sydney, NSW Government Printer.

Soil Conservation Service of NSW (1985) *Annual Report 1983–84*, Sydney, NSW Government Printer.

South Africa (1955) *Summary of the Report of the Commission for the Socioeconomic Development of the Bantu Areas within the Union of South Africa*, Pretoria, Government Printer.

Spate, O.H.K. (1959) *The Fijian People: Economic Problems and Prospects*, Council Paper no. 13 of 1959, Suva, Government Printer.

Spriggs, M.J.T. (1981) 'Vegetable kingdoms: Taro irrigation and Pacific prehistory', unpublished PhD thesis in Prehistory, Canberra, ANU.

Spriggs, M.J.T. (1986) 'Landscape, land use and political transformation in southern Melanesia', in Kirch, P.V. (ed.) *Island Societies: Archaeological Approaches to Evolution and Transformation*, Cambridge, Cambridge University Press.

Stadler, T. (1983) 'Nature conservation on rural land: attitudes of Tasmanian farmers to native vegetation retention – practice and policy', Master of Environmental Studies thesis, Hobart, University of Tasmania.

State Statistical Bureau (1984) *Statistical Yearbook of China 1984*, Hong Kong, Economic Information Agency.

Stavenhagen, R. (1969) 'Seven erroneous theses about Latin America', translated and revised by Stavenhagen, R. from an original article in Spanish, published in *El Dia* (newspaper, Mexico City) (1965), in Horowitz, I.L., de Castro, J. and Gerassi, J. (eds) *Latin American Radicalism: A Documentary Report on Left and Nationalist Movements*, New York, Random House, 102–17.

Stavenhagen, R. (ed.) (1970) *Agrarian Problems and Peasant Movements in Latin America*, New York, Doubleday Anchor.

Stebbing, E.P. (1935) *The Creeping Desert in the Sudan and Elsewhere in Africa*, Khartoum, Ministry of Agriculture.

Stebelsky, I. (1974) 'Soil erosion in the central Russian Black Earth region', in Volgyes, I. (ed.), *Environmental Deterioration in the Soviet Union and Eastern Europe*, New York, Praeger, 55–66.

Stewart, A.T.Q. (1972) *The Pagoda War*, London, Faber and Faber.

Stewart, P.J. (1975) *Algerian Peasantry at the Crossroads: Fight Erosion or Migrate*, IDS discussion paper no. 69, Brighton, Sussex, Institute of Development Studies.

Stiller, L.F. (1975) *The Rise of the House of Gorkha 1786–1816*, Kathmandu, Ratna Pustak Bhandar.

Stocking, M.A. (1978) 'Remarkable erosion in central Rhodesia', *Proceedings of the Geographical Association of Rhodesia*, 11, 42–56.

Stocking, M.A. (1983) 'Farming and environmental degradation in Zambia: the human dimension', School of Development Studies, University of East Anglia, Norwich, 41 pp, mimeo.

Stocking, M.A. (1984a) 'Erosion and soil productivity: a review', Consultants' working paper no. 1, Soil Conservation Programme, Land and Water Development Division, Rome, FAO.

Stocking, M.A. (1984b) 'Rates of erosion and sediment yield in the African environment', in Walling, D.E., Foster, S.S.D. and Wurzel, P. (eds) *Challenges in African Hydrology and Water Resources*, International Association of Hydrological Sciences publication no. 144, Wallingford, UK, IAHS, 285–93.

Stocking, M.A. (1984c) 'Soil conservation policy in colonial Africa and its continuation, effects and lessons', paper presented to the Symposium on the History of Soil and Water Conservation, 23–26 May, Missouri.

Stocking, M.A. and Peake, L. (1985) *Erosion-induced Loss in Soil Productivity: Trends in Research and International Cooperation*, paper presented to the IVth International Conference on Soil Conservation, Maracay, Venezuela, 3–9 November, undertaken under Soil Conservation Programme, Soil Resources, Management and Conservation Service, Land and Water Development Division, Rome, FAO.

Stocking, M.A. and Peake, L. (1986) 'Cropyield losses from the erosion of Alfisols', *Tropical Agriculture* (Trinidad), 63, 41–5.

Strakhov, N.M. (1967) *Principles of Lithogenesis*, vol. 1, Edinburgh, Oliver and Boyd.

Strathern, A, (1966) 'Despots and dictators in the New Guinea Highlands', *Man*, 1, 356–67.

Strathern, A. (1969) 'Finance and production: two strategies in New Guinea Highlands exchange systems', *Oceania*, 40, 42–67.

Strathern, A. (1971) *The Rope of Moka: Big-Men and Ceremonial Exchange in Mount Hagen, New Guinea*, Cambridge, Cambridge University Press.

Strathern, A. (1982) 'Tribesmen or peasants?', in Strathern, A. (ed.) *Inequality in New Guinea Highlands Societies*, Cambridge, Cambridge University Press, 137–57.

Ström, G.W. (1978) *Development and Dependence in Lesotho, the Enclave of South Africa*, Uppsala, Scandinavian Institute of African Studies.

Strout, A.N. (1983) 'How productive are soils of Java?', *Bulletin of Indonesian Economic Studies*, 19, 1, 32–52.

Suryatna, E.S. and McIntosh, J.C. (1980) 'Food crops production and control of *Imperata cylindrica* L. (Beauv.) on small farms', *Proceedings of the Biotropical Workshop on Alang-Alang*, Bogor, Biotrop special publication 5, 135–47.

Susman, P., O'Keefe, P. and Wisner, B. (1983) 'Global disasters, a radical interpretation', in Hewitt, K. (ed.) *Interpretations of Calamity from the Viewpoint of Human Ecology*, Boston, Allen and Unwin, 263–83.

Sutherland, D.M.G. (1981) 'Weather and the peasantry of Upper Brittany, 1780–1789', in Wigley, T.M.L., Ingram, M.J. and Farmer, G. (eds) *Climate and History: Studies in Past Climates and their Impact on Man*, Cambridge, Cambridge University Press, 434–49.

Sutton, K. (1977) 'Reclamation of wasteland during the eighteenth and nineteenth centuries', in Clout, H.D. (ed.) *Themes in the Historical Geography of France*, London, Academic Press, 247–300.

Suwardjo and Aibyamia, S. (1983) 'Crop residue mulch for conserving soil in upland agriculture', paper presented to Malama Ana 83, Honolulu, Hawaii.

Swartz, G.L. (compiler) (1974) *Land Conservation in Fiji*, report by an Australian Aid Consultant Team to the Australian Development Assistance Agency, Melbourne, Soil Conservation Authority of Victoria.

Tarrant, J.R. (1980) *Food Policies*, Chichester, John Wiley.

Taussig, M. (1978) 'Peasant economies and the development of capitalist agriculture of the Cauca Valley, Colombia', *Latin American Perspectives*, 18, 5, 3, 62–90.

Thompson, M. and Warburton, M. (1985a) 'Uncertainty on a Himalayan scale', *Mountain Research and Development* 5, 115–35.

Thompson, M. and Warburton, M. (1985b) 'Knowing where to hit: a conceptual framework for the sustainable development of the Himalayas', *Mountain Research and Development*, 5, 203–20.

Thomson, J.T. (1977) 'Ecological deteriorations: local-level rule-making and enforcement problems in Niger', in Glantz, M.H. (ed.) *Desertification and Degradation in and around Arid Lands*, Boulder, Colorado, Westview Press, 57–79.

Thornes, J.B. and Brunsden, D. (1977) *Geomorphology and Time*, London, Methuen.

Timmons, J.F. (1979) 'Agriculture's natural resource base: demand and supply interactions, problems and remedies', in Soil Conservation Society of America, *Soil Conservation Policies: an Assessment*, Ankeny, Iowa, Soil Conservation Society of America, 53–74.

Titow, J. (1960) 'Evidence of weather in the account rolls of the Bishopric of Winchester 1209–1350', *Economic History Review*, 12, 360–407.

Tomaselli, R. (1977) 'Degradation of the Mediterranean maquis', in UNESCO, *Mediterranean Forests and Maquis: Ecology, Conservation and Management*, MAB Technical Notes 2, Paris, UNESCO, 33–72.

Trapnell, C.G. and Clothier, J.N. (1937) *The Soils, Vegetation and Agricultural Systems of North-Western Rhodesia*, Lusaka, Government Printer.

Tregubov, P.S. (1981) 'Effective erosion control in the USSR', in Morgan, R.P.C. (ed.) *Soil Conservation: Problems and Prospects*, Chichester, John Wiley, 451–9.

Tricart, J. (1953) 'La géomorphologie et les hommes', *Revue de Géomorphologie Dynamique*, 4, 153–6.

Tucker, R.P. (1984) 'The historical context of social forestry in the Kumaon Himalayas', *Journal of Developing Areas*, 18, 3, 341–56.

Turner, B. (1984) 'Changing land-use patterns in the fadamas of northern Nigeria', in Scott, E.P. (ed.) *Life Before the Drought*, Boston and London, Allen and Unwin, 149–70.

Twyford, I.T. and Wright, A.S.C. (1965) *The Soil Resources of the Fiji Islands*, vol. I, Suva, Government Printer.

Unger, P.W. and McCalla, T.M. (1980) 'Conservation tillage systems', *Advances in Agronomy*, 33, 1–58.

Vail, L. (1983) 'The political economy of east-central Africa', in Birmingham, D. and Martin, P.M. (eds) *History of Central Africa*, London, Longman, 200–50.

Vanelslande, A., Rosseau, P., Lal, R., Gabriels, D. and Ghuman, B.S. (1984) 'Testing the applicability of a soil erodibility nomogram for some tropical soils', in Walling, D.E., Foster, S.S.D. and Wurzel, P. (eds) *Challenges in African Hydrology and Water Resources*, International Association of Hydrological Sciences publication no. 144, Wallingford, UK, IAHS, 463–73.

Virgo, K.J. and Munro, R.N. (1978) 'Soil and erosion features of the Central Plateau

Region of Tigrai, Ethiopia', *Geoderma*, 20, 131–57.

Vita-Finzi, C. (1969) *The Mediterranean Valleys: Geological Changes in Historical Times*, Cambridge, Cambridge University Press.

Vogeler, I. (1981) *The Myth of the Family Farm: Agribusiness Dominance of US Agriculture*, Boulder, Colorado, Westview Press.

Vogt, J. (1953) 'Erosion des sols et techniques de culture en climat tempéré maritime en transition (France et Allemagne)', *Revue de Géomorphologie Dynamique*, 4, 157–83.

Vogt, J. (1957a) 'La dégradation des terroirs lorrains au milieu du XVIIIe siècle', *Bulletin de la Section de Géographie, Comité des Travaux Historiques et Scientifiques (Actes du Congrès National des Sociétés Savantes, Bordeaux, 1957)*, 111–16.

Vogt, J. (1957b) 'Culture sur brulis et érosion des sols', *Bulletin de la Section de Géographie, Comité des Travaux Historiques et Scientifiques (Actes du Congrès National des Sociétés Savantes, Rouen, Caen, 1956)*, 337–42.

Vogt, J. (1958a) 'Zur Bodenerosion in Lippe: ein historisches Beitrag zur Erforschung der Bodenerosion', *Erdkunde*, 12, 132–4.

Vogt, J. (1958b) 'Zur historischen Bodenerosion in Mitteldeutschland', *Petermanns Geographische Mitteilungen*, 102, 199–203.

Vogt, J. (1960) 'Hardt et nord des Vosges au XVIIIe siècle: le déclin d'une moyenne montagne', *Bulletin de la Section de Géographie, Comité des Travaux Historiques et Scientifiques (Actes du Congrès National des Sociétés Savantes, Dijon, 1959)*, 201–4.

Vogt, J. (1966–74) 'Questions agraires alsaciennes (XVI–XVIIIe s.), IX. Temoignages sur l'érosion historique des sols au Sundgau (XVI–XVIIIe siècles)', *Revue d'Alsace*, 104, 70–3.

Vogt, J. (1967–71) 'L'érosion des sols par les eaux de fonte: l'exemple de la région de Saint-Dié en 1784', *Annales de la Société d'Emulation des Vosges*, 142–6, 279–84.

Vogt, J. (1968) 'Une texte remarquable sur l'érosion historique des sols au Sundgau', *Revue Géographique de l'Est*, 8, 361–2.

Vogt, J. (1970a) 'Un bel exemple d'érosion historique des sols dans le nord de la plaine d'Alsace', *Bulletin de l'Association Philomathique d'Alsace et de Lorraine*, 14, 53–5.

Vogt, J. (1970b) 'Aspects de l'érosion historique des sols en Bourgogne et dans les régions voisines', *Annales de Bourgogne*, 42, 30–50.

Vogt, J. (1972a) 'Coup d'oeil à l'érosion historique des sols dans le nord et l'est de la Haute-Marne (matériaux d'archives)', *Les Cahiers Haut-Marnais*, 65–82.

Vogt, J. (1972b) 'Les Riceys: formations liées à l'érosion anthropique', Service Géologique Nationale (Service Formations Superficielles), mimeo.

Vogt, J. (1972c) 'Notice Raucourt: formations liées à l'érosion anthropique', Service Géologique Nationale (Service Formations Superficielles), mimeo.

Vogt, J. (1972d) 'Un exemple d'érosion catastrophique des sols en montagne (Grendelbruch en 1774)', *Annuaire de la Société Historique et Archéologique Molsheim*, 11–13.

Vogt, J. (1975) 'Une texte remarquable sur l'érosion historique des sols en Beauvaisis au XVIIIe siècle, un des grands maux de l'état', *Etudes de la Région Parisienne*, NS, 45–6, 27–9.

Vogt, J. (1977) 'Archives et géologie appliquée: seismes, glissements, éboulements, érosion anthropique', *La Gazette des Archives* NS, 98, 131–6.

Vogt, J. (1982) 'Exemples d'érosion catastrophique des sols dans le vignoble alsacien au XVIIe siècle (Marlenheim et Mutzig)', *Annuaire de la Société Historique et*

Archéologique Molsheim, 95–6.

Vogt, J. (1983) 'Organisation des terroirs et évolution des cultures'; 'Les problèmes de tenure'; 'Les relations ville-campagnes', in Boehler, J-M., Lerch, D. and Vogt, J. (eds) *Histoire de l'Alsace rurale*, Strasbourg, Librairie Istra, 227–69.

Vogt, J. (n.d.) *'L'érosion historique des sols et sa prise de conscience dans l'est du bassin de Paris*, Paris, Bureau de Recherches Géologiques et Minières, Département Géologie, mimeo.

Vohra, B.B. (1984) 'The greening of India', New Delhi, Government of India, 30 pp, mimeo.

Waddell, E. (1972) *The Mound Builders: Agricultural Practices, Environment and Society in the Central Highlands of New Guinea*, American Ethnological Society Monograph 53, Seattle, University of Washington Press.

Wade, M.K. and Sanchez, P.A. (1983) 'Mulching and green manure applications for continuous crop production in the Amazon basin', *Agronomy Journal*, 75, 39–45.

Wagner, R. (1978) *Soil Conservation in New South Wales – 1938–1978*, Sydney, NSW Government Printer.

Walker, D. and Flenley, J.R. (1979) 'Late Quaternary vegetational history of Enga Province of upland Papua New Guinea', *Philosophical Transactions of the Royal Society of London. B. Biological Sciences*, 286, 265–344.

Wallace, I. (1985) 'Toward a geography of agribusiness', *Progress in Human Geography*, 9, 491–514.

Walling, D.E. (1978) 'Reliability considerations in the evaluation and analysis of river loads', *Zeitschrift fur Geomorphologie Suppl.*, 29, 29–42.

Walling, D.E. (1984) 'The sediment yield of African rivers', in Walling, D.E., Foster, S.S.D. and Wurzel, P. (eds) *Challenges in African Hydrology and Water Resources*, International Association of Hydrological Sciences publication no. 144, Wallingford, UK, IAHS, 265–83.

Walling, D.E. and Webb, B.W. (1981) 'The reliability of suspended sediment load data', in International Association of Hydrological Sciences *Erosion and Sediment Transport Measurement*, publication no. 133, Wallingford, UK, IAHS, 177–94.

Wang, Y.Y. and Zhang, Z.H. (eds) (1980) *Loess in China*, Xi'an, Shaanxi People's Art Publishing House.

Ward, R.G. (1965) *Land Use and Population in Fiji: A Geographical Study*, London, HMSO.

Ward, R.G. (1980) 'Plus ça change . . . plantations, tenants, proletarians or peasants in Fiji', in Jennings, J.N. and Linge, G.J.R., *Of Time and Place: Essays in Honour of O.H.K. Spate*, Canberra, Australian National University Press, 134–52.

Ward, R.G. (1985) 'Land, land use and land availability in Fiji', in Brookfield H.C., Ellis, F. and Ward, R.G. (eds) *Land, Cane and Coconuts: Papers on the Rural Economy of Fiji*, Department of Human Geography Publication 17, Canberra, Australian National University, 15–64.

Watling, D. (1984) 'Irrigated terrace cultivation of dalo at Nawaikama, Gau', *Domodomo* (Fiji Museum Quarterly), 2, 121–35.

Watson, J.W. (1939) 'Forest or bog: man the deciding factor', *Scottish Geographical Magazine*, 55, 148–61.

Watts, M. (1983) 'On the poverty of theory: natural hazards research in context', in K. Hewitt (ed.) *Interpretations of Calamity from the Viewpoint of Human Ecology*, Boston, Allen and Unwin, 231–62.

Weatherall, M. (1968) *Scientific Method*, London, English Universities Press.

Webb, W.P. (1931) *The Great Plains*, New York, Grosset and Dunlap.

Whitehead, C.E. (1954) 'Soil erosion and soil conservation in Fiji', *Agricultural Journal* (Department of Agriculture, Fiji), 25, 1–4.

Whitehead, C.E. (1955) *Soil Conservation and Its Control in Fiji*, Bulletin no. 28, Suva, Department of Agriculture.

Whyte, A.V.T. (1977) *Guidelines for Field Studies in Environmental Perception*, Programme on Man and the Biosphere, Technical Note 5, Paris, UNESCO.

Williams, C.H. (1980) 'Soil acidification under clover pasture', *Australian Journal of Experimental Agriculture and Animal Husbandry*, 23, 181–91.

Williams, G. (1976) 'Taking the part of the peasants: rural development in Nigeria and Tanzania', in Gutkind, P. and Wallerstein, I. (eds) *The Political Economy of Contemporary Africa*, London, Sage Publications, 131–54.

Williams, M. (1974) *The Making of the South Australian Landscape: A Study in the Historical Geography of Australia*, London, Academic Press.

Williams, M. (1978) 'Desertification and technological adjustment in the Murray mallee of South Australia', *Search*, 9, 265–8.

Williams, M. (1979) 'The perception of the hazard of soil degradation in South Australia: a review', in Heathcote, R.L. and Thom, B.L. (eds) *Natural Hazards in Australia*, Canberra, Australian Academy of Science, 275–89.

Williams, V. (1977) 'Neotectonic implications of the alluvial record in the Sapta Kosi Drainage Basin, Nepalese Himalayas', PhD thesis in Geology, University of Washington, Seattle.

Wills, I. (1984) 'Agricultural land use, environmental degradation, and alternative property rules', paper presented to the Conservation and the Economy Conference, Sydney, September, Melbourne, Economics Department, Monash University (unpublished).

Wilson, A.D. and Graetz, R. (1979) 'Management of the semi-arid and arid rangelands of Australia', in Walker, B.H. (ed.) *Management of Semi-arid Ecosystems*, Amsterdam, Elsevier, 83–111.

Winiger, M. (1983) 'Stability and instability of mountains ecosystems: definitions for evaluation of human systems', *Mountain Research and Development*, 3, 103–11.

Wischmeier, W.H. (1976) 'Use and misuse of the Universal Soil Loss Equation', *Journal of Soil and Water Conservation*, 31, 5–9.

Wischmeier, W.H. and Mannering, J.V. (1969) 'Relation of soil properties to its erodibility', *Proceedings of the Soil Science Society of America*, 23, 131–7.

Wischmeier, W.H. and Smith, D.D. (1978) 'Predicting rainfall erosion losses: a guide to conservation planning', *Agriculture Handbook No. 537*, Washington DC, US Department of Agriculture.

Wise, S.M., Thornes, J.B. and Gilman, A. (1982) 'How old are the badlands? a case study from south-east Spain', in Bryan, R. and Yair, A. (eds) *Badland Geomorphology and Piping*, Norwich, Geo Books, 259–77.

Wisner, B. (1976) 'Man-made famine in Eastern Kenya: the interrelationship of environment and development', Institute of Development Studies discussion paper no. 96, Brighton, IDS.

Wittfogel, K.A. (1957) *Oriental Despotism: A Comparative Study of Total Power*, New Haven, Yale University Press.

Wohlt, P.B. (1978) 'Ecology, agriculture and social organization: the dynamics of group composition in the highlands of New Guinea', PhD thesis in Anthropology, University of Minnesota (unpublished).

Wolff, W. (1950–51) 'Bodenerosion in Deutschland', *Die Erde*, 2, 215–28.

Wood, A.W. (1984) 'Land for tomorrow: subsistence agriculture, soil fertility and ecosystem stability in the New Guinea Highlands', PhD thesis in Geography, University of Papua New Guinea, (unpublished).

Woods, L.E. (1984) *Land Degradation in Australia*, Canberra, Australian Government Printing Service.

Wrigley, E.A. and Schofield, R.S. (1981) *The Population History of England 1541–1871. A Reconstruction*, London, Edward Arnold.

Yen, D.E. (1974) 'The sweet potato and Oceania: an essay in ethnobotany, *Bernice P. Bishop Museum Bulletin*, 236, Honolulu, Bishop Museum Press.

Yi Zhi (1981) 'We should control the use of cultivated land as strictly as we do population growth', *Hongqi (Red Flag)*, 20, 43–6.

Young, A. (1892) *Arthur Young's Travels in France During the Years 1787, 1788, 1789*, with an Introduction, etc., edited by Miss Betham-Edwards, London, Bell.

Young, A.T. (1978) 'Recent advances in the survey and evaluation of land resources', *Progress in Physical Geography*, 2, 462–79.

Young, K. and Moser, M. (eds) (1981) *Women and the Informal Sector, IDS Bulletin*, 12, 3.

Young, M. (1981) 'Arid land administration, management and tenure in Australia – recent inquiries and current policy issues', *Rural Marketing and Policy*, 11, 29–31.

Young, R.A. and Fosbrooke, H. (1960) *Land and Politics among the Luguru of Tanganyika*, London, Routledge and Kegan Paul.

Yusoff, Datuk, M. (1983) *Decades of Change (Malaysia – 1910s – 1970s)*, Kuala Lumpur, Art Printing Works.

Zhang Pinghua (1981) 'Conscientiously implement private woodlots and energetically raise firewood and charcoal forests', *Zhongguo Linye (China's Forestry)*, 3, 4–5.

Index of Authors and Countries

This index lists only first authors of joint publications referred to in the text. Substantive references to countries and major regions are listed. Incidental references are not indexed.

Index of Subject Matter

Bold numerals are used where a topic is defined or discussed in depth. Certain ubiquitous topics (e.g. decision-making/makers, farmers, degradation, small/peasant farmers) are not indexed except where they are defined (e.g. degradation 7 [*and ubiquitous*]). Cross-references are simplified by using capitals for key topics with more than four cross references (e.g. LAND), and other topics are referenced back to these topics only, except where principal pairs are also identified. Other related topics may therefore be traced from these key topics.

commercialization, *see* commercial farms
COMMON PROPERTY RESOURCES
(CPRs) (*see* Coase's theorem, feudalism,
FOREST, free-rider strategies, green-apple
syndrome, LAND, LIVESTOCK,
POLITICAL ECONOMY, private
property, STATE, stinting), 2, 65, 74,
76, 77, 103, 131, 132, **186–207**, 195–6,
240
communes (*see* SOCIALISM, state/collective
farms), 215
CONSERVATION (*see* aid, colonialism,
ECONOMICS, environmental
fundamentalism, FOREST, SOIL, Soil
Conservation Services, STATE,
VEGETATION), 34; strategies, 4, 30, 50,
105–6, 172, 183, 212; conservationist
consciousness, 164, 181–5, 222, 225, 229,
230, 241–2; legislation, (*see* STATE), 129,
220, 225, 229–30, 231, 233–4, 236, 237;
populist approaches, 243–5; World
Conservation Strategy, 241–3
contour cultivation, *see* terracing
corporate business, in agriculture (*see*
AGRICULTURE, POLITICAL
ECONOMY), 79–81, 227–8, 231
cost-benefit analysis (*see* ECONOMICS), 87,
90, 98, 239
cotton (*see* CASH CROPS), 105, 108
crops, *see* AGRICULTURE, CASH CROPS,
rice; introductions, 44, 136, 141, 145,
147–8; yields, decline of, 5, 9, 15, 17, 45,
60–2, 97, 131, 138, 152–4, 172, 205, 213,
224
'Culture System' (*see* CASH CROPS), 104

degradation, **7** [*and ubiquitous*]
denudation rates, *see* soil-loss rates
desertification (*see* EROSION,
VEGETATION), 44, 107, 110, 218–19,
232
development/underdevelopment, 43, 119
diminishing returns (*see* ECONOMICS), 19,
20, 61, 221
discounting rates (*see* ECONOMICS), 9, 73,
81, **87–8**
disease (*see* population, LIVESTOCK), 106,
138
'downstream effects' (*see* ECONOMICS,
EROSION, TRANSPORATION), 8, 64,
85, 95, 96, 98, 162, 164
drought (*see* famine, LIVESTOCK), 2, 20,
36, 57, 72, 110, 120, 123, 133, 167, 171,
236, 237, 247

ecocide (*see* poverty), 13, 240
ECONOMICS (*see* Coase's theorem,
CONSERVATION, cost-benefit analysis,
development, diminishing returns,
discounting rates, 'downstream effects',
'Landesque capital', margin, POLITICAL
ECONOMY, risk aversion, SOCIALISM,
'value' of land), 96, 239; classical, 2;
Marxian, 2, 6, 24, 82, 209–10; neo-
classical, 19; resource-depletion models, 6;
shadow wage, 95; of soil and water conservation, **84–91**
ecosystems, ecology, 33, 209
environmental fundamentalism (*see*
CONSERVATION, VEGETATION),
xviii, xx, 2, 26, 41, 84, 97, 164, 182, 185,
229
erodibility (*see* EROSION, SOIL), 12, 71,
129, 179
EROSION (*see* alluvial deposition, 'badlands',
Bubnoff units, colluvial deposition,
desertification, erodibility, erosivity,
frequency-magnitude problem, gully
erosion, high-concentration events, Loess,
mass wasting, measurement of degradation,
sedimentation, sheet/rill erosion, SOIL,
soil-loss rates, T values,
TRANSPORATION, Universal Soil Loss
Equation, wind erosion), 60
erosion rates, *see* soil-loss rates
erosivity (*see* EROSION), 12, 57, 71, 122,
128, 133, 135, 139, 140, 141, 179
Eupatorium (*see* FOREST, VEGETATION),
169–70, 173
export commodity production, *see* CASH
CROPS

fallow (*see* AGRICULTURE, COMMON
PROPERTY RESOURCES,
LIVESTOCK, pastoralism, three-field
system), 136, 160, 200–1, 205, 211–12, 213
famine (*see* drought), xviii, 107, 120, 138, 167,
216–17
fertilizers (*see* AGRICULTURE,
LIVESTOCK, pollution, SOIL), 29, 224;
inorganic, 1, 14, 15, 141, 171–3, 175, 208,
221, 224, 228, 230, 242; organic, 30, 40, 42,
46, 68, 75, 104, 108, 131, 135, 141, 150,
153, 159, 171–3, 179, 191, 221, 230
feudalism (*see* COMMON PROPERTY
RESOURCES, LAND, POLITICAL
ECONOMY), 7, 102, 140, 141, 192–3, 198,
241
fire (*see* FOREST, VEGETATION), 144,